LOGIC OF MATHEMATICS

PURE AND APPLIED MATHEMATICS

A Wiley-Interscience Series of Texts, Monographs, and Tracts

A complete list of the titles in this series appears at the end of this volume.

LOGIC OF MATHEMATICS

A Modern Course of Classical Logic

ZOFIA ADAMOWICZ
Institute of Mathematics of the Polish Academy of Sciences

PAWEŁ ZBIERSKI
Department of Mathematics, Warsaw University

A Wiley-Interscience Publication

JOHN WILEY & SONS, INC.

New York · Chichester · Weinheim · Brisbane · Singapore · Toronto

This text is printed on acid-free paper.

Library of Congress Cataloging in Publication Data:

Adamowicz, Zofia.
Logic of mathematics : a modern course of classical logic / Zofia Adamowicz, Paweł Zbierski.
p. cm. -- (Pure and applied mathematics)
"A Wiley-Interscience publication."
Includes bibliographical references (p. 254–256) and index.
ISBN 0-471-06026-7 (cloth : alk. paper)
1. Logic, Symbolic and mathematical. I. Zbierski, Paweł.
II. Title. III. Series: Pure and applied mathematics
(John Wiley & Sons : Unnumbered)
QA9.A24 1997
511.3--dc20 95-20818

Printed in the United States of America

10 9 8 7 6 5 4 3 2

PREFACE

In this textbook on mathematical logic, we take the position of a mathematician rather than a logician. We select and discuss the material referring directly to mathematical practice either by applications to other branches of mathematics or by explaining the nature of mathematical reasoning. In our approach, relational structures are given priority over logical languages. In the exposition we treat the subject as any part of mathematics as far the methods and the level of accuracy is concerned.

The book is addressed first of all to students of mathematics and to all mathematicians who want to have some familiarity with this beautiful domain of science. The technical difficulties do not exceed those used in any standard course of, say, abstract algebra. Nevertheless, to understand the book, some mathematical experience seems necessary.

Part I of the book (Chapters 1 through 17) is an introductory course at graduate level. In Chapters 1 to 4 we develop the theory of relational structures with a particular emphasis on Boolean algebras. In Chapters 5 to 7 we introduce and discuss formulas, the truth relation, theories, and models. Chapters 8 to 11, devoted to the notion of proof, culminate in Gödel's completeness theorem. In Chapters 12 to 17 we deal mostly with model theoretic topics such as definability, compactness, ultraproducts, realization, omitting of types, and so on.

Part II (Chapters 18 through 24) consists of famous theorems crucial in the development of mathematical logic. In Chapters 18 to 21 we present Gödel's theory, leading to his celebrated incompleteness theorems. Chapter 22 is devoted to the independence proof of Goodstein's theorem from Peano arithmetic. The next chapter contains Cohen's proof of Tarski's theorem on elimination of quantifiers for the theory of real closed fields. Finally, in Chapter 24 we present the Matiyasevich theorem on diophantine relations giving a solution of the tenth Hilbert problem. All the above theorems are provided with complete and rigorous proofs.

Each chapter ends with a number of exercises. Some of them are easy; those more difficult are supplied with hints. We advise the reader to solve them all.

As in any branch of mathematics, we make use of some set-theoretical apparatus. The introductory chapter contains the set-theoretical notions and theorems (without proofs) used throughout the book.

ZOFIA ADAMOWICZ
PAWEŁ ZBIERSKI

CONTENTS

PART II Selected Topics

LOGIC OF MATHEMATICS

INTRODUCTION

In this introductory chapter we set forth the logical and set-theoretical notation and theorems to be used throughout the book.

Elementary Logic

Mathematical statements (expressing properties of some objects) are called formulas. Given any formulas ϕ and ψ, we can form the following new ones: the negation $\neg\phi$ (not ϕ) the implication $\phi \to \psi$ (if ϕ then ψ), the disjunction $\phi \vee \psi$ (ϕ or ψ), the conjunction $\phi \wedge \psi$ (ϕ and ψ), and the equivalence $\phi \equiv \psi$ (ϕ if and only if ψ). The operational symbols $\neg$, $\to$, $\vee$, $\wedge$ and $\equiv$ are called the (logical) connectives (besides the above there are also other connectives but we shall not use them). Recall that

$\neg\phi$ is true if ϕ is false and $\neg\phi$ is false if ϕ is true;
$\phi \to \psi$ is true except when the antecedent ϕ is true and the consequent ψ is false;
$\phi \vee \psi$ is true if at least one of the factors ϕ, ψ is true, otherwise false;
$\phi \wedge \psi$ is true if both ϕ and ψ are true, otherwise false;
$\phi \equiv \psi$ is true if both ϕ and ψ are true or both false, otherwise false.

The symbols $\forall$ and $\exists$ denote the universal and the existential quantifiers, respectively. Thus, $\forall x\phi$ abbreviates "for every x, ϕ" and $\exists x\phi$ stands for "there exists an x, such that ϕ." The symbols $\forall x \in X\phi$ and $\exists x \in X\phi$ denote, respectively, $\forall x(x \in X \to \phi)$ and $\exists x(x \in X \wedge \phi)$.

Operations on Sets

The membership relation is denoted by $\in$. Thus, $x \in A$ means that an object x belongs to (is an element of) the set A, while $x \notin A$ means that x does not belong to A. For any sets A and B we can form the union $A \cup B$, the intersection $A \cap B$, and the difference $A \setminus B$. Thus, we have

$$x \in A \cup B \equiv x \in A \vee x \in B,$$
$$x \in A \cap B \equiv x \in A \wedge x \in B,$$
$$x \in A \setminus B \equiv x \in A \wedge x \notin B.$$

The empty set is denoted by $\emptyset$ and the set inclusion (containment) by $\subseteq$. Thus, we have

$$A \subseteq B \equiv \forall x(x \in A \rightarrow x \in B).$$

The proper inclusion is defined as follows:

$$A \subsetneq B \equiv A \subseteq B \wedge A \neq B.$$

The power set $P(A)$ of A is the family of all subsets of the set A, $P(A) = \{x\colon\ x \subseteq A\}$. An indexed family of sets is denoted by $\{A_i\colon\ i \in I\}$ and its union and intersection by $\bigcup\{A_i\colon\ i \in I\}$ and $\bigcap\{A_i\colon\ i \in I\}$, or $\bigcup_{i \in I} A_i$ and $\bigcap_{i \in I} A_i$, respectively.

Thus, we have

$$x \in \bigcup_{i \in I} A_i \equiv \exists i \in I(x \in A_i)$$

$$x \in \bigcap_{i \in I} A_i \equiv \forall i \in I(x \in A_i).$$

Sets A, B are said to be disjoint, if $A \cap B = \emptyset$, that is, A and B have no common element. A family $\mathcal{F}$ is called disjoint if it consists of nonempty sets and any two sets $A, B \in \mathcal{F}$ are disjoint.

Functions

The symbols $\mathrm{dom}(f)$ and $\mathrm{rng}(f)$ denote, respectively, the domain (the set of arguments) and the range (the set of values) of a function f. The expression

$$f\colon\ X \longrightarrow Y$$

means that f is a function defined on X, $\mathrm{dom}(f) = X$, and with values in Y, $\mathrm{rng}(f) \subseteq Y$. If $\mathrm{rng}(f) = Y$, we say that f is onto Y, and f is one-to-one, if $a \neq b$ implies $f(a) \neq f(b)$, for any $a, b \in X$.

The symbols $f[A]$ and $f^{-1}[B]$ denote the image of a set A and the counterimage (inverse image) of B, respectively.

Let $f\colon X \longrightarrow X$. A subset $A \subseteq X$ is said to be closed under f, if $f[A] \subseteq A$, that is, $f(a) \in A$ for all $a \in A$.

The composition of an f: $X \longrightarrow Y$ and a g: $Y \longrightarrow Z$ is denoted by gf or $g \circ f$. The restriction of f to a subset $A \subseteq \mathrm{dom}(f)$ is denoted by $f|A$. The symbol N (sometimes ω) denotes the set of natural numbers, that is, nonnegative integers. A k-element sequence, where $k \in N$, is a function defined on the set $\{1, \ldots, k\}$ (or on another k-element set). We write $a = \langle a_1, \ldots, a_k \rangle$, where $a_i = a(i)$ is the value of a at i. Functions F defined on the set N are sometimes called infinite sequences and one denotes F by $\langle F_i$: $i \in N \rangle$.

Products and Relations

The product $A \times B$ of sets A, B is defined as the set of all ordered pairs $\langle a, b \rangle$ with $a \in A$ and $b \in B$. A (binary) relation on A is any subset $r \subseteq A \times A$. It is customary to write $r(a, b)$ or arb instead of $\langle a, b \rangle \in r$ and not $r(a, b)$ or $a \not r b$ instead of $\langle a, b \rangle \notin r$. In the first case we say that r holds for a, b and in the latter that r does not hold for a, b.

The product $A_1 \times \cdots \times A_n$ of any finite number of sets is defined as the set of all n-element sequences $\langle a_1, \ldots, a_n \rangle$ with $a_i \in A_i$, for $i = 1, \ldots, n$. For $n = 2$ this definition is consistent with the previous one, since the two-element sequences can be identified with the ordered pairs in an obvious way. If $A_1 = \cdots = A_n = A$, then $A_1 \times \cdots \times A_n$ is denoted by A^n (and called the nth power of A).

Any subset $r \subseteq A_1 \times \cdots \times A_n$ is called an n-ary (n-argument) relation.

A binary relation $\simeq$ on a set A is called an equivalence, if $\simeq$ is reflexive $[\forall a \in A(a \simeq a)]$, symmetric $[\forall a, b \in A(a \simeq b \rightarrow b \simeq a)]$, and transitive $[\forall a, b, c \in A(a \simeq b \wedge b \simeq c \rightarrow a \simeq c)]$. The subset

$$[a] = \{x \in A\text{: } x \simeq a\}$$

is called the equivalence (or abstraction) class of a. The family $A/\simeq\, = \{[a]\text{: } a \in A\}$ of all equivalence classes is a partition of A; that is, distinct classes are disjoint and the union of all classes is A.

Orderings

A binary relation $\leq$ is called a partial ordering if it is reflexive $[\forall x \in X(x \leq x)]$, antisymmetric $[\forall x, y \in X(x \leq y \wedge y \leq x \rightarrow x = y)]$, and transitive $[\forall x, y \in X(x \leq y \wedge y \leq z \rightarrow x \leq z)]$. A partial ordering is linear if, in addition, it is connected: $\forall x, y \in X(x \leq y \vee y \leq x)$. Given orderings $(X, \leq^X)$ and $(Y, \leq^Y)$, a one-to-one function f: $X \longrightarrow Y$ is called an order embedding if it satisfies the condition

$$x \leq^X y \quad \text{if and only if} \quad f(x) \leq^Y f(y), \quad \text{for all } x, y \in X.$$

If, in addition, f is onto Y, then f is called an order isomorphism.

The Kuratowski–Zorn Principle

A subfamily $\mathcal{F}_0 \subseteq \mathcal{F}$ of a given family of sets $\mathcal{F}$ is called a chain if we have

$$A \subseteq B \quad \text{or} \quad B \subseteq A,$$

for any $A, B \in \mathcal{F}_0$.

The maximum principle of Kuratowski-Zorn states the following: If $\mathcal{F}$ is a family such that the union $\bigcup\{A_i\colon\ i \in I\}$ of any chain $\{A_i\colon\ i \in I\} \subseteq \mathcal{F}$ belongs to $\mathcal{F}$, then $\mathcal{F}$ has maximal elements (i.e. sets $A \in \mathcal{F}$ such that $A \subsetneq B$ holds for no $B \in \mathcal{F}$).

Now, let $\mathcal{F}$ be an arbitrary family of sets. A choice function for $\mathcal{F}$ is any function g defined on $\mathcal{F}$ such that $g(A) \in A$, for each $A \in \mathcal{F}$. The axiom of choice says that every family $\mathcal{F}$ of nonempty sets has a choice function. In particular, there is always a choice function for the family $P(A) \setminus \{\emptyset\}$, of nonempty subsets of A, where A is any nonempty set.

Definitions by Induction

Let $\mathcal{F}$ be a family of sets. For every function $G\colon\ \mathcal{F} \longrightarrow \mathcal{F}$ there is a family $\{G^n\colon\ n \in N\}$ of all the iterations $G^n = G \circ \cdots \circ G$ (n-times). It follows that for every operation $G\colon\ \mathcal{F} \longrightarrow \mathcal{F}$ and every set $B \in \mathcal{F}$ there exists exactly one sequence $F\colon\ N \longrightarrow \mathcal{F}$ such that $F_0 = B$ and $F_{i+1} = G(F_i)$, for each $i \in N$. Further, it can be proved that for every set B there is a family $\mathcal{F}$ such that $B \subseteq \mathcal{F}$ and $\mathcal{F}^n \subseteq \mathcal{F}$, for every $n \in N$ (the latter property means that $\mathcal{F}$ is closed under formation of finite sequences).

Assume that a sequence $F\colon\ N \longrightarrow \mathcal{F}$ has the following property: For some set D,

$$F_{i+1} \subseteq D \times \bigcup_{n \in N} F_i^n \quad \text{for all } i \in N,$$

that is, F_{i+1} consists of some sequences of the form $\langle d, a_1, \ldots, a_n \rangle$, where $d \in D$ and $a_1, \ldots, a_n \in F_i$. We shall often make use of the following theorem.

Theorem. For an arbitrary function $H\colon\ D \times \bigcup_{n \in N} W^n \longrightarrow W$ and any $h\colon\ F_0 \longrightarrow W$ there exists exactly one function $g\colon\ \bigcup_{i \in N} F_i \longrightarrow W$ such that $g|F_0 = h$ and

$$g(d, a_1, \ldots, a_n) = H(d, g(a_1), \ldots, g(a_n))$$

for each $\langle d, a_1, \ldots, a_n \rangle \in F_i$, $i > 0$. Here, W is an arbitrary set.

Cardinal Arithmetic

The cardinality (power) of a set A is denoted by card A. Thus, card $A =$ card B

means that there exists a one-to-one function f from A onto B and card $A \leq$ card B means that card $A =$ card B_0 for some subset $B_0 \subseteq B$ (which is equivalent to the existence of a one-to-one function $g\colon A \longrightarrow B$).

A set A is countable, card $A = \omega$, if card $A =$ card N; that is, A is of the same power as the set N of natural numbers. The following rules hold true, for any infinite sets $A_1, \ldots, A_n$:

$$\text{card } (A_1 \cup \cdots \cup A_n) = \text{card } (A_1 \times \cdots \times A_n) = \max\{\text{card } A_1, \ldots, \text{card } A_n\}.$$

If the sets A_i are all of the same power equal to card A for $i \in I$, then

$$\text{card} \bigcup_{i \in I} A_i = \max\{\text{card } I, \text{card } A\}.$$

It follows, for example, that card $\bigcup_{n \in N} A^n =$ card A, for any infinite set A. The Cantor–Bernstein theorem states the following:

If $A \subseteq B \subseteq C$ and card $A =$ card C, then card $B =$ card $A (=$ card $C)$.

Classes

Intuitively, a class is a collection of objects which is too large (i.e., it has too many elements) to be a set. For example, the collection of all sets is a class, as the Russell paradox shows. Some other examples are the class of all orderings, the class of all groups, the class of all compact spaces. It is convenient to regard sets as classes and classes that are not sets are called then proper classes. Each formula $\phi(x)$ determines the class

$$\mathbb{K}_\phi = \{x : \phi(x)\},$$

consisting of all the xs with the property ϕ. The use of classes of this form is inessential, that is, does not lead out of the ordinary set theory (since statements about $\mathbb{K}_\phi$ can be replaced by equivalent statements about sets only). The union $A \cup B$, the intersection $A \cap B$, and the difference $A \setminus B$ of classes A, B are defined in an obvious way. Notice that

$$\mathbb{K}_\phi \cup \mathbb{K}_\psi = \mathbb{K}_{\phi \vee \psi}, \qquad \mathbb{K}_\phi \cap \mathbb{K}_\psi = \mathbb{K}_{\phi \wedge \psi}, \qquad \mathbb{K}_\phi \setminus \mathbb{K}_\psi = \mathbb{K}_{\phi \wedge \neg \psi}.$$

For a systematic exposition of set theory we refer the reader to Monk [M4], Vaught [V2], and Hayden and Kennison [HK]. An axiomatic theory of classes is developed in Kelley [K1].

I

MATHEMATICAL STRUCTURES AND THEIR THEORIES

1

RELATIONAL SYSTEMS

Relational systems (called also relational structures) are in common use in mathematics. Among the most familiar examples of relational systems are groups, rings, fields, linear spaces, modules, and so on. Thus, in any branch of mathematics we are concerned with a particular kind of relational structures (the terms *relational structure* and *relational system* will be used interchangeably). We can even say that relational structures are the main subject of interest in mathematical research.

The main subject of mathematical logic is a connection between semantics and syntax. To put it more directly, mathematical logic investigates the relationship between relational systems and formulas (expressing properties of elements in the system). Hence, relational systems and formulas are two fundamental notions of mathematical logic.

First, we shall be dealing with relational systems. Formulas will be introduced in Chapter 5.

Now we pass to the precise definition. First, let us recall the general notion of a relation with a finite number of arguments.

Let A be an arbitrary nonempty set. For any integer $n \geq 1$ we can form the product

$$A^n = A \times \cdots \times A \quad n \text{ times.}$$

The set A^n consists of all n-termed sequences $\langle a_1, \ldots, a_n \rangle$, where $a_i \in A$ for $i = 1, \ldots, n$.

Every subset $r \subseteq A^n$ is called then an *n-ary relation* on the set A. According to generally accepted notation, we write

$$r(a_1, \ldots, a_n), (r \text{ holds for the } a_1, \ldots, a_n) \quad \text{if } \langle a_1, \ldots, a_n \rangle \in r.$$

and

$$\text{not } r(a_1, \ldots, a_n), (r \text{ does not hold for the } a_1, \ldots, a_n) \quad \textit{if} \langle a_1, \ldots, a_n \rangle \notin r.$$

In the case of $n = 1$ the set A^1 can be identified with the set A (identifying the sequence $\langle a \rangle$ with the element a). Consequently, unary relations r on A will be identified with subsets $r \subseteq A$.

We shall often write r^A to indicate that r is a relation on the set A. Every function

$$f\colon\ A^n \longrightarrow A$$

defined on the product A^n and assuming its values in the set A is called also an *n-ary operation* on the set A. Similarly, for relations we shall often write f^A to emphasize that f is an n-ary operation on the set A, for some $n \geq 1$.

A relational system is a nonempty set A jointly with some selected relations and operations on A and some elements of A. More exactly, a relational system $\mathbb{A}$ is given by

1.1 $$\mathbb{A} = \langle A,\ \mathcal{R},\ \mathcal{F},\ \mathcal{C} \rangle,$$

where A is a nonempty set called the *universe* of $\mathbb{A}$, $\mathcal{R}$ is a family of finitary relations on A, $\mathcal{F}$ is a family of operations on A, and $\mathcal{C} \subseteq A$ is a subset of A. The elements $a \in \mathcal{C}$ are called *distinguished elements* of the system $\mathbb{A}$.

As already mentioned, relational systems are also called relational structures. We shall use also shorter terms: *a system* or *a structure*. The definition is due to Tarski [T1].

Now, let us comment on this definition. We pose no limitations on the number of relations, operations, or distinguished elements of $\mathbb{A}$—the families $\mathcal{R}, \mathcal{F}, \mathcal{C}$ can be finite or infinite of arbitrary cardinality. It is often convenient to represent $\mathcal{R}, \mathcal{F}, \mathcal{C}$ as indexed families. In this case, consequently accepted throughout this book, the system 1.1 can be written in the form

1.2 $$\mathbb{A} = \langle A,\ \{r_i^A : i \in I\}, \{f_j^A : j \in J\}, \{c_k^A : k \in K\} \rangle,$$

for some sets I, J, K of indices. We do not assume that the above enumeration is one-to-one. If $\mathbb{A}$ has finitely many relations, operations, and distinguished elements and, for example, $I = \{1, \ldots, n\}$, $J = \{1, \ldots, m\}$, and $K = \{1, \ldots, l\}$, then we write

$$\mathbb{A} = \langle A,\ r_1^A, \ldots, r_n^A, f_1^A, \ldots, f_m^A, c_1^A, \ldots, c_l^A \rangle.$$

Of course, we shall omit the indices, if there is no fear of misunderstanding. We say that $\mathbb{A}$ is finite or infinite if the universe A is finite or infinite, respectively. More generally, by the cardinality of $\mathbb{A}$, card $\mathbb{A}$, we mean the cardinality of the universe A.

Let us note that some of the sets $\mathcal{R}, \mathcal{F}, \mathcal{C}$ can be empty. If, for instance, $\mathcal{F} = \mathcal{C} = \emptyset$, then $\mathbb{A}$ has relations only and in this case $\mathbb{A}$ is called a *pure* relational system. If $\mathcal{R} = \emptyset$, that is, $\mathbb{A}$ has operations and (possibly) distinguished elements, then $\mathbb{A}$ is said to be an *algebraic system* or in shortened form, an *algebra*).

Examples

The system $\langle N, \leq\rangle$, of nonnegative integers with the usual ordering is an example of a pure relational system. It belongs to the class $\mathbb{K}_0$ of all linear orderings (i.e., systems $\langle A, \leq^A\rangle$, where the relation $\leq^A$ linearly orders the set A). We have $\mathbb{K}_0 \subseteq \mathbb{K}_1$, where $\mathbb{K}_1$ consists of all systems $\langle X, r\rangle$, where r is a binary relation on X.

Any group $\langle G, \cdot, 1\rangle$ and any ring $\langle P, +, \cdot, 0, 1\rangle$ are examples of algebraic systems. The class of all groups is contained in the class of all algebras with one binary operation and one distinguished element.

Similarly, the class of all rings is a subclass of the class of all algebras with two binary operations and two distinguished elements.

The field $\mathbb{R} = \langle R, \leq, +, \cdot, 0, 1\rangle$ of real numbers with the usual ordering is an example of a general relational system. $\mathbb{R}$ belongs to the class of all ordered fields, that is, systems

$$\langle F, \leq, +, \cdot, 0, 1\rangle$$

such that the algebra $\langle F, +, \cdot, 0, 1\rangle$ is a field, the relation $\leq$ linearly orders F and is congruent with the field operations; the latter condition means that for any $a, b \in F$,

$$a \leq b \quad \text{implies} \quad a + x \leq b + x \qquad \text{for all } x \in F.$$

and

$$a \leq b \quad \text{implies} \quad a \cdot x \leq b \cdot x, \qquad \text{for all } x \geq 0.$$

1.3. The Type of a System

For any relation r on the set A we let $\arg(r)$ denote the number of arguments of r. Similarly, we define $\arg(f)$, for any operation f on A. Thus, for any integer $n \in \omega$, we have

$$\arg(r) = n \quad \text{if and only if} \quad r \subseteq A^n$$

and

$$\arg(f) = n \quad \text{if and only if} \quad \operatorname{dom}(f) = A^n.$$

Let $\mathbb{A}$ be a structure of the form 1.2. The type $\tau = \tau(\mathbb{A})$ of the structure $\mathbb{A}$ is defined as the triple

$$\tau = \langle\langle \arg(r_i^A) \colon i \in I\rangle, \langle \arg(f_j^A) \colon j \in J\rangle, K\rangle.$$

Thus, the type of $\mathbb{A}$ says what the arity (i.e., the number of arguments) of any relation and operation of $\mathbb{A}$ is. Also, τ says whether A has distinguished elements and how they are enumerated. For example, the type of a linear ordering can be described informally as "one binary relation," that of a ring as "two binary operations and two distinguished elements," and so on.

Let us assume now that $\mathbb{A}$ and $\mathbb{B}$ are of the same type τ, that is, $\tau(\mathbb{A}) = \tau(\mathbb{B})$. If $\mathbb{A}$ is such as in 1.2, then it follows that $\mathbb{B}$ has the form

$$\mathbb{B} = \langle B, \ \{r_i^B\colon \ i \in I\}, \{f_j^B\colon \ j \in J\}, \{c_k^B\colon \ k \in K\}\rangle$$

and the equalities

1.4 $$\arg(r_i^A) = \arg(r_i^B), \qquad \arg(f_j^A) = \arg(f_j^B)$$

hold for all $i \in I$ and $j \in J$, respectively. Thus, any two structures $\mathbb{A}$ and $\mathbb{B}$ of a common type are built up in a similar way, in the sense that 1.4 holds (obviously, besides this, $\mathbb{A}$ and $\mathbb{B}$ can be totally different).

Usually, in a given branch of mathematics we investigate one particular structure (e.g., the ordered ring $\mathbb{Z}$ of integers, the field $\mathbb{C}$ of complex numbers, the three-dimensional Euclidean space) or the whole class of structures having some common properties (e.g., the class of partial orderings, the class of abelian groups, the class of rings of polynomials, the class of Banach spaces). In the latter case the structures under consideration are of some common type τ; that is, they constitute a subclass of the class $\mathbb{K}(\tau)$ consisting of all structures of type τ. We shall see later that any class $\mathbb{K}(\tau)$ has its logical language $L(\tau)$ (defined in Chapter 5), so that the formulas of $L(\tau)$ are interpretable in any structure $\mathbb{A} \in \mathbb{K}(\tau)$.

EXERCISES

1.1. Let τ be an arbitrary type. Show that the class $\mathbb{K}(\tau)$ (of all structures of type τ) contains structures of every (finite or infinite) cardinality ≥ 1. More generally, for every set $A \neq \emptyset$, there is a structure $\mathbb{A} \in \mathbb{K}(\tau)$ with the universe A.

1.2. Let A be a finite set, card $A = n$, and fix an integer $m \geq 1$.

(a) What is the number of systems of the form $\langle A, r\rangle$, with $\arg(r) \leq m$.

(b) What is the number of systems of the form $\langle A, f\rangle$, with $\arg(f) \leq m$.

2

BOOLEAN ALGEBRAS

The notion of a Boolean algebra is strictly connected with the logical calculus. It was introduced by Boole in the mid-ninteenth century and defined in full generality by Huntington in 1904; see [H3]. Later Boolean algebras were studied by Stone in the 1930s; see [S6], [S7]. The study of Boolean algebras is inspired not only by logic but also by other branches of mathematics, for example, set theory, measure theory, algebra, and topology. The main examples of Boolean operations are logical connectives (disjunction, conjunction, negation) and set theoretical operations (union, intersection, complementation).

Let us consider an algebraic system

$$\mathbb{A} = \langle A,\ +,\ \cdot,\ -,\ \mathbb{0},\ \mathbb{1} \rangle$$

in which the operations "$+$" and "$\cdot$" are binary, the operation "$-$" is unary, and the distinguished elements $\mathbb{0}$ and $\mathbb{1}$ are assumed to be distinct. A system of this form is called a *Boolean algebra* if the following conditions hold (for any $a, b, c \in A$):

2.1.
$$\begin{aligned} a + b &= b + a, & a \cdot b &= b \cdot a, \\ (a + b) + c &= a + (b + c), & (a \cdot b) \cdot c &= a \cdot (b \cdot c), \\ (a + b) \cdot c &= (a \cdot c) + (b \cdot c), & (a \cdot b) + c &= (a + c) \cdot (b + c), \\ a + \mathbb{0} &= a, & a \cdot \mathbb{1} &= a, \\ a + (-a) &= \mathbb{1}, & a \cdot (-a) &= \mathbb{0}. \end{aligned}$$

The elements $a + b$ and $a \cdot b$ are called the (*Boolean*) sum and product of the elements a, b, respectively, $-a$ is called the *Boolean complementation* of the element a, and $\mathbb{0}$ and $\mathbb{1}$ are called the *zero* and *unit*.

Example. (The power-set algebra). Let $A = P(X)$ be the family of all the subsets of a nonempty set X. The Boolean operations are defined as the usual set theoretical operations

$$a + b = a \cup b, \qquad a \cdot b = a \cap b \quad \text{and} \quad -a = X \setminus a.$$

Moreover, we put $\mathbb{0} = \emptyset$ (the empty set) and $\mathbb{1} = X$. By well-known laws of the algebra of sets, we infer that the so defined power-set algebra $P(X)$ is a Boolean algebra. More generally, any field of sets, that is, a family $R \subseteq P(X)$ containing $\emptyset$ and X and closed under union, intersection, and complementation (with respect to X) is a Boolean algebra. In particular, the two-element family $R = \{\emptyset, X\}$ is a field of sets.

Now we shall derive from the axioms 2.1 some laws that are true in any Boolean algebra.

2.2 $$a + a = a \quad \text{and} \quad a \cdot a = a.$$

Proof. We have

$$a = a + \mathbb{0} = a + [a \cdot (-a)] = (a + a)[a + (-a)] = (a + a) \cdot \mathbb{1} = a + a.$$

Similarly,

$$a = a \cdot \mathbb{1} = a \cdot [a + (-a)] = a \cdot a + a \cdot (-a) = a \cdot a. \quad \square$$

Let us note that from 2.2 we get

$$a + \mathbb{1} = \mathbb{1} \quad \text{and} \quad a \cdot \mathbb{0} = \mathbb{0}.$$

Because

$$a + \mathbb{1} = a + [a + (-a)] = (a + a) + (-a) = a + (-a) = \mathbb{1}$$

and

$$a \cdot \mathbb{0} = a \cdot (a \cdot (-a)) = (a \cdot a) \cdot (-a) = a \cdot (-a) = \mathbb{0}.$$

For arbitrary sets a, b, the conditions $a \cup b = b$ and $a \cap b = a$ are equivalent and characterize the inclusion $a \subseteq b$. Similarly, in an arbitrary Boolean algebra we have

2.3 *The conditions $a + b = b$ and $a \cdot b = a$ are equivalent.*

Proof. Multiplying both sides of the equality $a + b = b$ by a we get

$$a \cdot b = a \cdot (a + b) = (a \cdot a) + (a \cdot b) = a + (a \cdot b) = a \cdot (\mathbb{1} + b) = a \cdot \mathbb{1} = a.$$

Conversely, by adding b to both sides of $a \cdot b = a$, we obtain

$$a + b = a \cdot b + b = (a + \mathbb{1}) \cdot b = \mathbb{1} \cdot b = b. \quad \square$$

The above remarks suggest the following definition.

Definition. In any Boolean algebra we define a binary relation $\leq$ as follows:

$$a \leq b \quad \text{if and only if} \quad a + b = b.$$

By 2.3 we can write

$$a \leq b \;\text{ if and only if }\; a \cdot b = a.$$

If the algebra is a field of sets, then the relation $\leq$ coincides with the inclusion. In the general case

2.4 *the relation $\leq$ is a partial ordering.*

Proof. We have $a \leq a$, since always $a + a = a$, by 2.2. Assume that $a \leq b$ and $b \leq a$, that is, $a + b = b$ and $b + a = a$. Hence, we get immediately $a = b$. If $a \leq b$ and $b \leq c$, then

$$a + c = a + (b + c) = a + b + c = b + c = c,$$

and thus $a \leq c$. □

2.5 *From $a \leq b$ it follows that $a + x \leq b + x$ and $a \cdot x \leq b \cdot x$, for all x.*

Proof.

$$(a + x) + (b + x) = (a + b) + (x + x) = b + x$$

and

$$(a \cdot x) \cdot (b \cdot x) = (a \cdot b) \cdot (x \cdot x) = a \cdot x. \quad \square$$

From the already known laws $\mathbb{0} + a = a$ and $a + \mathbb{1} = \mathbb{1}$, we infer

$$\mathbb{0} \leq a \leq 1 \quad \text{for any } a,$$

that is, $\mathbb{0}$ is the least and $\mathbb{1}$ is the greatest element of the algebra.

Now, we prove the "lattice" property of $\leq$.

2.6 $a + b = \sup\{a, b\}$ *and* $a \cdot b = \inf\{a, b\}$.

Proof. Since $a + (a + b) = (a + a) + b = a + b$, we have $a \leq a + b$. Similarly, $b \leq a + b$. If x is such that $x \geq a$ and $x \geq b$, then

$$(a + b) + x = a + (b + x) = a + x = x,$$

that is, $a+b \leq x$. Hence, $a+b$ is the least upper bound of the set $\{a, b\}$, that is, $a+b=\sup\{a, b\}$. The other equality can be proved similarly. □

From 2.6, by an easy induction, we infer

$$a_1 + \cdots + a_n = \sup\{a_1, \ldots, a_n\}$$

$$a_1 \cdot \cdots \cdot a_n = \sup\{a_1, \ldots, a_n\}$$

for any elements $a_1, \ldots, a_n \in A$, while for infinite sets $Z \subseteq A$, the bounds $\sup Z$ and $\inf Z$ need not exist.

The following theorem characterizes the Boolean complement.

2.7 *If* $a+x=\mathbb{1}$ *and* $a \cdot x=\mathbb{0}$ *then* $x=-a$.

Proof. Using the assumption $a+x=\mathbb{1}$ we get

$$\begin{aligned}(-a)+x &= [(-a)+x] \cdot (a+x) = (-a) \cdot a + x \cdot a + (-a) \cdot x + x \cdot x \\ &= (a+(-a)) \cdot x + x = x + x = x,\end{aligned}$$

that is, $-a \leq x$. On the other hand, using $a \cdot x = \mathbb{0}$,

$$x \cdot (-a) = x \cdot (-a) + x \cdot a = x \cdot ((-a)+a) = x \cdot \mathbb{1} = x,$$

that is $x \leq -a$. Thus $x=-a$. □

Double complementation acts as identity; that is,

2.8 $$-(-a) = a.$$

Proof. In 2.7 we substitute $-a$ for a and put $x = a$ □

From 2.7 we also get $-\mathbb{1} = \mathbb{0}$ and $-\mathbb{0} = \mathbb{1}$.

The De Morgan rules known from elementary logic or set theory can be stated in Boolean terms as follows:

2.9 $-(a+b) = (-a)(-b)$ *and* $-(ab) = (-a)+(-b)$.

Proof. To obtain the first equality, substitute $x=(-a)(-b)$ in 2.8, replacing a by $a+b$. The other equality can be proved in a similar way. □

The De Morgan rules 2.9 can be generalized (by an obvious induction) to an arbitrary finite number of elements,

$$-(a_1 + \cdots + a_n) = (-a_1) \cdot \cdots \cdot (-a_n)$$

and

$$-(a_1 \cdot \cdots \cdot a_n) = (-a_1) + \cdots + (-a_n).$$

The Boolean complementation is related to the Boolean ordering in the following way:

2.10 $$a \le b \quad \textit{if and only if} \quad -b \le -a.$$

Proof. The implication to the right follows from 2.9 and the definition of $\le$. The converse implication can be obtained from the first and 2.8. □

An element $a > \mathbb{0}$ is called an *atom*, if the following is satisfied:

for every x, if $x \le a$, then $x = \mathbb{0}$ or $x = a$.

In other words, atoms are the minimal elements among nonzero elements.

2.11 *If $a > \mathbb{0}$ is an atom, then for any x we have $a \le x$ or $a \cdot x = \mathbb{0}$.*

Proof. We have $a \cdot x \le a$, and thus $a \cdot x = \mathbb{0}$ or $a \cdot x = a$, that is, $a \le x$. □

Example. In a power-set algebra $P(X)$ the atoms coincide with one-element sets $\{x\}$, where $x \in X$.

A Boolean algebra is called *atomic*, if for every element $x > \mathbb{0}$ there is an atom a, $a \le x$. For example, any power-set algebra $P(X)$ is atomic.

There are Boolean algebras without any atoms (*atomless* algebras).

EXERCISES

The Boolean difference $a - b$ is defined as the element $a \cdot (-b)$. Thus, in a field of sets $R \subseteq P(X)$ we have $a - b = a \cdot (-b) = a \cap (X \setminus b) = a \setminus b$ (the set difference).

2.1. Show that

$$a \le b \quad \text{if and only if} \quad a - b = \mathbb{0}.$$

2.2. Show that if $\mathbb{0} < a < \mathbb{1}$, then also $\mathbb{0} < -a < \mathbb{1}$.

An algebra in which every subset E of the universe has a least upper bound $\sup E$ (and a greatest lower bound $\inf E$) is called *complete*.

2.3. Show that any power-set algebra is complete and

$$\sup\{a_s: s \in S\} = \bigcup\{a_s: s \in S\},$$

$$\inf\{a_s: s \in S\} = \bigcap\{a_s: s \in S\}.$$

2.4. Let X be an infinite set. The subfamily $FC(X) \subseteq P(X)$ consisting of the subsets $a \subseteq X$ that are finite or for which the complement $X \setminus a$ is finite is a field of sets. Show that $FC(X) \neq P(X)$ and that $FC(X)$ is not a complete algebra.

2.5. Prove the De Morgan laws (in a complete algebra);

$$\sup\{a_s\colon\ s \in S\} = -\inf\{-a_s\colon\ s \in S\}$$

and

$$-\inf\{a_s\colon\ s \in S\} = \sup\{-a_s\colon\ s \in S\}.$$

Instead of completeness, one may assume the existence of bounds at the left-hand sides.

2.6. If in a given algebra (not necessarily complete) there exists the bound $\sup\{a_s\colon\ s \in S\}$, then there exist also the bounds $\sup\{a + a_s\colon\ s \in S\}$, $\sup\{a \cdot a_s\colon\ s \in S\}$, and

$$\sup\{a + a_s\colon\ s \in S\} = a + \sup\{a_s\colon\ s \in S\}$$

and

$$\sup\{a \cdot a_s\colon\ s \in S\} = a \cdot \sup\{a_s\colon\ s \in S\}.$$

Analogous rules hold for the infima.

For an arbitrary metric space (more generally, topological space) X, the family $B(X)$ of open–closed sets (i.e., the sets that are both open and closed) is a field of sets. The equality $B(X) = \{\emptyset, X\}$ holds for connected spaces and $B(X) = P(X)$ for discrete spaces. For an infinite set S let $X = \{0,1\}^S$; that is, X consists of all the zero–one functions defined on the set S. Any finite sequence $\varepsilon = \langle \varepsilon(s_1), \ldots, \varepsilon(s_n)\rangle$ of zeros and ones determines the set $U_\varepsilon = \{x \in X\colon\ \varepsilon \subseteq x\}$ consisting of all the $x \in X$, for which $x(s_1) = \varepsilon(s_1), \ldots, x(s_n) = \varepsilon(s_n)$. The sets U_ε form a basis of a topology on X in which they are open–closed. The space $X = X(S)$ defined in this way is called the Cantor cube of the weight card S.

2.7. Show that the field $B(X)$ for a Cantor cube X has no atoms.

2.8. Show that the field $B(X)$ is incomplete. (*Hint:* for any infinite $I \subseteq S$, the set $\{x \in X\colon\ \forall i \in I(x(i) = 1)\}$ is nowhere dense in X).

2.9. Prove the converse to 2.11.

2.10. Prove that every finite Boolean algebra is atomic.

2.11. The family of finite unions of intervals of the form

$$[a, b) = \{x \in \mathbb{R}\colon\ a \leq x < b\},$$

where a, b are real numbers (including $a = -\infty, b = +\infty$) is a noncomplete atomless field of sets.

3

SUBSYSTEMS AND HOMOMORPHISMS

Relational structures of a common type τ can be related to each other in some way. Fundamental relationships of this kind are that of containment and homomorphism. Both notions are well known in the case of such structures as groups or rings. Here, we generalize them to arbitrary relational systems.

Let $\mathbb{A}$ and $\mathbb{B}$ be relational systems of the same type

$$\mathbb{A} = \langle A, \{r_i^A: i \in I\}, \{f_j^A: j \in J\}, \{c_k^A: k \in K\}\rangle,$$

$$\mathbb{B} = \langle B, \{r_i^B: i \in I\}, \{f_j^B: j \in J\}, \{c_k^B: k \in K\}\rangle.$$

3.1. *Definition.* We say that $\mathbb{A}$ is a *subsystem* of $\mathbb{B}$, $\mathbb{A} \subseteq \mathbb{B}$, if $A \subseteq B$ and

$$r_i^A = r_i^B \cap A^{\arg(r_i)}, \quad \text{for every } i \in I,$$

that is

$$r_i^A(a_1, \ldots, a_{n_i}) \quad \text{if and only if} \quad r_i^B(a_1, \ldots, a_{n_i})$$

for any $a_1, \ldots, a_{n_i} \in A$, where $n_i = \arg(r_i)$;

$$f_j^A = f_j^B | A, \quad \text{for } j \in J,$$

that is,

$$f_j^A(a_1, \ldots, a_{m_j}) = f_j^B(a_1, \ldots, a_{m_j})$$

for any $a_1, \ldots, a_{m_j} \in A$, where $m_j = \arg(f_j)$;

$$c_k^A = c_k^B \quad \text{for every } k \in K.$$

From the definition it follows that if $\mathbb{A} \subseteq \mathbb{B}$, then the set A is closed under all the operations f^B of the system $\mathbb{B}$ and contains all the distinguished elements. Every subset $B_0 \subseteq B$ of the universe of $\mathbb{B}$ that has these properties determines, in a natural way, a subsystem $\mathbb{B}_0 \subseteq \mathbb{B}$, where

$$\mathbb{B}_0 = \langle B_0,\ \{r_i^B \cap B_0^{\arg(i)}\colon i \in I\}, \{f_j^B | B_0\colon\ j \in J\}, \{c_k^B\colon\ k \in K\}\rangle.$$

3.2. Generating a Subsystem

Let $E \subseteq A$ be an arbitrary subset of the universe A. We define by induction a sequence of sets $E_0 \subseteq E_1 \subseteq, \ldots$, as follows:

$$E_0 = E \cup \{c_k^A\colon\ k \in K\},$$

$$E_{n+1} = E_n \cup \{f_j^A(a_1, \ldots, a_{m_j})\colon\ a_1, \ldots, a_{m_j} \in E_n,\ j \in J\},$$

where, as usual, $m_j = \arg(f_j^A)$. Then the set $E^* = \bigcup\{E_n\colon\ n \in N\}$ is the smallest set such that $E \subseteq E^*$, $c_k^A \in E^*$, for $k \in K$, and E^* is closed under all the operations f_j^A. Since, if $a_1, \ldots, a_{m_j}$ are in E^* then $a_1, \ldots, a_{m_j} \in E_n$, for some n, and $f_j^A(a_1, \ldots, a_{m_j}) \in E_{n+1} \subseteq E^*$. Therefore E^* determines a subsystem $\mathbb{A}[E] \subseteq \mathbb{A}$, whose universe is E^* and which is called the subsystem *generated* by E.

3.3. Homomorphism

Let $\mathbb{A}$, $\mathbb{B}$ be relational systems of a common type. A function $h\colon A \longrightarrow B$ is called a *homomorphism* (of the system $\mathbb{A}$ into the system $\mathbb{B}$) if the following conditions hold:

$$\text{if } r_i^A(a_1, \ldots, a_n) \quad \text{then} \quad r_i^B(h(a_1), \ldots, h(a_n)),$$

for every $i \in I$ and $a_1, \ldots, a_n \in A$, $n = \arg(r_i^A)$;

$$h(f_j^A(a_1, \ldots, a_m)) = f_j^B(h(a_1), \ldots, h(a_m)),$$

for every $j \in J$ and $a_1, \ldots, a_m \in A$, $m = \arg(f_j^A)$;

$$h(c_k^A) = c_k^B \quad \text{for every } k \in K.$$

This is a generalization of the well-known notion of a homomorphism of groups or rings to the case of arbitrary relational systems.

Let $h\colon \mathbb{A} \longrightarrow \mathbb{B}$ be a homomorphism. The image $h[A] \subseteq B$ contains all the distinguished elements c_k^B, since $c_k^B = h(c_k^A)$. It is also closed under the operations f_j^B. Since, if $b_1, \ldots, b_m \in h[A]$, where $m = \arg(f_j^B)$, then there are $a_1, \ldots, a_m \in A$ such that

$$h(a_1) = b_1,\ \ldots,\ h(a_m) = b_m.$$

Consequently,

$$f_j^B(b_1, \ldots, b_m) = f_j^B(h(a_1), \ldots, h(a_m)) = h(f_j^A(a_1, \ldots, a_m)) \in h[A].$$

Thus, $h[A]$ determines a subsystem $h[\mathbb{A}] \subseteq \mathbb{B}$, which is called the *image* of the system $\mathbb{A}$.

Under a homomorphism h: $\mathbb{A} \longrightarrow \mathbb{B}$, the properties of the relations may change drastically. Therefore, the notion of a homomorphism is applied mostly to algebraic systems. In the general case some stronger notions are sometimes introduced. For example, call a given homomorphism h: $\mathbb{A} \longrightarrow \mathbb{B}$ *strong* if it satisfies the following additional condition:

$$\textit{if } r_i^B(h(a_1), \ldots, h(a_n)), \textit{ then there are } a_1', \ldots, a_n' \in A \textit{ such that}$$
$$h(a_1') = h(a_1), \ldots, h(a_n') = h(a_n) \quad \text{and} \quad r_i^A(a_1', \ldots, a_n').$$

Of course, in the case of algebraic systems, both notions coincide.

A one-to-one strong homomorphism is called a *monomorphism* or an *embedding*. Thus, a one-to-one homomorphism h: $\mathbb{A} \longrightarrow \mathbb{B}$ is an embedding if the implication in the first condition of the definition in Section 3.3 is strengthened to the equivalence

$$r^A(a_1, \ldots, a_n) \quad \text{if and only if} \quad r^B(h(a_1), \ldots, h(a_n)).$$

If, additionally, $h[A] = B$, that is, if h is a monomorphism onto $\mathbb{B}$, then h is called an *isomorphism*. It follows that a monomorphism h: $\mathbb{A} \longrightarrow \mathbb{B}$ is an isomorphism from $\mathbb{A}$ onto the image $h[\mathbb{A}] \subseteq \mathbb{B}$.

3.4. Filters

Let $\mathbb{A}$ be a Boolean algebra. A proper subset $F \subseteq A$ is called a *filter* if the following two conditions hold:

$$\text{if } a \in F \text{ and } b \geq a, \text{ then } b \in F,$$

and

$$\text{if } a, b \in F, \text{ then also } a \cdot b \in F.$$

Example. Let h: $\mathbb{A} \longrightarrow \mathbb{B}$ be a homomorphism of Boolean algebras. Then

$$F(h) = \{a \in A\colon\ h(a) = \mathbb{1}\}$$

is a filter. In fact, if $a \leq b$, that is, $a + b = b$, then

$$h(b) = h(a + b) = h(a) + h(b), \quad \text{that is} \quad h(a) \leq h(b).$$

Hence, if $a \in F(h)$ and $a \leq b$, then we have $h(b) \geq h(a) = \mathbb{1}$, whence $h(b) = \mathbb{1}$, that is, $b \in F(h)$. If $a, b \in F(h)$, then $h(a \cdot b) = h(a) \cdot h(b) = \mathbb{1} \cdot \mathbb{1} = \mathbb{1}$, that is, $a \cdot b \in F(h)$. Finally, $\mathbb{O} \notin F(h)$, since $h(\mathbb{O}) = \mathbb{O} \neq \mathbb{1}$.

Every element $a > \mathbb{O}$ generates a filter

$$F_a = \{b \in A\colon\ b \geq a\}.$$

A filter of this form is called *principal*.

The family R of all the filters in a given algebra is partially ordered by inclusion. If $R_0 \subseteq R$ is a linearly ordered subfamily, then $F_0 = \bigcup R_0$ is a filter. Since, if $a \in F_0$ and $b \geq a$, then we have $a \in F$ for some $F \in R_0$, and thus $b \in F$, whence $b \in F_0$. Similarly, if $a, b \in F_0$ then there are $F_1, F_2 \in R_0$ such that $a \in F_1$ and $b \in F_2$. Since $F_1 \subseteq F_2$ or $F_2 \subseteq F_1$ we infer that both a and b belong to the greater one. Then $a \cdot b$ also belongs to that filter and hence to F_0. Obviously, the filter F_0 bounds the family R_0 from above. In view of the Kuratowski-Zorn lemma we obtain the following theorem.

3.5. Theorem. *Every filter can be extended to a maximal filter.*

Maximal filters are called also *ultrafilters*. Note the following fact:

3.6. *If $a > \mathbb{O}$, then there is an ultrafilter p such that $a \in p$.*

Proof. We have $a \in F_a$ and in view of Theorem 3.5, the filter F_a can be extended to an ultrafilter p. □

EXERCISES

3.1. A subsystem of a Boolean algebra is a subalgebra, that is, it is a Boolean algebra.

3.2. A subsystem of a group (ring) is not necessarily a subgroup (subring). Expand the type by adding a unary operation so that a subsystem of a group is a group. Expand similarly for rings.

3.3. If the sets X, Y are of the same cardinality, then the algebras $P(X)$, $P(Y)$ are isomorphic. Every two-element Boolean algebra is isomorphic to $P(X)$, where X is a one-element set.

3.4. Let card $X \geq$ card Y. Thus there is a mapping g of the set X onto Y. Show that the function $h(a) = g^{-1}[a]$, for $a \subseteq Y$, is a monomorphism of the algebra $P(Y)$ into $P(X)$.

3.5. Let $Y \subseteq X$. Show that the mapping

$$h(a) = a \cap Y, \quad \text{for } a \subseteq X,$$

is a homomorphism of the algebra $P(X)$ into $P(Y)$. Find the filter $F(h)$.

3.6. Let $e > \mathbb{0}$ be a fixed element of the universe of a Boolean algebra. Define a new Boolean algebra

$$\mathbb{A}(e) = \langle A(e),\ +_e, \cdot_e, -_e, \mathbb{0}, e \rangle,$$

where $A(e) = \{a \in A: a \leq e\}$, the operations "$+_e$" and "$\cdot_e$" are the restrictions of the operations "$+$", "$\cdot$" of the algebra $\mathbb{A}$ to the subset $A(e)$, and $-_e a = e - a = e \cdot (-a)$. If $e = \mathbb{1}$, then $\mathbb{A}(e) = \mathbb{A}$. If $\mathbb{A} = P(X)$, then $\mathbb{A}(e) = P(e)$. Show that the function

$$h(a) = a \cdot e, \quad \text{for } a \in A,$$

is a homomorphism of $\mathbb{A}$ onto $\mathbb{A}(e)$. Find the filter $F(h)$.

In a Boolean algebra we define the symmetric difference as

$$a \triangle b = (a - b) + (b - a).$$

3.7. Show that,

$$a = b \quad \text{if and only if} \quad a \triangle b = \mathbb{0}.$$

3.8. Prove that a homomorphism $h\colon \mathbb{A} \longrightarrow \mathbb{B}$ is an embedding if and only if $F(h) = \{\mathbb{1}\}$.

3.9. A family $R \subseteq A$ is said to be *centered*, if for every finite set $\{a_1, \ldots, a_n\} \subseteq R$, $a_1 \cdot \cdots \cdot a_n > \mathbb{0}$. Show that every centered family can be extended to a filter.

3.10. A filter p is maximal if and only if for every $a \in A$ we have either $a \in p$ or $-a \in p$.

3.11. A principal filter $F_a = \{b \in A\colon\ b \geq a\}$ is maximal if and only if the generator a is an atom.

3.12. Let $h\colon \mathbb{A} \longrightarrow \mathbb{B}$ be a homomorphism of Boolean algebras. Show that the filter $F(h) = \{a\colon\ h(a) = \mathbb{1}\}$ is maximal if and only if $h[A] = \{\mathbb{0}, \mathbb{1}\}$; that is, the image $h[\mathbb{A}]$ is a two-element algebra.

Prime filters. We say that a filter F in $\mathbb{A}$ is *prime* if it satisfies the condition

$$a + b \in F \quad \text{implies} \quad a \in F \quad \text{or} \quad b \in F \quad \text{for any } a, b \in A.$$

3.13. Prove that a filter F is prime if and only if F is an ultrafilter. (*Hint:* Use De Morgan rules).

A nonempty subset $I \subseteq A$ of a Boolean algebra $\mathbb{A}$ is called an *ideal* if it has the following properties:

(a) $a \in I$ and $b \leq a$ imply $b \in I$

(b) $a, b \in I$ implies $a + b \in I$.

3.14. If h: $\mathbb{A} \longrightarrow \mathbb{B}$ is a homomorphism (of Boolean algebras $\mathbb{A}$ and $\mathbb{B}$) then $I(h) = \{a \in A\colon\ h(a) = \mathbb{O}\}$ is an ideal.

3.15. If I is an ideal, then the set $-I(=\{-a\colon\ a \in I\})$ is a filter and conversely, if F is a filter then $-F$ is an ideal.

3.16. The family J of finite subsets of a given infinite set X is an ideal in the power-set algebra $P(X)$.

4

OPERATIONS ON RELATIONAL SYSTEMS

Let τ be a fixed type and $\mathbb{K}(\tau)$ be the class of all the relational systems of type τ. In this chapter we shall describe a few constructions, which applied to a system (or to a family of systems) of the class $\mathbb{K}(\tau)$ yield as a result a new system of the same class.

4.1. Congruences

An equivalence relation $\simeq$ on the universe A of $\mathbb{A}$ is called a *congruence* if the following holds

$$(*)\qquad \begin{array}{l} \text{if } a_1 \simeq a'_1, \ldots, a_m \simeq a'_m \text{ then } f_j^A(a_1, \ldots, a_m) \simeq f_j^A(a'_1, \ldots, a'_m) \\ \text{for every } j \in J \text{ and arbitrary } a_1, a'_1, \ldots, a_m, \ a'_m \in A, \ m = arg(f_j^A). \end{array}$$

Example. Let H be a normal subgroup of a group G, that is, $aHa^{-1} = H$, for $a \in G$. A simple checking shows that the relation

$$a =_H b \ \equiv \ a \cdot b^{-1} \in H \ \equiv \ aH = bH$$

is a congruence in G. Similarly, if I is an ideal in a ring, then the relation $=_I$ defined by the equivalence

$$a =_I b \ \equiv \ a - b \in I$$

is a congruence.

Let $\simeq$ be a congruence in $\mathbb{A}$. We construct the factor system $\mathbb{B} = \mathbb{A}/\simeq$. Let us denote

$$[a] = \{b \in A\colon \ b \simeq a\} \quad \text{(the equivalence class of } \simeq).$$

The new universe B consists of all the equivalence classes

$$B = A/\simeq = \{[a]\colon\ a \in A\}.$$

The relations r_i^B and operations f_j^B are defined as follows:

$$r_i^B([a_1], \ldots, [a_n]) \equiv r_i^A(a_1', \ldots, a_n'), \quad \text{for some } a_1' \simeq a_1, \ldots, a_n' \simeq a_n,$$

$$f_j^B([a_1], \ldots, [a_n]) = [f_j^A(a_1, \ldots, a_n)].$$

The condition $(*)$ ensures that the f^Bs are well defined (it does not depend on the choice of the representatives $a_1, \ldots, a_n$).

The distinguished elements c_k^B are defined as the equivalence classes

$$c_k^B = [c_k^A], \quad \text{for } k \in K.$$

We check easily that the mapping $h(a) = [a]$, for $a \in A$, is a (strong) homomorphism from $\mathbb{A}$ onto $\mathbb{B}$.

Now we shall apply this construction to Boolean algebras. Let F be a filter in an algebra $\mathbb{A}$. We set

4.2 $\qquad a =_F b$ if and only if $a \cdot x = b \cdot x \quad$ for some $x \in F$.

The relation $=_F$ is a congruence; $a =_F a$ is obtained from 4.2 with $x = \mathbb{1}$, the symmetry is evident, and from $a =_F b$ and $b =_F c$ we infer $a \cdot x = b \cdot x$ and $b \cdot y = c \cdot y$ for some $x, y \in F$. Thus, $a \cdot z = c \cdot z$ for $z = x \cdot y$, that is, $a =_F c$. If $a =_F a'$ and $b =_F b'$ then we have $a + b =_F a' + b'$ since from $a \cdot x = a' \cdot x$ and $b \cdot y = b' \cdot y$ for some $x, y \in F$, we obtain $a \cdot z = a' \cdot z$ and $b \cdot z = b' \cdot z$ for $z = x \cdot y$, whence $(a + b) \cdot z = (a' + b') \cdot z$ [and also $(a \cdot b) \cdot z = (a' \cdot b') \cdot z$].

Also the condition $a =_F a'$ implies $-a =_F -a'$, since from $a \cdot x = a' \cdot x$ for an $x \in F$ and from the De Morgan laws we infer $(-a) + (-x) = (-a') + (-x)$; multiplying the last equality by x we get $(-a) \cdot x = (-a') \cdot x$.

The factor algebra $\mathbb{B} = \mathbb{A}/{=_F}$ is denoted in a short form by $\mathbb{A}/F$ (or $\mathbb{A}$ mod F).

4.3. Products

For two systems $\mathbb{A}, \mathbb{B}$ the *product* of $\mathbb{A}, \mathbb{B}$ is defined as follows:

$$\mathbb{A} \times \mathbb{B} = \langle A \times B;\ \{r^{A \times B}\}, \{f^{A \times B}\}, \{c^{A \times B}\}\rangle,$$

where

$$r_i^{A \times B}(\langle a_1, b_1\rangle, \ldots, \langle a_n, b_n\rangle) \equiv (r_i^A(a_1, \ldots, a_n) \text{ and } r_i^B(b_1, \ldots, b_n)),$$

$$f_j^{A \times B}(\langle a_1, b_1\rangle, \ldots, \langle a_n, b_n\rangle) = \langle f_j^A(a_1, \ldots, a_n), f_j^B(b_1, \ldots, b_n)\rangle,$$

$$c_k^{A \times B} = \langle c_k^A, c_k^B\rangle.$$

It is easy to generalize this definition to the case of any finite or infinite family of systems of the class $\mathbb{K}(\tau)$. Let $\{\mathbb{A}_s\colon\ s \in S\}$ be an indexed family of systems of type τ. We construct a product $\mathbb{A} = \prod\{\mathbb{A}_s\colon\ s \in S\}$ as follows: $A = \prod\{A_s\colon\ s \in S\}$, that is, A consists of all functions $a\colon\ S \longrightarrow \bigcup\{A_s\colon\ s \in S\}$, for which $a(s) \in A_s$ for every $s \in S$, and the relations r_i^A, operations f_j^A, and distinguished elements c_k^A are defined as follows:

$$r_i^A(a_1, \ldots, a_n) \equiv r_i^{A_s}(a_1(s), \ldots, a_n(s)), \quad \text{for every } s \in S,$$

$$f_j^A(a_1, \ldots, a_n)(s) = f_j^{A_s}(a_1(s), \ldots, a_n(s)), \quad \text{for each } s \in S,$$

$$c_k^A(s) = c_k^{A_s}, \quad \text{for each } s \in S.$$

Any product of groups (rings) is a group (ring). We check easily that a product $\mathbb{A} = \prod\{\mathbb{A}_s\colon\ s \in S\}$ of Boolean algebras is a Boolean algebra.

4.4. Directed Sum

Let S be a set linearly ordered by the relation $\leq$. Assume that $\{\mathbb{A}_s\colon\ s \in S\}$ is a family of systems of type τ satisfying the condition

$$\textit{if } s \leq t, \textit{ then } \mathbb{A}_s \subseteq \mathbb{A}_t, \quad \textit{for any } s, t \in S.$$

In this case we define a directed sum $\mathbb{A} = \bigcup\{\mathbb{A}_s\colon\ s \in S\}$. The universe A is the union of the universes, $A = \bigcup\{A_s\colon\ s \in S\}$, and the relations r_i^A, operations f_j^A and distinguished elements c_k^A are defined as follows:

$$r_i^A(a_1, \ldots, a_n) \equiv r_i^{A_s}(a_1, \ldots, a_n),$$

for any s such that $a_1, \ldots, a_n \in A_s$,

$$f_j^A(a_1, \ldots, a_m) = f_j^{A_s}(a_1, \ldots, a_m),$$

for any s such that $a_1, \ldots, a_m \in A_s$,

$$c_k^A = c_k^{A_s} \quad \text{for any } s.$$

We check now that all these objects are well defined. Let $a_1, \ldots, a_n \in A$. Since $A = \bigcup\{A_s\colon\ s \in S\}$, for some $s_1, \ldots, s_n \in S$ we have $a_1 \in A_{s_1}, \ldots, a_n \in A_{s_n}$. Let s be the largest among the indices $s_1, \ldots, s_n$. Then $\mathbb{A}_{s_1}, \mathbb{A}_{s_2}, \ldots, \mathbb{A}_{s_n} \subseteq \mathbb{A}_s$, by the assumption, and thus $a_1 \ldots, a_n \in A_s$. So, every finite collection $a_1, \ldots, a_n$ of elements of the universe of $\mathbb{A}$ lies in the universe of some $\mathbb{A}_s$. If $\mathbb{A}_s$, $\mathbb{A}_t$ are two systems with that property, then $s \leq t$ or $t \leq s$, whence $\mathbb{A}_s \subseteq \mathbb{A}_t$ or $\mathbb{A}_t \subseteq \mathbb{A}_s$, respectively. In either case,

$$r_i^{A_s}(a_1, \ldots, a_n) \ \text{ if and only if } \ r_i^{A_t}(a_1, \ldots, a_n) \quad \text{for } i \in I$$

and

$$f_j^{A_s}(a_1, \ldots, a_m) = f_j^{A_t}(a_1, \ldots, a_m) \quad \text{for } j \in J.$$

Thus, $r_i^{A_s}$ holds (or does not hold) for the elements $a_1, \ldots, a_n$ simultaneously for all s for which $a_1, \ldots, a_n \in A_s$. Similarly, the value of the function $f_j^{A_s}(a_1, \ldots, a_m)$ is the same in all the systems $\mathbb{A}_s$ for which $a_1, \ldots, a_m \in A_s$. Also the corresponding distinguished elements are the same; $c_k^{A_s} = c_k^{A_t}$ for all $s, t \in S$.

Directly from the definition it follows that all the systems $\mathbb{A}_s$ are subsystems of the sum $\mathbb{A} = \bigcup\{\mathbb{A}_s\colon\ s \in S\}$;

$$\mathbb{A}_s \subseteq \mathbb{A}, \qquad \text{for every } s \in S.$$

We define the directed sum $\mathbb{A} = \bigcup\{\mathbb{A}_s\colon\ s \in S\}$ also in a more general situation. Notice that in the construction just given it suffices to assume that the set S is directed, that is, that the relation $\leq$ partially orders S in such a way that for every finite collection $s_1, \ldots, s_n \in S$ there exists an $s \in S$ such that $s_1, \ldots, s_n \leq s$. The assumption

$$\textit{if } s \leq t, \textit{ then } \mathbb{A}_s \subseteq \mathbb{A}_t \quad \textit{for all } s \leq t,$$

remains unchanged as well as the definition of the system $\mathbb{A} = \bigcup\{\mathbb{A}_s\colon\ s \in S\}$. Since the set S is directed, it follows that any finite collection $a_1, \ldots, a_n \in A$ lies in some A_s, and if s_1, s_2 are two indices with that property, then for any $s \geq s_1, s_2$ we have the equivalence

$$r_i^{A_{s1}}(a_1, \ldots, a_n) \equiv r_i^{A_s}(a_1, \ldots, a_n),$$

and also

$$r_i^{A_{s2}}(a_1, \ldots, a_n) \equiv r_i^{A_s}(a_1, \ldots, a_n),$$

because $\mathbb{A}_{s_1}, \mathbb{A}_{s_2} \subseteq \mathbb{A}_s$. Therefore

$$r_i^{A_{s1}}(a_1, \ldots, a_n) \equiv r_i^{A_{s2}}(a_1, \ldots, a_n),$$

which proves that the relations r^A are well defined. Also, the values of the operations $f^{(A_{s_1})}, f^{(A_{s_2})}$ for the arguments $a_1, \ldots, a_m$ are the same since they are equal to $f^{A_s}(a_1, \ldots, a_m)$. Finally, for arbitrary s_1, s_2 we have $c^{A_{s_1}} = c^{A_{s_2}}$, since taking $s \geq s_1, s_2$ we find that $c^{A_{s_1}} = c^{A_s}$ and $c^{A_{s_2}} = c^{A_s}$. This proves that the sum is well defined in the case of a directed set S. Obviously also here we have $\mathbb{A}_s \subseteq \mathbb{A}$, for every $s \in S$.

EXERCISES

4.1. Let H be a normal subgroup of a group G. Check that the relation $a =_H b \equiv a \cdot b^{-1} \in H$ is a congruence and that $G \bmod =_H$ coincides with the factor group G/H (in the usual sense). Check similarly for rings (see Example in Section 4.1).

4.2. Let F be a filter in a Boolean algebra $\mathbb{A}$. Check that the unit $[\mathbb{1}]$ of the factor algebra $\mathbb{A}/F$ is identical with F. Show that

$$[a] \leq [b] \quad \text{in} \quad \mathbb{A}/F \quad \text{if and only if} \quad -a+b \in F.$$

4.3. Prove that every finite Boolean algebra $\mathbb{A}$ is isomorphic to a power-set algebra $P(X)$. (*Hint*: Let X be the set of all atoms of $\mathbb{A}$; cf. Exercise 2.10, and put $h(x) = \{a \in X\colon\ a \leq x\}$).

4.4. Any two finite Boolean algebras with the same number of elements are isomorphic.

4.5. Prove that any finite Boolean algebra can be embedded into any infinite one.

4.6. Let $h\colon\ \mathbb{A} \longrightarrow \mathbb{B}$ be a homomorphism of the relational system $\mathbb{A}$ onto $\mathbb{B}$. Check that the relation $a =_h b \equiv h(a) = h(b)$ is a congruence. If h is a strong homomorphism then the factor relational system $\mathbb{A}/{=_h}$ is isomorphic with $\mathbb{B}$.

4.7. Let $\mathbb{A} = \prod\{A_s\colon\ s \in S\}$ be a product of relational systems. Define the projections $\pi_s\colon\ \mathbb{A} \longrightarrow \mathbb{A}_s$ by the equality $\pi_s(a) = a(s)$ for $a \in A$. Check that the projection π_s is a homomorphism from $\mathbb{A}$ onto $\mathbb{A}_s$. If $r^{A_s} \neq \emptyset$ for every $s \in S$ and every relation r of the systems A_s then the homomorphism π_s is strong.

4.8. The power-set Boolean algebra $P(S)$ is isomorphic with the product $\prod\{\mathbb{A}_s\colon\ s \in S\}$, where $\mathbb{A}_s = \{\mathbb{0}, \mathbb{1}\}$ for every $s \in S$.

4.9. If $\mathbb{A}$ is a Boolean algebra and $e_1, \ldots, e_n$ are such that $\mathbb{1} = e_1 + \cdots + e_n$ and $e_i \cdot e_j = \mathbb{0}$ for $i, j = 1, \ldots, n$ then $\mathbb{A}$ is isomorphic to the product $A(e_n) \times \cdots \times A(e_n)$ (see Exercise 3.6).

4.10. For a given relational system $\mathbb{A}$, let S be the family of all finite subsets of the universe A. The set S is directed by the inclusion. For $s \in S$ let $\mathbb{A}[s] \subseteq \mathbb{A}$ be the subsystem generated by the set s. Check that $\mathbb{A}$ is the directed sum $\mathbb{A} = \bigcup\{\mathbb{A}[s]\colon\ s \in S\}$.

4.11. Let $\mathbb{A}$ be a directed sum, $\mathbb{A} = \bigcup\{\mathbb{A}_s\colon\ s \in S\}$. Let us construct the product $\mathbb{A}^* = \prod\{\mathbb{A}_s\colon\ s \in S\}$. A function $a \in A^*$ is called *eventually constant* if the following holds:

$$\exists t \forall s \geq t [a(s) = a(t)].$$

The subset $B \subseteq A^*$ of eventually constant functions contains the distinguished elements and is closed under the operations of the product $\mathbb{A}^*$—thus, it determines a subsystem $\mathbb{B} \subseteq \mathbb{A}^*$. Find a homomorphism from $\mathbb{B}$ onto $\mathbb{A}$.

5

TERMS AND FORMULAS

In this chapter we define the basic notions of syntax—a logical (or a formal) language—its terms and formulas. Further, we discuss the truth relation introduced by Tarski in the 1930s; see [T2]. The Tarski truth relation is a bridge connecting semantics and syntax. Let us mention also that the notion of a formal language comes essentially from Peano, who defined and used such a language for the theory of integers in the 1880s.

5.1. Logical Language

Mathematical practice shows that properties of specified objects under consideration (numbers, vectors, points of a space, etc.) can always be expressed in a uniform way. Compound properties can be obtained from simpler ones by means of such syntactical operations as connectives and quantifiers. Therefore, all the properties will be fully determined, once the *atomic* (or *primitive*) properties are chosen.

Given a relational structure $\mathbb{A} = \langle A, \{r_i^A\colon i \in I\}, \{f_j^A\colon j \in J\}, \{c_k^A\colon k \in K\}\rangle$ of some type τ (in which the universe A consists of our objects of consideration), it is natural to choose as the primitive properties the simplest relational and operational connections in $\mathbb{A}$. That is to say, the atomic properties (of $\mathbb{A}$) are roughly of the form $r_i^A(a_1, \ldots, a_n)$ and $f_j^A(a_1, \ldots, a_m) = a$. This observation allows us to define a logical language referring to $\mathbb{A}$ as well as to any structure of type τ.

Definition. Let $\mathbb{K} = \mathbb{K}(\tau)$ be the class of all relational structures of a given type τ. The language $L = L(\tau)$ of type τ [or of the class $\mathbb{K}(\tau)$] is defined as a sequence of arbitrary pairwise disjoint sets

$$L(\tau) = \langle R, F, C, X, S\rangle,$$

in which the sets R, F, C are enumerated according to τ,

$$R = \{r_i\colon i \in I\}, \qquad F = \{f_j\colon j \in J\}, \qquad C = \{c_k\colon k \in K\};$$

the set X is countably infinite and

$$S = \{=, \neg, \rightarrow, \forall\}$$

is a four-element set.

The elements of the sets R, F, C are called, respectively, the *relation symbols*, *operation (function) symbols*, and *constant symbols*. The elements of X are called *variables*. The set S is the set of *logical signs*. We assume that the above enumerations are one-to-one, that is, $r_{i'} \neq r_{i''}$, whenever $i' \neq i''$ and similarly for the fs and the cs.

Our intention is that the relation symbol r_i denote the relation r_i^A, for any $\mathbb{A} \in \mathbb{K}(\tau)$ and $i \in I$. Similarly, the operation symbol f_j will to denote the operation f_j^A, and the constant symbol c_k denotes the distinguished element c_k^A. On the other hand, any variable $x \in X$ may denote an arbitrary element of the universe of an $\mathbb{A} \in \mathbb{K}$. Finally, the meaning of the elements of S is clear from their notation: $=$ is always interpreted as the identity, $\neg$ as the negation, $\rightarrow$ as the implication, and $\forall$ as the universal quantifier. Although some of the symbols such as $=$ or $\forall$ may be used in the current text as the usual abbreviations, there will be no misunderstandings.

Why have we omitted the existential quantifier $\exists$ and the other connectives such as the disjunction $\vee$, the conjunction $\wedge$, and the equivalence $\equiv$? Because this will make the forthcoming definitions and proof shorter—there will be fewer cases of inductive proofs. Note that we do not lose anything, since the missing symbols are definable from those already included in S.

We do not specify what exactly the sets R, F, C, X, S are. The nature of their elements is completely inessential. Replacing, for example, the set R by a set R' of the same cardinality we obtain another copy of the same language. Note that the cardinalities of the sets R, F, C are determined by the type τ. The sets X and S are independent of τ—we may assume that they are the same in every language L.

The language $L = L(\tau)$, as defined here, consists solely of the symbols from which the expressions of L can be built. In subsequent sections we define two kinds of such expressions: terms (to denote elements of a universe) and formulas (to express properties of elements).

5.2. Terms

As mentioned in the foregoing, by *terms* of the language $L = L(\tau)$ we mean those expressions which can be interpreted as elements of the universes [of the systems of the class $\mathbb{K}(\tau)$]. Thus, the variables and the constants are terms. If there are any operation symbols f in L, then the finite sequence $\langle f, t_1, \ldots, t_n\rangle$, where $t_1, \ldots, t_n$ are terms, can be interpreted in a system $\mathbb{A}$ as the result of applying the operation f^A to the elements corresponding to the terms $t_1, \ldots, t_n$, and hence $\langle f, t_1, \ldots, t_n\rangle$ can be regarded as a term.

More exactly, the set $\mathrm{Tm} = \mathrm{Tm}(L)$ for language L is defined by induction as follows:

$$\mathrm{Tm} = \bigcup\{\mathrm{Tm}_l\colon\ l \in N\},$$

where $\mathrm{Tm}_0 = X \cup C$ and

$$\mathrm{Tm}_{l+1} = \mathrm{Tm}_l \cup \{\langle f_j,\ t_1, \ldots, t_m \rangle:\ j \in J,\ t_1, \ldots, t_m \in \mathrm{Tm}_l\}.$$

Obviously, $m = \arg(f_j)$ in this definition. Thus, the terms are obtained from the variables and the constants by applying repeatedly the operation of forming finite sequences of the form $\langle f_j,\ t_1, \ldots, t_m \rangle$. According to the intended meaning we shall use a more suggestive notation: the sequence $\langle f,\ t_1, \ldots, t_m \rangle$ will be denoted by $f(t_1, \ldots, t_m)$.

5.3. Theorem (On induction for terms). *If $Z \subseteq \mathrm{Tm}$ is a set satisfying the conditions $X \cup C \subseteq Z$ and for every $j \in J$, if $t_1, \ldots, t_m \in Z$, $m = \arg(j)$, then $f_j(t_1, \ldots, t_m) \in Z$, then $Z = \mathrm{Tm}$.*

Proof. We apply induction. The first assumption reads $\mathrm{Tm}_0 \subseteq Z$. Assume that $\mathrm{Tm}_l \subseteq Z$. Then the second assumption implies $\mathrm{Tm}_{l+1} \subseteq Z$. Thus, $\mathrm{Tm}_l \subseteq Z$ for every $l \in N$, whence $\mathrm{Tm} = \bigcup\{\mathrm{Tm}_l:\ l \in N\} = Z$. □

From Theorem 5.3 it follows that in order to prove that every term has a given property, it suffices to prove that the variables and the constants have this property and that it holds for $f(t_1, \ldots, t_m)$ whenever it holds for $t_1, \ldots, t_m$.

It follows also that if we want to define by induction a function g: $\mathrm{Tm} \longrightarrow W$ (for some set W), it is sufficient to define the values $g(x)$ and $g(c)$, for all $x \in X$ and $c \in C$, and to express the value $g(f(t_1, \ldots, t_m))$ as a function of $f, g(t_1), \ldots, g(t_m)$. Precisely, the latter condition means that

$$g(f(t_1, \ldots, t_m)) = H(f, \langle g(t_1), \ldots, g(t_m) \rangle),$$

for some function H: $F \times \bigcup_{n \in N} W^n \longrightarrow W$.

A few examples will clarify the use of this kind of induction.

5.4. Variables of a Term

We define by induction a function V on the set of terms as follows:

$$V(x) = \{x\} \text{ for every variable } x, \quad V(c) = \emptyset \quad \text{for every constant } c,$$

$$V(f(t_1, \ldots, t_m)) = V(t_1) \cup \cdots \cup V(t_m).$$

From this definition and Theorem 5.3 it follows that for every $t \in \mathrm{Tm}$, $V(t) \subseteq X$ is a finite set.

In fact, this property holds in the initial step, that is, for $V(x)$ and $V(c)$, and if it holds for $V(t_1), \ldots, V(t_m)$, then it holds also for $V(f(t_1, \ldots, t_m))$, since it is preserved under finite unions.

The elements $x \in V(t)$ are called *variables of the term t* (or *occurring in t*).

Similarly, we can define the set of *constants of a term*:

$$\mathrm{C}(x) = \emptyset \quad \text{for every variable } x, \qquad \mathrm{C}(c) = \{c\} \quad \text{for every constant c,}$$

$$\mathrm{C}(f(t_1, \ldots, t_m)) = \mathrm{C}(t_1) \cup \cdots \cup \mathrm{C}(t_m).$$

Also, in a similar way we can define the set of function symbols occurring in a term.

5.5. The Value of a Term

Let $Y \subseteq X$ be an arbitrary set of variables. Denote

$$\mathrm{Tm}(Y) = \{t \in \mathrm{Tm}\colon\ V(t) \subseteq Y\}.$$

Obviously, we have $\mathrm{Tm}(Y) = \bigcup_{l \in N} \mathrm{Tm}_l(Y)$, where the sets $\mathrm{Tm}_l(Y)$ are defined by induction exactly as the sets Tm_l with the exception that now $\mathrm{Tm}_0(Y) = Y \cup C$. Therefore, we can apply the induction principle and the inductive definition for $\mathrm{Tm}(Y)$ exactly as for $\mathrm{Tm} = \mathrm{Tm}(X)$.

Let $\mathbb{A} \in \mathbb{K}(\tau)$ be a structure of type τ. Any function $p\colon Y \longrightarrow A$ is said to be an assignment in $\mathbb{A}$. The value $t[p]$, of a term $t \in \mathrm{Tm}(Y)$, is defined by induction as follows:

$$x[p] = p(x) \quad \text{for any variable } \ x \in Y$$

$$c[p] = c^A \quad \text{for any constant } \ c \in C$$

$$f(t_1, \ldots, t_m)[p] = f^A(t_1[p], \ldots, t_m[p]).$$

Thus, the value of a variable is given directly by the assignment, and the value of a constant is independent of the assignment and always equal to the corresponding distinguished element of $\mathbb{A}$. The value of $f(t_1, \ldots, t_m)$ is the result of the corresponding operation f^A on the values $t_1[p], \ldots, t_m[p]$. This agrees with the intuitive remarks of the previous section.

We write, if necessary, $t^{\mathbb{A}}[p]$ to indicate that the value $t[p]$ is calculated in $\mathbb{A}$. By definition, the value $t[p]$ can be calculated for those assignments p which are defined at least on $V(t)$. On the other hand, it seems that the elements $p(y)$, for $y \notin V(t)$, are inessential. This is proved in the following lemma.

5.6. Lemma *If $p|V(t) = q|V(t)$, then $t[p] = t[q]$.*

Proof. The proof is by induction on t. If t is a constant c_k, then for any assignments p, q we have $p(c_k) = c_k^A = q(c_k)$. For a variable, the equality holds by the assumption. Assuming the equality $t_1[p] = t_1[q], \ldots, t_m[p] = t_m[q]$

we have

$$f_j(t_1, \ldots, t_m)[p] = f_j^A(t_1[p], \ldots, t_m[p])$$

$$= f_j^A(t_1[q], \ldots, t_m[q]) = f_j^A(t_1, \ldots, t_m)[q]. \quad \square$$

According to the lemma, the value $t[p]$ depends only on $p|V(t)$, that is, on the values $p(x)$ for $x \in V(t)$ of the assignment p.

If Y is a finite set of variables, $Y = \{y_1, \ldots, y_n\}$, then we write customarily $t(y_1, \ldots, y_n)$ for the elements $t \in \mathrm{Tm}(Y)$. In other words, the symbol $t(y_1, \ldots, y_n)$ bears the information that all variables occurring in t are among the $y_1, \ldots, y_n$ (but possibly $V(t) \subsetneq Y$). In this case, any finite sequence $\langle a_1, \ldots, a_n \rangle$ of elements of the universe A, can be identified with the assignment p: $Y \longrightarrow A$, for which $p(y_1) = a_1, \ldots, p(y_n) = a_n$. Consequently, the value $t[p]$ will be denoted by

$$t[a_1/y_1, \ldots, a_n/y_n]$$

or in shorter form,

$$t[a_1, \ldots, a_n],$$

if it is clear for which variables the elements $a_1, \ldots, a_n$ are substituted. If t is a constant term, that is, $V(t) = \emptyset$, then $t^{\mathbb{A}}[p]$ does not depend on p and we write $t^{\mathbb{A}}$ or t^A for the value of t in $\mathbb{A}$.

To illustrate the above notions consider the language L of groups containing one binary operation symbol $\circ$ and a constant c. In the additive group $\mathbb{Z} = \langle Z, +, 0 \rangle$ of integers $\circ$ is interpreted as the addition $+$, that is, $\circ^Z$ is $+$ and c^Z is 0, while in the multiplicative group $\mathbb{Z}^* = \langle Z^*, \cdot, 1 \rangle$ (here $Z^* = Z \setminus \{0\}$) the operation $\circ^{Z^*}$ is the multiplication $\cdot$ and c^{Z^*} is the number 1.

For arbitrary terms $t, s \in \mathrm{Tm}$, the sequence $\langle \circ, t, s \rangle$ is also a term. By our convention it can be denoted by $\circ(t, s)$, but following the usual mathematical notation, we shall write $t \circ s$ in this case.

For a fixed variable x we define the terms $t_n(x)$ by induction on $n \in N$ as follows:

$$t_0(x) \text{ is } c$$

$$t_{n+1}(x) \text{ is } t_n(x) \circ x.$$

Clearly, $t_n^{\mathbb{Z}}[a/x] = n \cdot a$ for each $a \in Z$.

Because, $t_0^{\mathbb{Z}}[a] = c^Z = 0 = 0 \cdot a$ and assuming $t_n^{\mathbb{Z}}[a] = n \cdot a$, we infer

$$t_{n+1}^{\mathbb{Z}}[a] = (t_n(x) \circ x)^{\mathbb{Z}}[a] = t_n^{\mathbb{Z}}[a] + x^{\mathbb{Z}}[a] = n \cdot a + a = (n+1) \cdot a.$$

Similarly, $t_n^{\mathbb{Z}^*}[a] = a^n$, for each $a \in Z^*$.

More generally, for an arbitrary group $\mathbb{G} = \langle G, \circ^G, c^G \rangle$ we have

$$t_n^{\mathbb{G}}[a] = a^n (= \underbrace{a \circ^G \ldots \circ^G a}_{n \text{ times}})$$

for each $a \in G$. Therefore, the term $t_n(x)$ is usually denoted by

$$x^n (\text{or } \underbrace{x \circ \cdots \circ x}_{n \text{ times}})$$

as the ordinary power. Abelian groups are often presented as additive groups $\mathbb{G} = \langle G, +, 0 \rangle$ and then $t_n(x)$ is usually denoted by $n \cdot x$ (or $x + \cdots + x$).

5.7. Termal Operations

Let $Y = \{x_1, \ldots, x_n\}$ be a finite set of variables. Each term $t(x_1, \ldots, x_n)$ defines, in any structure $\mathbb{A} \in \mathbb{K}(\tau)$, an n-ary operation f_t^A, as follows:

$$f_t^A(a_1, \ldots, a_n) = t^A[a_1/x_1, \ldots, a_n/x_n] \quad \text{for } a_1, \ldots, a_n \in A.$$

For example, if t is a variable x_r, where $1 \leq r \leq n$, then f_t^A is the projection

$$f_t^A(a_1, \ldots, a_n) = a_r \quad \text{for arbitrary } a_1, \ldots, a_n \in A,$$

and if t is a constant symbol c_k, for some $k \in K$, then f_t^A is the constant function with the value c_k^A,

$$f_t^A(a_1, \ldots, a_n) = c_k^A \quad \text{for all } a_1, \ldots, a_n \in A.$$

Of course, if t is $f_j(x_1, \ldots, x_n)$, where $n = \arg(f_j)$, then f_t^A coincides with the operation f_j^A, $j \in J$. It is not difficult to notice that the family $I_n(\mathbb{A})$, of all n-ary *termal* operations f_t^A, where $t \in \mathrm{Tm}(Y) = \mathrm{Tm}(x_1, \ldots, x_n)$, is closed under compositions (superpositions);

$$f_t^A(f_{t_1}^A, \ldots, f_{t_n}^A) \text{ is in } \mathrm{In}(\mathbb{A}) \text{ for any } t_1, \ldots, t_n \in \mathrm{Tm}(x_1, \ldots, x_n).$$

Actually, $I_n(\mathbb{A})$ is the smallest family of n-ary operations on A containing projections and constant operations (with the value c_k^A) and closed under compositions.

5.8. ***Example.*** Consider the algebraic rings with subtraction. The corresponding language L has three binary operation symbols $+, -, \cdot$ (for addition, subtraction, and multiplication, respectively), and also two constants c_0, c_1 (for the zero and the unit, respectively). Define terms kc_1 (the sum of k many ones), by induction on k: let $0c_1$ be c_0 and let $(k+1)c_1$ be $kc_1 + c_1$. Also for negative $k \in Z$ let kc_1 be

$c_0 - (-k)c_1$ (the first minus is a symbol of L, while the second is the subtraction in $\mathbb{Z}$). It follows immediately that

$$(kc_1)^{\mathbb{Z}} = k$$

for each $k \in Z$.

Generally, the mapping $k \longrightarrow (kc_1)^R$, where R is an arbitrary ring, is a homomorphism of $\mathbb{Z}$ into R and hence $R_0 = \{(kc_1)^R\colon\ k \in \mathbb{Z}\}$ is a subring of R. We claim that the termal operations f_t^R for $t \in \mathrm{Tm}(x_1, \ldots, x_n)$ coincide with the polynomial functions of n indeterminates with coefficients in R_0. First, by induction on t we show that for any term $t \in \mathrm{Tm}(x_1, \ldots, x_n)$ there is a polynomial $\varphi_t \in R_0[\xi_1, \ldots, \xi_n]$, such that

5.9 $$f_t^R(a_1, \ldots, a_n) = \varphi_t(a_1, \ldots, a_n) \quad \text{for all } a_1, \ldots, a_n \in R.$$

To see this, let φ_{x_r} be the polynomial ξ_r for $r = 1, \ldots, n$ and φ_{c_0}, φ_{c_1} be the null and unit polynomials, respectively. If 5.9 holds for terms t and s, then it suffices to put $\varphi_{t \circ s} = \varphi_t \circ \varphi_s$, where $\circ$ is any of $+, -$, or $\cdot$. On the other hand using the exponential terms x^n (see the remarks before Section 5.7), for a given polynomial $\varphi \in R_0[\xi_1, \ldots, \xi_n]$,

$$\varphi(\xi_1, \ldots, \xi_n) = \sum_{k_1, \ldots, k_n} a_{k_1, \ldots, k_n} \xi_1^{k_1} \cdot \ \cdots \ \cdot \xi_n^{k_n}$$

we construct in an obvious way a term t, such that 5.9 holds with φ in place of φ_t. This completes the proof of the claim.

In view of Section 5.7 and Example 5.8, we can say that terms are generalizations of polynomials. In fact, in a relational structure (of an arbitrary type) terms play a similar role to polynomials in an algebraic ring.

5.10. Formulas

In this section we define the other kind of expressions—the formulas of a given language $L = L(\tau)$. The definition is similar to that of terms—by induction with respect to the formation of finite sequences. The set $\mathrm{Fm} = \mathrm{Fm}(L)$ of the *formulas* of L is defined as the union

$$\mathrm{Fm} = \bigcup\{\mathrm{Fm}_l\colon\ l \in N\},$$

where

$$\begin{aligned}\mathrm{Fm}_0 = &\{\langle t, =, s\rangle\colon\ t, s \in \mathrm{Tm}\}\\ &\cup \{\langle r_i, t_1, \ldots, t_n\rangle\colon i \in I \text{ and } n = \arg(r_i) \text{ and } t_1, \ldots, t_n \in \mathrm{Tm}\},\end{aligned}$$

and

$$\begin{aligned}\mathrm{Fm}_{l+1} = &\mathrm{Fm}_l \cup \{\langle \neg, F\rangle: F \in \mathrm{Fm}_l\}\\ &\cup \{\langle F, \rightarrow, G\rangle: F, G \in \mathrm{Fm}_l\}\\ &\cup \{\forall, x, F\rangle: F \in \mathrm{Fm}_l \text{ and } x \in X\}.\end{aligned}$$

Let us comment on this definition and introduce a more suggestive notation as we did in the case of terms. Formulas F on the zero level, that is, $F \in \mathrm{Fm}_0$ are called *atomic*. We shall write $t = s$ and $r(t_1, \ldots, t_n)$ in place of $\langle t, =, s\rangle$ and $\langle r, t_1, \ldots, t_n\rangle$, respectively. Thus, the set Fm_0 of atomic formulas consists of all *equalities* $t = s$ and *relational expressions* $r(t_1, \ldots, t_n)$, with the obvious intended meaning.

Also, we shall write, respectively, $\neg F$, $F \rightarrow G$, and $\forall x\ F$ instead of $\langle \neg, F\rangle$, $\langle F, \rightarrow, G\rangle$, and $\langle \forall, x, F\rangle$. Thus, the $(l+1)$th level Fm_{l+1} consists of all the negations $\neg F$, implications $F \rightarrow G$, and quantifications (generalizations) $\forall x\ F$, where F, G run over Fm_l and x is in X. Of course, the signs $\neg$, $\rightarrow$, and $\forall$ are understood here as symbols of language L (the elements of S) and hence as some mathematical objects. Similarly, the equality sign $=$ in a formula $t = s$ is a symbol of L. Generally, whenever a logical connective, a quantifier, or the equality sign occurs in a formula of L, then it is clearly a symbol of L, and we shall use (or try to use) the connectives and quantifiers in such positions only. The sign $=$ is often used in the text in its normal meaning—as a symbol of L it occurs only in an equality $t = s$, where t and s are terms of L. Parentheses will be also in use as in ordinary logical notation. For example, $\langle\langle F, \rightarrow G\rangle, \rightarrow, H\rangle$ will be denoted by $(F \rightarrow G) \rightarrow H$ and $\langle F, \rightarrow, \langle G, \rightarrow, H\rangle\rangle$ by $F \rightarrow (G \rightarrow H)$. Notice the formal correspondence between the parentheses (,) and the sequence formation $\langle\ ,\ \rangle$.

According to the definition, if F is a formula, then $\forall x\ F$ is also a formula, for each variable x. Thus, we can quantify any variable, even one that does not occur at all in F. This makes the definition considerably simpler. Of course, we shall ensure that in this case F and $\forall x\ F$ are equivalent.

The induction principle for formulas looks very similar to that for terms; see Theorem 5.3. In order to prove that a given property P holds for all $F \in \mathrm{Fm} = \bigcup_{l \in N} \mathrm{Fm}_l$ it is sufficient to show the following.

1. The property P holds for all atomic formulas $F \in \mathrm{Fm}_0$.
2. If P holds for all $F \in \mathrm{Fm}_l$, then it holds also for all $F \in \mathrm{Fm}_{l+1}$.

The proof of this induction principle is then the same as for Theorem 5.3.

Let us note that the case stated in item 2 usually reduces to the proof that P holds for $\neg F$, $F \rightarrow G$, and $\forall x\ F$ under the assumption that it holds for F and G.

Analogously, to define by induction a function g on the set Fm it is sufficient to define the values $g(F)$, for atomic F, and to express the values $g(\neg F)$, $g(F \rightarrow G)$, and $g(\forall x\, F)$ as some functions depending possibly on $g(F)$ and $g(G)$ (see the remarks following Theorem 5.3 for a more precise formulation).

Recall that L has two connectives and one quantifier only (in order to shorten the inductive definitions and proofs). We define the other useful connectives and the existential quantifier in a well-known way, as follows:

the disjunction $F \vee G$	is the formula	$\neg F \to G$,
the conjunction $F \wedge G$	is the formula	$\neg(F \to \neg G)$,
the equivalence $F \equiv G$	is the formula	$(F \to G) \wedge (G \to F)$,
the existential quantification $\exists x F$	is the formula	$\neg\forall x \neg F$.

5.11. Free and Bound Variables

It is clear from the definition that a variable may occur in a formula in two ways: as free or bound (by a quantifier). First, we define the set $V_f(F)$ of *free* variables in F by induction, as follows. For atomic formulas let

$$V_f(t = s) = V(t) \cup V(s), \quad V_f(r_i(t_1, \ldots, t_n)) = V(t_1) \cup \cdots \cup V(t_n)$$

and (the inductive conditions)

$$V_f(\neg F) = V_f(F), \quad V_f(F \to G) = V_f(F) \cup V_f(G), \quad V_f(\forall x\, F) = V_f(F) \setminus \{x\}.$$

Thus, a variable stops being free whenever it falls into the range of a quantifier.

An easy induction shows that $V_f(F)$ is always a finite set of variables. For atomic F this property follows from the result of Section 5.4 and it is preserved under the inductive operations: finite union and subtraction.

For an arbitrary set $Y \subseteq X$ of variables, we denote

$$\mathrm{Fm}(Y) = \{F \in \mathrm{Fm}\colon\ V_f(F) \subseteq Y\},$$

that is, $\mathrm{Fm}(Y)$ consists of formulas whose free variables all belong to Y.

The following notation is often used: $F(x_1, \ldots, x_n)$ means that F is a formula from $\mathrm{Fm}(x_1, \ldots, x_n)$, that is, $V_f(F) \subseteq \{x_1, \ldots, x_n\}$.

Now we define the set $V_b(F)$ of *bound* variables in F by induction, as follows:

$$V_b(F) = \emptyset \quad \text{for all atomic } F,$$

$$V_b(\neg F) = V_b(F), \ V_b(F \to G) = V_b(F) \cup V_b(G), \ V_b(\forall x\, F) = V_b(F) \cup \{x\}.$$

Thus, every new quantifier adds one bound variable more. As in the case of V_f, it is simple to check that $V_b(F)$ is a finite set of variables, for all $F \in \mathrm{Fm}$.

Let us note that a variable can be simultaneously free and bound in a single formula. For example, if F is the equality $x = x$ and G is $\forall x\, F$, then in the implication $F \to G$, which can be written as

$$(x = x) \to \forall x(x = x),$$

the variable x is both free and bound.

A formula for which $V_f(F) = \emptyset$ (i.e., it has no free variables) is called a *sentence*. Intuitively, a formula $F(x_1, \ldots, x_n)$ encodes some property of the objects denoted by $x_1, \ldots, x_n$. A sentence expresses a property that does not depend on parameters.

5.12. Other Symbols and Subformulas

It is easy to define the set of any kind of symbols occurring in a given formula. For example, the function C, defined for terms in Section 5.4, can be extended to the whole of Fm (note that $\mathrm{Tm} \cap \mathrm{Fm} = \emptyset$), as shown.

For atomic formulas let $\mathrm{C}(t = s) = \mathrm{C}(t) \cup \mathrm{C}(s)$, and $\mathrm{C}(r_i(t_1, \ldots, t_n)) = \mathrm{C}(t_1) \cup \cdots \cup \mathrm{C}(t_n)$.

The inductive conditions are $\mathrm{C}(\neg F) = \mathrm{C}(F)$, $\mathrm{C}(F \to G) = \mathrm{C}(F) \cup \mathrm{C}(G)$, and $\mathrm{C}(\forall x\ F) = \mathrm{C}(F)$.

Obviously, $\mathrm{C}(F)$ is always a finite set of constant symbols.

In a similar way we can define the set of relation (function) symbols occurring in F.

Finally, consider the notion of a subformula. First, define a binary relation $<_0$ on Fm as follows. $F <_0 G$ if and only if one of the following cases hold:

G is $\neg F$

G is an implication with antecedent or consequent F

G is $\forall x\ F$ for some variable x.

Now, let us define a relation $F \leq G$ (F is a *subformula* of G) by the following rules.

$F \leq G$ if and only if $F = G$ or there is a finite sequence $H_1, \ldots, H_n$ such that $H_1 <_0 \cdots <_0 H_n$, $F = H_1$, and $H_n = G$.

Of course, if $F <_0 G$, then also $F \leq G$. It is easy to check that the relation $\leq$ is a partial ordering on the set Fm. Intuitively, $F \leq G$ means that F appears as a part of G in the course of the construction.

5.13. The Truth Relation

Let $\mathbb{K}(\tau)$ be the class of all relational systems of a given type τ and $L = L(\tau)$ be the corresponding language. For every system $\mathbb{A} \in \mathbb{K}(\tau)$ we define the truth relation

$$\mathbb{A} \models F[p]$$

(read: *F is true in* $\mathbb{A}$ *at p* or *p satisfies F in* $\mathbb{A}$) between formulas $F \in \mathrm{Fm}(L)$ and assignments p: $\mathrm{dom}(p) \longrightarrow A$, defined at least on $V_f(F)$, (i.e., $V_f(F) \subseteq \mathrm{dom}(p)$). The definition is by induction on $F \in \mathrm{Fm}$. For atomic formulas, let

$$\mathbb{A} \models (t = s)[p] \quad \text{iff} \quad t[p] = s[p],$$

and

$$\mathbb{A} \models r(t_1, \ldots, t_n)[p] \quad \text{iff} \quad r^A(t_1[p], \ldots, t_n[p]),$$

for any relation symbol r and $n = \arg(r)$.

Assuming that $\mathbb{A} \models F[p]$ is already defined for all $F \in \mathrm{Fm}_l$ and all suitable assignments p, let

$$\mathbb{A} \models \neg F[p] \quad \text{iff} \quad \mathbb{A} \not\models F[p],$$

$$\mathbb{A} \models (F \to G)[p] \quad \text{iff} \quad \text{either } \mathbb{A} \not\models F[p], \text{ or } \mathbb{A} \models G[p],$$

and

$$\mathbb{A} \models \forall x\, F[p] \quad \text{iff} \quad \text{for every } a \in A,\ \mathbb{A} \models F[p(a/x)].$$

In the last condition we use the assignment $p(a/x)$ which is defined as follows:

$$\mathrm{dom}(p(a/x) = \mathrm{dom}(p) \cup \{x\} \quad \text{and} \quad p(a/x)(y) = \begin{cases} p(y), & \text{if } y \neq x \\ a, & \text{if } y = x. \end{cases}$$

Therefore, at the right-hand side of the equivalence of the quantifier case the values of the $p(a/x)$s at x run over all the elements of the universe A.

After a short reflection we become convinced that the truth definition is such as it should be. It consists of the expected interpretation of the atomic formulas, while the inductive conditions are dictated simply by the normal understanding of the connectives and the quantifier. Actually, its real significance is to show that the truth is able to be defined [in the case of the relational structures of $\mathbb{K}(\tau)$ and the formulas of $L(\tau)$].

Since any formula F expresses some property of the interpretations of its free variables, it seems obvious that the relation $\mathbb{A} \models F[p]$ depends on the values $p(x)$, for $x \in V_f(F)$ only. Formally, this is proved in the next lemma.

5.14. Lemma. *If* $p|V_f(F) = q|V_f(F)$, *then*

$$(*) \qquad \mathbb{A} \models F[p] \text{ if and only if } \mathbb{A} \models F[q].$$

Proof. (By induction on F). For atomic F the conclusion follows immediately from Lemma 5.6. Assume that the lemma holds for an F and suppose that $p|V_f(\neg F) = q|V_f(\neg F)$. Since $V_f(\neg F) = V_f(F)$ we see that $(*)$ holds for F and, negating both the sides, we infer that it holds for $\neg F$ as well. Similarly, assuming the lemma for F and G, we prove that it holds also for the implication $F \to G$

[because $V_f(F), V_f(G) \subseteq V_f(F \to G)$]. Finally, consider the quantifier case: assume the lemma for F and let $p|V_f(\forall x\, F) = q|V_f(\forall x\, F)$. We know that $V_f(F) = V_f(\forall x\, F)$, if $x \notin V_f(F)$, or $V_f(F) = V_f(\forall x\, F) \cup \{x\}$, otherwise. In either case $p(a/x)|V_f(F) = q(a/x)|V_f(F)$, for all $a \in A$. Thus, the equivalence $(*)$ holds for F and the assignments $p(a/x)$, $q(a/x)$, where a is an arbitrary element of the universe A. This implies that $(*)$ holds for the formula $\forall x\, F$ and the assignments p, q. □

Let us note the following remark on redundant quantifiers: if $x \notin V_f(F)$, then we have

$$\mathbb{A} \models \forall x\, F[p] \quad \text{if and only if} \quad \mathbb{A} \models F[p].$$

The implication on the right-hand side follows from the truth definition and the converse follows from the lemma.

A few words on the notation. For a formula $F(x_1, \ldots, x_n)$, that is, such that $V_f(F) \subseteq \{x_1, \ldots, x_n\}$, we shall write more explicitly

$$\mathbb{A} \models F[a_1/x_1, \ldots, a_n/x_n] \quad \text{or even shorter,} \quad \mathbb{A} \models F[a_1, \ldots, a_n]$$

instead of $\mathbb{A} \models F[p]$, where p is any assignment with $p(x_1) = a_1, \ldots, p(x_n) = a_n$. This is justified by the lemma—only the values $p(x_1), \ldots, p(x_n)$ can matter. In other words, the sequence $\langle a_1, \ldots, a_n \rangle$ can be understood as an assignment p satisfying $p(x_1) = a_1, \ldots, p(x_n) = a_n$.

Now let F be a sentence, that is, $V_f(F) = \emptyset$. In this case the relation $\mathbb{A} \models F[p]$ is meaningful for all the assignments including $p = \emptyset$, the empty function. It follows from the lemma that either we have

$$\mathbb{A} \models F[p] \quad \text{for all} \quad p,$$

or else

$$\mathbb{A} \models F[p] \quad \text{for no} \quad p.$$

In the former case we write $\mathbb{A} \models F$ and we say "F is true (or F holds) in $\mathbb{A}$" and in the latter we write $\mathbb{A} \not\models F$ and we say "F is false in $\mathbb{A}$."

Directly from the truth definition, there follow analogous conditions concerning other connectives and the existential quantifier. Thus, we have the following equivalences:

$$\begin{array}{lll}
\mathbb{A} \models (F \vee G)[p] & \text{iff} & \mathbb{A} \models F[p] \text{ or } \mathbb{A} \models G[p], \\
\mathbb{A} \models (F \wedge G)[p] & \text{iff} & \mathbb{A} \models F[p] \text{ and } \mathbb{A} \models G[p], \\
\mathbb{A} \models (F \equiv G)[p] & \text{iff} & \mathbb{A} \models F[p] \text{ if and only if } \mathbb{A} \models G[p], \\
\mathbb{A} \models \exists x\, F[p] & \text{iff} & \text{for some } a \in A,\ \mathbb{A} \models F[p(a/x)].
\end{array}$$

5.15. *Example*. Let $\mathbb{K}$ be the class of relational systems of the form

$$\mathbb{A} = \langle A, \leq^A \rangle,$$

where $\leq^A$ is a two-argument relation. Thus, the corresponding language L has a binary relation symbol denoted by $\leq$. Given a variable x we form the sentence S_1: $\forall x(x \leq x)$. Clearly, $x \leq x$ stands for $\leq (x, x)$, (cf. the convention at the end of Lemma 5.6). From the truth definition and the foregoing remarks we obtain

$$\mathbb{A} \models S_1 \quad \text{iff} \quad \mathbb{A} \models S_1[\emptyset] \quad \text{iff} \quad \mathbb{A} \models \forall x(x \leq x)[\emptyset]$$

$$\text{iff} \quad \text{for each } a \in A,\ \mathbb{A} \models (x \leq x)[\emptyset(a/x)]$$

$$\text{iff} \quad \text{for each } a \in A,\ x[a/x] \leq^A x[a/x]$$

$$\text{iff} \quad \text{for each } a \in A,\ a \leq^A a.$$

Hence, $\mathbb{A} \models S_1$ if and only if the relation $\leq^A$ is reflexive.

Choose another variable y and let S_2 be the sentence

$$\forall x \forall y(x \leq y \wedge y \leq x \rightarrow x = y).$$

Similarly, we obtain

$$\mathbb{A} \models S_2 \quad \text{iff} \quad \text{for all } a, b \in A\{\text{if } \mathbb{A} \models (x \leq y)[a/x, b/y]$$

$$\text{and} \quad \mathbb{A} \models (y \leq x)[a/x, b/y], \text{then } \mathbb{A} \models (x = y)[a/x, b/y]\}$$

$$\text{iff} \quad \text{for all } a, b \in A\{\text{if } a \leq^A b \text{ and } b \leq^A a, \text{ then } a = b\},$$

and thus, $\mathbb{A} \models S_2$ if and only if the relation $\leq^A$ is antisymmetric.

Finally, let S_3 be the sentence

$$\forall x, y, z(x \leq y \wedge y \leq z \rightarrow x \leq z).$$

(A chain of quantifiers $\forall x_1, \ldots, \forall x_n$ is denoted more briefly by $\forall x_1, \ldots, x_n$). We have

$$\mathbb{A} \models S_3 \ \text{iff} \ \text{for all } a, b, c \in A\{\text{if } \mathbb{A} \models (x \leq y)[a/x, b/y, c/z]$$

$$\text{and} \quad \mathbb{A} \models (y \leq z)[a/x, b/y, c/z], \text{then } \mathbb{A} \models (x \leq z)[a/x, b/y, c/z]\}$$

$$\text{iff} \ \text{for all } a, b, c \in A\{\text{if } a \leq^A b \text{ and } b \leq^A c, \text{ then } a \leq^A c\},$$

Hence, $\mathbb{A} \models S_3$ if and only if the relation $\leq^A$ is transitive.

Consequently, $\mathbb{A} \models (S_1 \wedge S_2 \wedge S_3)$ if and only if the relation $\leq^A$ is a partial ordering on A.

Now consider another example. Let $\mathbb{K}$ be the class of algebraic systems of the form $\mathbb{A} = \langle A, \circ^A, c^A \rangle$, with one binary operation and one distinguished element. Thus, the corresponding language L contains one binary operation symbol $\circ$ and one constant c. Given two variables x and y, we build up the terms x, $x \circ y$, $x \circ c$ [cf. the remarks at the end of Lemma 5.6] and we calculate the following values in an arbitrary system $\mathbb{A} \in \mathbb{K}$:

$$x^{\mathbb{A}}[a/x] = a, \qquad (x \circ c)^A[a/x] = a \circ^A c^A, \qquad (x \circ y)^A[a/x, b/y] = a \circ^A b.$$

Hence

$$\mathbb{A} \models \forall x(x \circ c = x) \quad \text{iff} \quad \text{for each } a \in A \ (a \circ^A c^A = a)$$
$$\text{iff} \quad c^A \text{ is a (right) unit,}$$

and

$$\mathbb{A} \models \forall x \exists y(x \circ y = c) \quad \text{iff} \quad \text{for every } a \in A \text{ there is a b}\in\text{A}$$
$$\text{such that } a \circ^A b = c^A \quad \text{iff} \quad \text{every element has a (right) inverse in A.}$$

To conclude, let us make the following remark.

5.16. Remark

Let us agree to use (temporarily) the signs $\neg$, $\rightarrow$, and $\forall$ not only as symbols of a language L but also as the ordinary abbreviations. Now the connective conditions of the truth definition can be rewritten as follows:

$$\mathbb{A} \models \neg F[p] \quad \text{if and only if} \quad \neg\mathbb{A} \models F[p]$$

and

$$\mathbb{A} \models (F \rightarrow G)[p] \quad \text{if and only if} \quad \mathbb{A} \models F[p] \rightarrow \mathbb{A} \models G[p].$$

Thus, on the left-hand side of the above equivalences the signs $\neg$ and $\rightarrow$ are objects of language L, while on the right they are expressions of the ordinary colloquial language used in everyday mathematical practice (called a metalanguage in this situation).

By the above equivalences we can say that the connectives $\neg$, $\rightarrow$ commute with the truth sign $\models$. For the quantifier $\forall$ we have

$$\mathbb{A} \models \forall x F[p] \quad \text{if and only if} \quad \forall a \in A \ \mathbb{A} \models F[p(a/x)],$$

that is, the quantifier $\forall$ also commutes with $\models$, provided its range is limited to the universe A. Of course, this kind of commutation holds also for the remaining connectives and the quantifier $\exists$. Applying the above rules to a given formula F, we see that the relation $\mathbb{A} \models F[p]$ is equivalent to a logical combination (with the same position of the connectives and quantifiers as in F) of the relations $\mathbb{A} \models H[p]$, where H runs over all the atomic subformulas of F. Replacing the latter by their equivalents from the truth definition we come to the following conclusion.

The relation $\mathbb{A} \models F[p]$ is equivalent to the property obtained from F by replacing each atomic subformula H of F by the equality $t^{\mathbb{A}}[p] = s^{\mathbb{A}}[p]$ or $r^A(t_1^{\mathbb{A}}[p], \ldots, t_n^{\mathbb{A}}[p])$ depending on whether H is $t = s$ or $r(t_1, \ldots, t_n)$, respectively, and relativizing all the quantifiers to the universe of $\mathbb{A}$.

This clarifies (and justifies) the double role of the signs $\neg, \rightarrow, \forall$. Actually, there is no need for a strict distinction—the interpretation of the connectives and quantifiers as symbols of L under the truth relation is standard (that is, it coincides with the ordinary one).

It should be emphasized that not every property of the systems under consideration can be expressed by a formula (or a set of formulas) of the corresponding L. In fact, only some strictly definite properties have their counterparts in L. For example, the variables of L, free or bound, always refer to elements of any universe and atomic properties of elements are determined by the choice of primitive relations and operations. Of course, one can define other, nonclassical languages. For example, variables may denote also subsets of a universe, one can form infinite disjunctions and conjunctions, one can consider different kinds of quantifiers, and so on. Nonclassical languages are a special part of logic and will not be considered in this book.

EXERCISES

5.1. If $\mathbb{A} \subseteq \mathbb{B}$, that is, $\mathbb{A}$ is a subsystem of $\mathbb{B}$, then $t^{\mathbb{A}}[p] = t^{\mathbb{B}}[p]$ for every term t and every assignment p in $\mathbb{A}$.

5.2. We say that the equality $t = s$ is universally valid in $\mathbb{A}$ if

$$\mathbb{A} \models \forall x_1, \ldots, x_n(t = s),$$

where $V(t) \cup V(s) \subseteq \{x_1, \ldots, x_n\}$, that is, all the free variables of t and s are among $x_1, \ldots, x_n$.

Prove the following properties:

(a) $t = s$ is valid in $\mathbb{A}$ if and only if $\mathbb{A} \models (t = s)[p]$ for any assignment p,

(b) if $t = s$ is valid in $\mathbb{A}$, then it is valid in any subsystem $\mathbb{B}$ of $\mathbb{A}$,

(c) if $t = s$ is valid in $\mathbb{A}$, then it is valid in any homomorphic image $\mathbb{B}$ of $\mathbb{A}$,

(d) if $t = s$ is valid in each factor $\mathbb{A}_i$ of a product $\mathbb{A} = \prod_{i \in I} \mathbb{A}_i$, then it is also valid in the product $\mathbb{A}$ as well,

(e) prove the same as in (d) for directed sums.

5.3. *Algebra of Terms.* The family Tm of all the terms of a given algebraic language $L = L(\tau)$ can be considered as an algebra of type τ in the following way. Put $T = \mathrm{Tm}$ and let

$$f_j^T(t_1, \ldots, t_m) = f_j(t_1, \ldots, t_m) \quad \text{for } t_1, \ldots, t_m \in T,\ m = \arg(f_j),$$

$$\text{for each } j \in J,$$

and

$$c_k^T = c_k \quad \text{for each } k \in K.$$

Clearly, the structure

$$\mathbb{T} = \langle T,\ \{f_j^T\}_{j \in J}, \{c_k^T\}_{k \in K} \rangle$$

is an algebra of type τ. It is called the *algebra of terms* (of $L(\tau)$). Check that $\mathbb{T} = \mathbb{T}[X]$, that is, the set X of the variables generates the algebra $\mathbb{T}$. Actually, the variables are *free generators* of $\mathbb{T}$ with respect to the class $\mathbb{K}(\tau)$ (of all the algebras of type τ); that is, for an arbitrary algebra $\mathbb{B} \in \mathbb{K}(\tau)$ each mapping p: $X \longrightarrow B$ extends uniquely to a homomorphism h: $\mathbb{T} \longrightarrow B$. Prove that $h(t) = t^{\mathbb{B}}[p]$.

5.4. In the language of the class of algebraic rings there are terms t_n of the form

$$t_n(y_0, \ldots, y_n, x) = y_0 + y_1 \cdot x + \cdots + y_n \cdot x^n, \qquad n \geq 1.$$

Verify that the condition $\mathbb{A} \models \forall y_0, \ldots, y_n \exists x (t_n = 0)$ is equivalent to the statement: every nonconstant polynomial of degree $\leq n$ and coefficients in A has a root in $\mathbb{A}$.

5.5. In an arbitrary language $L(\tau)$ construct sentences F_n and G_n, $n \geq 1$, such that

$$\mathbb{A} \models F_n \text{ if and only if the universe } A \text{ has at least } n \text{ elements,}$$

and

$$\mathbb{A} \models G_n \text{ if and only if the universe } A \text{ has exactly } n \text{ elements.}$$

5.6. A formula F is called *open* if it has no quantifiers. Thus, open formulas arise from atomic ones by adjoining $\neg$ and of connecting with $\rightarrow$. Prove that whenever $\mathbb{A} \subseteq \mathbb{B}$, then for every open F we have

$$\mathbb{A} \models F[p] \quad \text{if and only if} \quad \mathbb{B} \models F[p]$$

for any assignment p in $\mathbb{A}$.

5.7. A formula of the form $\exists x_1, \ldots, x_n\ F$, where F is open (see Exercise 5.6) and $x_1, \ldots, x_n$ are arbitrary variables, is called *existential*. Similarly, a formula of the form $\forall x_1, \ldots, x_n\ F$, with an open F, is called *universal*. Prove that whenever $\mathbb{A} \subseteq \mathbb{B}$, then for every assignment p in $\mathbb{A}$ we have:

(a) if $\mathbb{A} \models G[p]$, then $\mathbb{B} \models G[p]$ for each existential formula G

(b) if $\mathbb{B} \models H[p]$, then $\mathbb{A} \models H[p]$ for each universal H.

6

THEORIES AND MODELS

Now we can introduce the notion of a theory and of a model based on the truth definition. Also, we describe briefly the axiom system of the Zermelo–Fraenkel set theory—a theory which can be regarded as the foundation of mathematics, since each classical mathematical theory can be treated as part of it.

The notion of a model dates back to the nineteenth century when Beltrami and Klein constructed their models of a non-euclidean geometry. The precise definition in full generality was given by Hilbert and Tarski in the late 1920s and early 1930s.

Let T be an arbitrary set of sentences of a given language $L(\tau)$. A system $\mathbb{A}$ of type τ is called a *model* of the set T, if

$$\mathbb{A} \models F \quad \text{for every } F \in T,$$

that is, if every sentence F of the set T is true in $\mathbb{A}$.

The class of all models of the set T will be denoted by $\mathrm{Mod}(T)$.

Let $\mathbb{K} \subseteq \mathbb{K}(\tau)$ be a class of some systems of the type τ. We say that the class $\mathbb{K}$ is *axiomatizable*, if $\mathbb{K} = \mathrm{Mod}(T)$ holds for some set of sentences T of the language $L(\tau)$. Every such set T is called an *axiom system* for the class $\mathbb{K}$. By the *theory* of the given class $\mathbb{K}$ we mean the set of all the sentences true in every $\mathbb{A} \in \mathbb{K}$;

$$\mathrm{Th}(\mathbb{K}) = \{F\colon \ \mathbb{A} \models F, \quad \text{for every } \mathbb{A} \in \mathbb{K}\}.$$

We write shortly $\mathbb{K} \models F$, if for every $\mathbb{A} \in \mathbb{K}$, $\mathbb{A} \models F$. Directly from the definition we have $\mathbb{K} \subseteq \mathrm{Mod}(\mathrm{Th}(\mathbb{K}))$. Clearly, the equality $\mathbb{K} = \mathrm{Mod}(\mathrm{Th}(\mathbb{K}))$ holds precisely for axiomatizable classes.

Example. The class of all groups (rings, fields) is axiomatizable by means of the well-known system of axioms of group theory (ring theory, field theory). The class of all Boolean algebras is axiomatizable by the system of axioms given in Chapter 2.

We can also consider the notion of a theory and of a model in a more general situation, where T consists of formulas with free variables. Let $T \subseteq \mathrm{Fm}(L(\tau))$ be some set of formulas and let p be an assignment in the system $\mathbb{A}$, of type τ, defined at least on $V(T) = \bigcup\{V(F)\colon\ F \in T\}$. The pair $\langle \mathbb{A}, p\rangle$ is called a *model of the set* T, if $\mathbb{A} \models F[p]$ holds for every formula F belonging to T. If $V(T) \subseteq \{x_1, \ldots, x_n\}$, then we use a slightly different notation, by a *model of the set* T we mean every sequence $\langle \mathbb{A}, a_1, \ldots, a_n\rangle$ for which

$$\mathbb{A} \models F[a_1/x_1, \ldots, a_n/x_n], \quad \text{for every formula } F \in T.$$

Example. Let $T = T(x)$ consist of the axioms of group theory and of the formulas $\neg(x^n = 1)$ for $n \in N, n \geq 1$. Every model of the set $T(x)$ is of the form $\langle \mathbb{G}, a\rangle$, where $\mathbb{G}$ is a group and $a \in G$ is an element of infinite order. Thus, the class of the groups with a distinguished element of infinite order is axiomatizable.

6.1. Equivalence of Models

First, we shall prove the following theorem.

Theorem. *If h: $\mathbb{A} \longrightarrow \mathbb{B}$ is an isomorphism, then we have*

6.2 $\mathbb{A} \models F[a_1/x_1, \ldots, a_n/x_n]$ if and only if $\mathbb{B} \models F[h(a_1)/x_1, \ldots, h(a_n)/x_n]$,

for every formula $F = F(x_1, \ldots, x_n)$ and arbitrary elements $a_1, \ldots, a_n \in A$.

Proof. First, we check that for any term t the following equality holds:

$$(*) \qquad h(t^{\mathbb{A}}[a_1, \ldots, a_n]) = t^{\mathbb{B}}[h(a_1), \ldots, h(a_n)].$$

We apply induction. For any variable x we have

$$h(x[a/x]) = h(a) = x[h(a)/x],$$

and for any constant c_k we always have $h(c_k^A) = c_k^B$. Assuming the equality $(*)$ for $t_1, \ldots, t_m$, we infer

$$\begin{aligned} h\big(f_j(t_1, \ldots, t_m)[a_1, \ldots, a_n]\big) &= h\big(f_j^A(t_1^{\mathbb{A}}[a_1, \ldots, a_n], \ldots, t_m^{\mathbb{A}}[a_1, \ldots, a_n])\big) \\ &= f_j^B\big(h(t_1^{\mathbb{A}}[a_1, \ldots, a_n]), \ldots, h(t_m^{\mathbb{A}}[a_1, \ldots, a_n])\big) \\ &= f_j^B\big(t_1^{\mathbb{B}}[h(a_1), \ldots, h(a_n)], \ldots, t_m^{\mathbb{B}}[h(a_1), \ldots, h(a_n)]\big) \\ &= f_j(t_1, \ldots, t_m)[h(a_1), \ldots, h(a_n)]. \end{aligned}$$

Now, we prove 6.2 for atomic formulas.

$$\mathbb{A} \models (t = s)[a_1, \ldots, a_n] \quad \text{iff} \quad t[a_1, \ldots, a_n] = s[a_1, \ldots, a_n]$$

$$\text{iff} \quad h(t[a_1, \ldots, a_n]) = h(s[a_1, \ldots, a_n])$$

$$\text{iff} \quad t[h(a_1), \ldots, h(a_n)] = s[h(a_1), \ldots, h(a_n)]$$

$$\text{iff} \quad \mathbb{B} \models (t = s)[h(a_1), \ldots, h(a_n)].$$

$$\mathbb{A} \models r_i(t_1, \ldots, t_m)[a_1, \ldots, a_n]$$

$$\text{iff} \quad r_i^A(t_1[a_1, \ldots, a_n], \ldots, t_m[a_1, \ldots, a_n])$$

$$\text{iff} \quad r_i^B\big(h(t_1[a_1, \ldots, a_n]), \ldots, h(t_m[a_1, \ldots, a_n])\big)$$

$$\text{iff} \quad r_i^B(t_1[h(a_1), \ldots, h(a_n)], \ldots, t_m[h(a_1), \ldots, h(a_n)])$$

$$\text{iff} \quad \mathbb{B} \models r_i^B([h(a_1), \ldots, h(a_n)]).$$

Assuming 6.2 for F and negating both the sides, we obtain 6.2 for $\neg F$. Now, assume 6.2 for formulas F and G. Then we have

$$\mathbb{A} \models (F \to G)[a_1, \ldots, a_n] \quad \text{iff} \quad (\mathbb{A} \models F[a_1, \ldots, a_n] \to \mathbb{A} \models G[a_1, \ldots, a_n])$$

$$\text{iff} \quad (\mathbb{B} \models F[h(a_1), \ldots, h(a_n)] \to \mathbb{B} \models G[h(a_1), \ldots, h(a_n)])$$

$$\text{iff} \quad \mathbb{B} \models (F \to G)[h(a_1), \ldots, h(a_n)].$$

Finally, assuming 6.2 for F we check the case of the quantifier;

$$\mathbb{A} \models \forall x\, F[a_1, \ldots, a_n] \quad \text{iff} \quad \forall a \in A\ \mathbb{A} \models F[a_1, \ldots, a_n, a/x])$$

$$\text{iff} \quad \forall a \in A\ \mathbb{B} \models F[h(a_1), \ldots, h(a_n), h(a)/x]$$

$$\text{iff} \quad \forall b \in B\ \mathbb{B} \models F[h(a_1), \ldots, h(a_n), b/x]$$

$$\text{iff} \quad \mathbb{B} \models \forall x\ F[h(a_1), \ldots, h(a_n)],$$

which completes the proof. □

Systems $\mathbb{A}, \mathbb{B}$ are called *equivalent* (or *equivalent in the language* $L(\tau)$ or $L(\tau)$-*equivalent*), if we have

(∗) $\quad \mathbb{A} \models F$ iff $\mathbb{B} \models F$, $\quad$ for every sentence F of language $L(\tau)$.

The equivalence of structures is usually denoted by $\mathbb{A} \equiv \mathbb{B}$ or $\mathbb{A} \equiv_{L(\tau)} \mathbb{B}$ to indicate that the equivalence $(*)$ holds for all formulas of $L(\tau)$. This condition can also be stated as

$$\mathrm{Th}(\mathbb{A}) = \mathrm{Th}(\mathbb{B}).$$

From the theorem proved in this section, it follows immediately that isomorphic structures are equivalent. The class of models $\mathrm{Mod}(T)$, of a given theory T, is closed under the equivalence:

$$\textit{if } \mathbb{A} \in \mathrm{Mod}(T) \text{ and } \mathbb{A} \equiv \mathbb{B}, \textit{ then also } \mathbb{B} \in \mathrm{Mod}(T).$$

(We also write $\mathbb{A} \models T$ if $\mathbb{A} \in \mathrm{Mod}(T)$).

6.3. Set Theory

The primitive notion of set theory is a "set" and the primitive relation is the relation of "membership" denoted by $\in$. Thus, the language L of set theory contains one binary relation symbol $\in$. In this approach, elements of the sets under consideration are also sets. Below we give the system of axioms established by Zermelo in 1908; see [Z].

1. *The Axiom of Extensionality.* Sets having the same elements are identical. In language L this is expressed by the formula

$$\forall x, y[\forall z(z \in x \equiv z \in y) \to x = y].$$

To simplify the writing of formulas of L we use all the logical connectives and quantifiers.

2. *The Axiom of the Empty Set.* There exists a set without elements (such a set is usually denoted by the symbol $\emptyset$). This property is expressed by the formula

$$\exists x \forall y(\neg y \in x).$$

3. *The Axiom of Pair.* For any sets a, b there exists the pair $\{a, b\}$. The corresponding formula is

$$\forall x, y \exists p \forall z(z \in p \equiv z = x \vee z = y).$$

4. *The Axiom of Union.* For any set x there exists the union $s = \bigcup x$ (we recall that all elements of sets are sets). This axiom can be expressed in L as follows:

$$\forall x \exists s \forall z[z \in s \equiv \exists y(y \in x \wedge z \in y)].$$

5. *The Power-Set Axiom.* For every set x there exists the set $P(x)$, consisting of all the subsets of the set x;

$$\forall x \exists y \forall z(z \in y \doteq z \subseteq x),$$

where $z \subseteq x$ abbreviates the formula $\forall u(u \in z \to u \in x)$.

6. *The Axiom of Infinity*. There exists an infinite set.

To express this property in L let us denote $s(x) = x \cup \{x\}$. Then $s(\emptyset) = \{\emptyset\}$, $s(s(\emptyset)) = \{\emptyset, \{\emptyset\}\}, \ldots$ and so on. Iterating this operation n times we obtain the set $s^n(\emptyset)$, which obviously has n elements. Thus, any set w having $\emptyset$ as an element and closed under the operation s must be infinite. Hence, the axiom of infinity can be expressed by the following sentence of L:

$$\exists w[\emptyset \in w \wedge \forall x(x \in w \rightarrow s(x) \in w)].$$

It is easy to show that there exists a unique smallest set w such that $\emptyset \in w$ and w is closed under the operation s. The set of nonnegative integers can be defined (and usually is) in set theory as the set w.

7. *The Axiom of Choice*. A nonempty set x all of whose elements are also nonempty and pairwise disjoint has a selector, that is, a set intersecting every element of the set x in exactly one point.

We find without difficulty a sentence of L expressing this property.

8. *The Foundation Axiom*. Every nonempty set x has an element disjoint with x:

$$\forall x[x \neq \emptyset \rightarrow \exists y[y \in x \wedge \forall z(z \in y \rightarrow z \notin x)]].$$

This axiom eliminates objects which are not sets, for example, objects x for which $x \in x$.

9. *The Axiom of Subsets*. For every set a there exists the subset $\{x \in a\colon H(x, y_1, \ldots, y_n)\}$, of all the elements $x \in a$ having the property H, where H is an arbitrary formula of L.

This axiom can be formulated in language L in the form of a (so-called) scheme. Every formula H and every variable x determines a sentence $S_{H,x}$:

$$\forall y_1, \ldots, y_n \forall a \exists y \forall x[x \in y \equiv x \in a \wedge H(x, y_1, \ldots, y_n)].$$

Now, each $S_{H,x}$ is assumed as an axiom.

On the basis of the Zermelo theory we can develop the algebra of sets, the theory of functions and relations, and the theory of ordering and power to an extent sufficient for mathematical practice. In particular we can define the set of natural numbers, the set of integers, and the set of rationals and of reals with the ordinary arithmetical operations. We can also build the well-known function and topological spaces, and so on. However, a stronger system of a set theory is more popular namely the so-called Zermelo–Fraenkel theory, [F2]. We obtain that theory if instead of the axiom of subsets we assume the following axiom.

10. *The Replacement Axiom*. If a formula $F(x, y)$ defines a mapping (i.e., to every x there corresponds exactly one y such that $F(x, y)$), then the image of any set under F is a set.

We express this axiom in the form of a scheme similar to that in paragraph 9.

$$\forall x \exists! y F(x, y) \rightarrow \forall u \exists v \forall y[y \in v \equiv \exists x(x \in u \wedge F(x, y))],$$

and we precede the above formula with the quantifiers $\forall z_1, \ldots, \forall z_n$, where $z_1, \ldots, z_n$ are all the free variables of F distinct from x, y.

The symbol $\exists! y F(x, y)$ is here an abbreviation of the formula ($\exists!$ reads "there exists exactly one")

$$\exists y F(x, y) \wedge \forall y', y'' \big(F(x, y') \wedge F(x, y'') \rightarrow y' = y''\big).$$

As above, each formula F and each pair x, y of variables determines one instance of the axiom scheme.

The set A is called *transitive* if $z \subseteq A$ follows from $z \in A$. In the Zermelo–Fraenkel theory, we define by induction the following sets $R(\alpha)$, where α runs over ordinal numbers:

$$R(0) = \emptyset, \qquad R(\alpha + 1) = P(R(\alpha));$$

$$R(\alpha) = \bigcup\{R(\beta)\colon\ \beta < \alpha\}, \quad \text{for limit ordinals } \alpha.$$

It is easy to show that all the sets $R(\alpha)$ are transitive and that the condition $\alpha \leq \beta$ implies $R(\alpha) \subseteq R(\beta)$. The axiom of foundation is equivalent to the assertion that every set belongs to some $R(\alpha)$.

To the language of set theory there corresponds relational systems of the form $\langle A; \in^A \rangle$, where $\in^A$ is a binary relation on A. Of particular interest are transitive systems, that is, systems of the form $\langle A; \in^A \rangle$, where A is a transitive set and $\in^A$ is the usual membership relation restricted to A [that is, such that $\forall a, b \in A(a \in^A b \ \equiv \ a \in b)$].

Example. For any transitive system $\mathbb{A} = \langle A; \in^A \rangle$ we have

6.4 $\qquad \mathbb{A} \models \forall y(y \notin x)[a]$ if and only if $a = \emptyset$, since $a \subseteq A$.

Note that $\emptyset \in A$, for every transitive A. Thus, the set $\emptyset^A$ (the empty set of $\mathbb{A}$), that is, the unique a satisfying the left-hand side of 6.4, coincides with the ordinary set $\emptyset$. This need not be true in the case of a nontransitive $\mathbb{A}$.

EXERCISES

6.1. Show that the following classes of relational structures are axiomatizable:

(a) dense linear orderings without end points;

(b) discrete orderings, that is, orderings in which every element has an immediate successor and an immediate predecessor (except possibly the least one);

(c) groups in which every element has order $\leq n$ for a fixed $n \geq 2$;

(d) torsion free groups; every element $\neq \mathbb{1}$ has an infinite order;

(e) fields of a given characteristic > 0;
(f) fields of characteristic zero;
(g) algebraically closed fields;
(h) ordered fields;
(i) real closed fields (Exercises 17.7, 17.8);
(j) atomless Boolean algebras.

6.2. Show that if a class $\mathbb{K}$ is axiomatizable, then the subclass consisting of infinite systems of the class $\mathbb{K}$ is also axiomatizable. (*Hint*: Find a sentence s_n such that $\mathbb{A} \models s_n \equiv \text{card } A \geq n$, for $n \in N$).

6.3. Generalization of Exercise 5.3: Let $\mathbb{K}(\tau)$ be the class of all the algebras of a given type τ. For an arbitrary set $X \neq \emptyset$ we define the family of terms $\text{Tm}(X)$ in the usual way, treating the elements $x \in X$ as variables. The definition of an assignment and of the value $t[p]$ remains unchanged. Make sure that the algebra of terms

$$\mathbb{T}(X) = \langle \text{Tm}(X);\ \{f_j^{T(X)}\}, \{c_k^{T(X)}\}\rangle$$

is still a free algebra; the variables $x \in X$ are free generators of the algebra $\mathbb{T}(X)$ with respect to the class $\mathbb{K}(\tau)$. Thus, the cardinality of X is inessential here.

6.4. Let $\mathbb{K} \subseteq \mathbb{K}(\tau)$ be a class of algebras of type τ and let $\mathbb{A} \in \mathbb{K}$, card $\mathbb{A} \geq 2$. Show that the relation (cf. Exercise 5.2)

$$t \simeq_{\mathbb{K}} s \quad \text{if and only if} \quad \mathbb{K} \models t = s$$

$$\text{(i.e., } \mathbb{A} \models t = s \text{ holds, for every algebra } \mathbb{A} \in \mathbb{K})$$

is a congruence in $\mathbb{T}(X)$, for any set of variables X (cf. Exercise 6.3), and that the factor algebra (the free algebra of $\mathbb{K}$);

$$\mathbb{F}(X) = \mathbb{T}(X)/\simeq_{\mathbb{K}}$$

has card$X$-many free generators with respect to the class $\mathbb{K}$. [$\mathbb{F}(X)$ need not to belong to $\mathbb{K}$].

6.5. Let R be the family of all the congruences of the algebra $\mathbb{T}(X)$. If $\mathbb{K}$ is closed under isomorphism and under taking subalgebras, then we have

(a) $\mathbb{K} \models t = s$ if and only if $\forall r \in R[\mathbb{T}(X)/r \in \mathbb{K} \rightarrow [t]_r = [s]_r]$ (See Exericise 4.3.)

(b) Let $R_0 = \{r \in R\colon\ \mathbb{T}(X)/r \in \mathbb{K}\}$. Show that the algebra $\mathbb{F}(X)$ (See Exercise 6.4) can be embedded into the product $\prod_r\{\mathbb{T}(X)/r\colon\ r \in R_0\}$.

Hence, if the class $\mathbb{K}$ is closed under isomorphism, subalgebras, and products, then the free algebra $\mathbb{F}(X)$ belongs to $\mathbb{K}$.

6.6. Birkhoff's theorem (cf. [B2]): *If the class* $\mathbb{K}$ *is closed under isomorphism,* *subalgebras, products and homomorphic images, then* $\mathbb{K}$ *is axiomatizable by equalities* (cf. Exercise 5.2).

To show this, let E denote the set of all the equalities $t = s$ that are true in $\mathbb{K}$. Let $\mathbb{A} \in \mathrm{Mod}(E)$. Prove that $\mathbb{A}$ is a homomorphic image of the free algebra $\mathbb{F}(X)$ for sufficiently large X and apply Exercise 6.5.

6.7. For every set A there exists a transitive set B such that $A \subseteq B$.

6.8. Show that in any transitive system $\langle A, \in^A \rangle$ the following axioms are true: the axiom of the empty set, the extensionality axiom, and the foundation axiom.

6.9. Let $\mathbb{A} = \langle A, \in^A \rangle$ be a transitive system. Show that

$$\mathbb{A} \models \forall z(z \in u \;\equiv\; z = x \vee z = y)[a/x, b/y, c/u] \text{ if and only if } c = \{a, b\}.$$

Hence, the validity in $\mathbb{A}$ of the pairing axiom is equivalent with the condition $\forall a, b \in A(\{a, b\} \in A)$, that is, with the property that A is closed under the operation of pair.

6.10. Show that the validity in a transitive system $\mathbb{A}$ of the axiom of unions is equivalent with the property that A is closed under unions (cf. Exercise 6.9).

6.11. Prove that for every set A there exists a transitive set M containing A and such that in the system $\mathbb{M} = \langle M, \in^M \rangle$ besides the axioms of the empty set, extensionality, and foundation (cf. Exercise 6.8), also the axioms of pair and unions are also true. (Define inductively an increasing chain $A = M_0 \subseteq M_1 \subseteq \ldots$, applying at each stage the results of Exercises 6.7, 6.9, and 6.10).

7

SUBSTITUTION OF TERMS

Replacing a variable by another variable or by a composed term is a common syntactical operation widely used in mathematical practice. We prove the important Substitution Lemma 7.1 and apply it to show how to eliminate variables from the axioms of a theory.

Let L be a fixed language. An arbitrary function q: $Y \longrightarrow \mathrm{Tm}$, where $Y \subseteq X$ is a set of variables of L, will be called a *term assignment* or a *substitution* (in fact, q is an assignment in the algebra $\mathbb{T}$ of terms of L; cf. Exercise 5.3). Intuitively, any such assignment q determines a substitution operation; in any formula F, substitute the term $q_y = q(y)$ for the variable y, for all $y \in Y \cap V_f(F)$, simultaneously. The resulting formula will be denoted by $F(q)$ or $F(\{q_y/y\}_{y \in Y})$ or $F(t_1/y_1, \ldots, t_n/y_n)$, if $Y = \{y_1, \ldots, y_n\}$ and $t_1 = q_{y_1}, \ldots, t_n = q_{y_n}$. Clearly, such a substitution operation must commute with the connectives and quantifiers and will change, in general, the free variables.

We pass to the precise definition. The formula $F(q)$ is defined by induction on $F \in \mathrm{Fm}$. First, we define $t(q)$, for $t \in \mathrm{Tm}$ also by induction, as follows:

$$x(q) = \begin{cases} q_x, & x \in Y, \\ x, & x \notin Y, \end{cases}$$

for any variable $x \in X$, $c(q) = c$ for any constant c, and

$$f(t_1, \ldots, t_m)(q) = f(t_1(q), \ldots, t_m(q)).$$

Now, for atomic formulas define

$$(t = s)(q) \quad \text{as} \quad t(q) = s(q)$$

and

$$r(t_1, \ldots, t_n)(q) \quad \text{as} \quad r(t_1(q), \ldots, t_n(q)).$$

Finally, assume that

$$(\neg F)(q) \quad \text{is} \quad \neg(F(q)), \qquad (F \rightarrow G)(q) \quad \text{is} \quad F(q) \rightarrow G(q)$$

and

$$(\forall x\ F)(q) \quad \text{is} \quad \forall x\ F(q|Y \setminus \{x\}).$$

Let us comment on the quantifier case. Why do we not substitute to F just q but its restriction $q|Y \setminus \{x\}$; that is, why do we ignore the substitution q_x for x [of course, in the case where $x \in Y \cap V_f(F)$]? Just because we want to substitute for free variables of $\forall xF$ only and we want the bound variable x to remain bound after the substitution. Actually, we require something more. Substituting t for y in $F(y)$ we expect that t will have the same property that F imposes on y. To explain what we have in mind, let us fix a structure $\mathbb{A}$ and a formula $F(y)$. The set

$$D_F = \{b \in A\colon\ \mathbb{A} \models F[b/y]\}$$

consists of all the elements b of A having the property encoded by $F(y)$. Let $t(x_1, \ldots, x_n)$ be a term and let $f_t\colon\ A \longrightarrow A$ be the termal operation corresponding to t (Section 5.7). Thus,

$$f_t(a_1, \ldots, a_n) = t^{\mathbb{A}}[a_1, \ldots, a_n]$$

for $a_1, \ldots, a_n \in A$. Now, $F(t(x_1, \ldots, x_n)/y)$ should say "$x_1, \ldots, x_n$ are such that $t(x_1, \ldots, x_n)$ fulfills $F(y)$", that is,

$$\langle a_1, \ldots, a_n\rangle \in D_{F(t/x)} \ \equiv\ f_t(a_1, \ldots, a_n) \in D_F$$

or, in other words, $D_{F(t/x)} = f_t^{-1}[D_F]$.

This leads to the concept of a *proper substitution*. The idea is that no variable of substituted terms falls into the range of a quantifier in F.

Definition (Induction on $F \in \mathrm{Fm}$). For an atomic F each substitution is proper. A substitution q is proper for $\neg F$ if it is proper for F, and it is proper for $F \rightarrow G$ if it is proper for both F and G. Finally, a substitution $q\colon\ Y \longrightarrow \mathrm{Tm}$ is proper for $\forall x\ F$, if $q|Y \setminus \{x\}$ is proper for F and $x \notin V(q_y)$ for $y \in V_f(\forall xF) \cap Y$.

Example. In the ring of integers $\mathbb{Z}$ the formula

$$\exists y(x = y^2)$$

defines the set of squares. Substituting the term $x \cdot y$ for x (a nonproper substitution) we get the formula

$$\exists y(xy = y^2),$$

which is satisfied by all the integers. Thus,

$$D_{F(t/x)} = Z \times Z, \ \text{while}\ f_t^{-1}[D_F] = \{\langle a, b\rangle\colon\ a = b\}.$$

Now, we prove that proper substitutions meet our requirements.

7.1. Substitution Lemma. *Let $\mathbb{A}$ be a structure and $p\colon X \longrightarrow A$ an assignment in $\mathbb{A}$. If $q\colon Y \longrightarrow \mathrm{Tm}$ is a substitution, then for an arbitrary term t,*

(a) $$(t(q))^{\mathbb{A}}[p] = t^{\mathbb{A}}[p(\{q_y^{\mathbb{A}}[p]/y\}_{y\in Y})].$$

For an arbitrary formula F,

(b) $$\mathbb{A} \models F(q)[p] \;\equiv\; \mathbb{A} \models F[p(\{q_y^{\mathbb{A}}[p]/y\}_{y\in Y})],$$

provided that q is proper for F.

In particular, for any $s \in \mathrm{Tm}$ and $y \in X$,

$$(t(s/y))^{\mathbb{A}}[p] = t^{\mathbb{A}}[p(s^{\mathbb{A}}[p]/y)],$$

for $t \in \mathrm{Tm}$; and for any $F \in \mathrm{Fm}$,

$$\mathbb{A} \models F(s/y)[p] \;\equiv\; \mathbb{A} \models F[p(s^{\mathbb{A}}[p]/y)],$$

provided the substitution is proper.

Proof (By induction). Denote by $\bar{p}$ the assignment $p(\{q_y^{\mathbb{A}}[p]/y\}_{y\in Y})$. In the case of a variable $x \notin Y$, we have

$$x(q)^{\mathbb{A}}[p] = x^{\mathbb{A}}[p] = p(x) = x^{\mathbb{A}}[\bar{p}],$$

since $\bar{p}(x) = p(x)$.

If $x \in Y$, then

$$(x(q))^{\mathbb{A}}[p] = q_x^{\mathbb{A}}[p] = x^{\mathbb{A}}[\bar{p}],$$

since $\bar{p}(x) = q_x^{\mathbb{A}}[p]$ in this case.

For any constant symbol c,

$$(c(q))^{\mathbb{A}}[p] = c^{\mathbb{A}}[p] = c^{A} = c^{\mathbb{A}}[\bar{p}].$$

Now, the inductive step follows

$$\begin{aligned}\big(f(t_1,\ldots,t_m)(q)\big)^{\mathbb{A}}[p] &= f\big(t_1(q),\ldots,t_m(q)\big)^{\mathbb{A}}[p]\\ &= f^{A}\big((t_1(q))^{\mathbb{A}}[p],\ldots,(t_m(q))^{\mathbb{A}}[p]\big)\\ &= f^{A}\big(t_1^{\mathbb{A}}[\bar{p}],\ldots,t_m^{\mathbb{A}}[\bar{p}]\big) = \big(f(t_1,\ldots,t_m)\big)^{\mathbb{A}}[\bar{p}].\end{aligned}$$

Next, we prove the equivalence (b) using the just proved equality (a). For atomic formulas we have

$$\begin{aligned}\mathbb{A} \models ((t = s)(q))[p] \;&\equiv\; t^{\mathbb{A}}(q)[p] = s^{\mathbb{A}}(q)[p]\\ &\equiv\; t^{\mathbb{A}}[\bar{p}] = s^{\mathbb{A}}[\bar{p}] \;\equiv\; \mathbb{A} \models (t = s)[\bar{p}]\end{aligned}$$

and

$$\mathbb{A} \models \big(r(t_1,\ldots,t_n)(q)\big)[p] \;\equiv\; r^A\big(t^{\mathbb{A}}(q)[p],\ldots,t_n(q)^{\mathbb{A}}[p]\big)$$

$$\equiv\; r^A\big(t_1^{\mathbb{A}}[\bar{p}],\ldots,t_n^{\mathbb{A}}[\bar{p}]\big) \;\equiv\; \mathbb{A} \models r(t_1,\ldots,t_n)[\bar{p}].$$

The inductive step for the negation and implication is obvious:

$$\mathbb{A} \models \neg F(q)[p] \;\equiv\; \mathbb{A} \not\models F(q)[p]$$

$$\equiv\; \mathbb{A} \not\models F[\bar{p}] \;\equiv\; \mathbb{A} \models \neg F[\bar{p}]$$

and

$$\mathbb{A} \models (F \to G)(q)[p] \;\equiv\; \mathbb{A} \models \big(F(q) \to G(q)\big)[p]$$

$$\equiv\; \mathbb{A} \not\models F(q)[p] \text{ or } \mathbb{A} \models G(q)[p]$$

$$\equiv\; \mathbb{A} \not\models F[\bar{p}] \text{ or } \mathbb{A} \models G[\bar{p}]$$

$$\equiv\; \mathbb{A} \models (F \to G)[\bar{p}],$$

assuming the properness of q in F and $F \to G$, respectively.

Finally, we consider the quantifier case. Assume that q is proper for $\forall xF$. Thus, $x \notin V(q_y)$ for each $y \in \big(Y \setminus \{x\}\big) \cap V_f(F)$.

Hence, $p|V(q_y) = p(a/x)|V(q_y)$ for an arbitrary $a \in A$, whence

$$q_y^{\mathbb{A}}[p(a/x)] = q_y^{\mathbb{A}}[p]$$

for $y \in (Y \setminus \{x\}) \cap V_f(F)$ (cf. Lemma 5.6).

Note also that $p(a/x)\big(\{b_z/z\}_{z\in Z}\big)$ denotes the assignment that takes the value a at x and b_z at z, for $z \in Z$ (and $p(y)$ at the remaining $y \in Y$).

Using this and the inductive assumption we obtain

$$\mathbb{A} \models (\forall xF)(q)[p] \;\equiv\; \mathbb{A} \models \forall x\big(F(q|Y \setminus \{x\})\big)[p]$$

$$\equiv\; \forall a \in A \;\; \mathbb{A} \models F(q|Y \setminus \{x\})[p(a/x)]$$

$$\equiv\; \forall a \in A \;\; \mathbb{A} \models F[p(a/x)(\{q^{\mathbb{A}}[p(a/x)]/y\}_{y\in Y\setminus\{x\}})]$$

$$\equiv\; \forall a \in A \;\; \mathbb{A} \models F\big[p\big(\{q^{\mathbb{A}}[p(a/x)]/y\}_{y\in(Y\setminus\{x\})\cap V_f(F)}\big)(a/x)\big]$$

$$\equiv\; \forall a \in A \;\; \mathbb{A} \models F\big[p\big(\{q^{\mathbb{A}}[p]/y\}_{y\in(Y\setminus\{x\})\cap V_f(F)}\big)(a/x)\big]$$

$$\equiv\; \mathbb{A} \models \forall x\; F\big[p\big(\{q^{\mathbb{A}}[p]/y\}_{y\in(Y\setminus\{x\})\cap V_f(F)}\big)\big]$$

$$\equiv\; \mathbb{A} \models \forall x\; F[\bar{p}].$$

The last equivalence follows from Lemma 5.14. $\square$

Throughout the book it is tacitly assumed that all the occurring substitutions are proper.

We conclude this section with some additional remarks on $F(q)$. The proofs by induction are easy and will be left to the reader. First, let us notice that $F(q)$ depends only on the q_y's for $y \in V_f(F)$. More precisely, we have the following lemma.

7.2. Lemma. *If $q'|V_f(F) = q''|V_f(F)$, then $F(q')$ coincides with $F(q'')$.*
In particular, $F(q)$ is identical with $F(q|V_f(F))$.
[Prove first that $t(q')$ is $t(q'')$, for all $t \in \mathrm{Tm}$].

It follows that if the domain of q contains no free variables of F, that is, if $\mathrm{dom}(q) \cap V_f(F) = \emptyset$, then $F(q)$ is equal to F. This is because $F(q)$ is $F(\emptyset)$, by the lemma, and $F(\emptyset)$ is F. [The rigorous proof of this is by induction preceded by a proof of $t(\emptyset) = t$].

In particular, $F(q)$ is F for each sentence F and each substitution q.

Note also that an identical substitution makes no change in F. More generally, if $q_y = y$, for some variable y, then $F(q)$ is $F(q|Y \setminus \{y\})$.

7.3. The Replacement of Constants by Terms

We consider also substitutions q: $D \to \mathrm{Tm}$, where $D \subseteq C$ is a set of constants of the language L. Only the initial conditions are different than before in the definition of $t(q)$: $x(q) = x$ for every variable x and $c(q) = q(c)$ for any constant c (we assume, of course, that q is defined at least on the set of constants of a term or of a formula). In the definition of $F(q)$ the last condition can be simplified—we have $(\forall xF)(q) = \forall x(F(q))$. The definition of the properness of a substitution remains unchanged.

7.4. Expansions and Reducts

Assume that we are given two types,

$$\tau_1 = \langle\langle \arg(r_i): i \in I_1\rangle, \langle \arg(f_j): j \in J_1\rangle, K_1\rangle$$

and

$$\tau_2 = \langle\langle \arg(r_i): i \in I_2\rangle, \langle \arg(f_j): j \in J_2\rangle, K_2\rangle.$$

Type τ_2 is called an *expansion* of type τ_1 if the function $\langle \arg(r_i): i \in I_2\rangle$ is an extension of $\langle \arg(r_i): i \in I_1\rangle$, and the function $\langle \arg(f_j): j \in J_2\rangle$ is an extension of $\langle \arg(f_j): j \in J_1\rangle$ and $K_1 \subseteq K_2$. Informally, type τ_2 has more relations, functions, and distinguished elements than type τ_1. Then, if $\mathbb{A}$ is a system of type τ_2, we obtain system $\mathbb{A}|\tau_1$ by means of rejecting the relations, functions, and distinguished elements that are not in τ_1. System $\mathbb{A}|\tau_1$ is called the *reduct* of system $\mathbb{A}$ to type τ_1. Conversely, every system $\mathbb{A}$ of type τ_1 can be expanded (in general, in many ways) to a system of type τ_2. Also, the language $L(\tau_2)$ is called an *expansion* of the language $L(\tau_1)$. We may assume that the symbols corresponding to type τ_1 are identical in both languages. Thus, if τ_2 is an expansion of τ_1, then $\mathrm{Tm}(L(\tau_1)) \subseteq \mathrm{Tm}(L(\tau_2))$ and $\mathrm{Fm}(L(\tau_1)) \subseteq \mathrm{Fm}(L(\tau_2))$. In this situation the truth of the formulas F of the language $L(\tau_1)$ can be considered both for systems $\mathbb{A}$ of type τ_2 and for their reducts $\mathbb{A}|\tau_1$. As expected, by a straightforward induction on t and F we easily obtain the following lemma:

7.5. Lemma. *For an arbitrary term $t \in \mathrm{Tm}(L(\tau_1))$, a formula $F \in \mathrm{Fm}(L(\tau_1))$ and any assignment p in $\mathbb{A}$ (where $\mathbb{A}$ is of type τ_2) we have*

$$t^{\mathbb{A}}[p] = t^{\mathbb{A}|\tau_1}[p] \quad \text{and} \quad \mathbb{A} \models F[p] \quad \text{if and only if} \quad \mathbb{A}|\tau_1 \models F[p].$$

7.6. *Example*. For a fixed subset $Y \subseteq X$, let the language L^* be an expansion of L by new constants d_y, $y \in Y$. Then any system $\mathbb{A}$ (for the language L) and any assignment p in $\mathbb{A}$ defined for all the variables $y \in Y$ determine a system $\mathbb{A}^*$ (for language L^*) obtained by adding new distinguished elements $a_y = p(y)$, for $y \in Y$. This can be written as $\mathbb{A}^* = \langle \mathbb{A}, \{a_y\}_{y \in Y} \rangle$ and is called an expansion of a system by new distinguished elements.

For an arbitrary formula $F = F(y_1, \ldots, y_n)$ of language L we have

$$\mathbb{A} \models F[p] \quad \text{if and only} \quad \mathbb{A}^* \models F(d_{y_1}/y_1, \ldots, d_{y_n}/y_n).$$

Of course, the constant d_y is interpreted in $\mathbb{A}^*$ as $a_y = p(y)$. In fact, in view of Lemmas 7.1 and 7.5 we have:

$$\mathbb{A}^* \models F(d_{y_1}/y_1, \ldots, d_{y_n}/y_n) \quad \text{iff} \quad \mathbb{A}^* \models F[a_{y_1}/y_1, \ldots, a_{y_n}/y_n]$$

$$\text{iff} \quad \mathbb{A} \models F[a_{y_1}/y_1, \ldots, a_{y_n}/y_n] \quad \text{iff} \quad \mathbb{A} \models F[p].$$

Thus, if $T \subseteq \mathrm{Fm}(L)$ is any set of formulas and T^* is a set of sentences of language L^* of the form $F(d_{y_1}/y_1, \ldots, d_{y_n}/y_n)$ for $F \in T$, then we have

$$\langle \mathbb{A}, p \rangle \in \mathrm{Mod}(T) \quad \text{if and only if} \quad \mathbb{A}^* \in \mathrm{Mod}(T^*).$$

Therefore, any axiom system consisting of formulas (possibly with free variables) can be replaced, by means of adding new constants, by an axiom system consisting of sentences exclusively.

Let us note also that we often do not need to distinguish strictly between reducts and expansions, since the type—in view of Lemma 7.5—has no influence on the truth of formulas.

EXERCISES

7.1. Every substitution $q: X \longrightarrow \mathrm{Tm}$ can be extended uniquely to a homomorphism of the algebra of terms $\mathbb{T}$ into itself. Conversely, every homomorphism $h: \mathbb{T} \longrightarrow \mathbb{T}$ is of this form.

7.2. If $V(F) \subseteq \{x_1, \ldots, x_n\}$, then the formula

$$\forall x_1, \ldots, x_n F \rightarrow F(s_1/x_1, \ldots, s_n/x_n)$$

is true in every system $\mathbb{A}$, under every assignment p in $\mathbb{A}$.

7.3. Let $F(x)$ be the formula $\exists y(x + y = 0)$. Thus, in the field of reals we have $\mathbb{R} \models F[a/x]$ for every $a \in R$. However, the sentence $F(y^2 + 1/x)$, arising from an improper substitution, is false in $\mathbb{R}$. Find another example of this type.

7.4. Define the quantifier $\exists! xF$ (there exists exactly one x satisfying F) by means of a change of variables.

7.5. By the *length* $l(F)$ of a formula F we mean the least number l' for which $F \in \mathrm{Fm}_{l'}$. Show that $l(F(q)) = l(F)$ for any substitution q of terms. Thus, in inductive proofs with respect to F, one may apply the inductive assumption also to $F(q)$ for any q.

7.6. The algebras belonging to a given class $\mathbb{K}$ and having the same number of free generators with respect to the class $\mathbb{K}$ are isomorphic (cf. Exercises 6.4 and 6.5.)

7.7. Consider the algebraic rings with subtraction. Show that the ring $\mathbb{Z}[X]$ of polynomials with variables $x \in X$, for an arbitrary X, and integer coefficients is free with respect to the class of all rings.

7.8. Find a free group and an abelian free group with a fixed number of generators.

7.9. Every four-element Boolean algebra is isomorphic with the power-set algebra $P(X)$, where X is any two-element set. Show that this is a free algebra with one generator.

7.10. Prove the generalized distributivity laws for Boolean algebras:

$$\sum_{i=1}^{n} a_{i1} \cdot \cdots \cdot a_{im} = \prod_{j_1, \ldots, j_n = 1}^{m} (a_{1j_1} + \cdots + a_{nj_n})$$

and

$$\prod_{i=1}^{n} (a_{i1} + \cdots + a_{im}) = \sum_{j_1, \ldots, j_n = 1}^{m} a_{1j_1} \cdot \cdots \cdot a_{nj_n}$$

for an arbitrary matrix $\{a_{ij}\colon\ i = 1, \ldots, n,\ j = 1, \ldots, m\}$ of elements of the algebra.

7.11. Show that the subalgebra generated by a subset E of a Boolean algebra consists of finite sums of elements of the form

$$\varepsilon_1 a_1 \cdot \cdots \cdot \varepsilon_n a_n, \quad a_i \in E \quad \text{and} \quad \varepsilon_i = \pm 1, \quad i = 1, \ldots, n,$$

where $\varepsilon a = a$ for $\varepsilon = +1$ and $\varepsilon a = -a$ for $\varepsilon = -1$.

8

THEOREMS AND PROOFS

It is one of the most fascinating achievements of logic that the notion of a mathematical proof (and hence of a theorem) can be defined in a fully satisfactory way. The definition is in fact very simple but its justification requires the completeness theorem which is not proven until Chapter 11.

As already mentioned an arbitrary set T of formulas of a given language L can be regarded as a system of axioms of some mathematical theory T^*. Any theorem of the theory T^* is derived from the axioms $F \in T$ or from theorems already proved with the help of logical reasoning. Selected elementary steps of the logical reasonings are called rules of inference. Here is an example, the rule of *modus ponens* or *detachment* (explicitly stated by Russell):

$$\text{from } F \to G \text{ and } F \text{ infer } G.$$

Also, in the proofs we can make use of some selected universally true formulas (that is, true in every structure $\mathbb{A}$ under every assignment p), called the axioms of logic (of course, the choice of the logical axioms does not depend on T).

The first axiom system of logic was introduced and developed by Hilbert and Ackerman [HA]. In our approach we follow Enderton [E1].

8.1. The Axioms of Logic

By a *(universal) closure* with respect to the variables $x_1, \ldots, x_n$ of a formula F we mean the formula $\forall x_1, \ldots, x_n F$. The axioms of logic are closures with respect to every finite sequence of variables of all the formulas occurring in the following groups.

1. *Axioms of the Propositional Calculus*. For any formulas F, G, H:

$$F \to (G \to F),$$

$$[F \to (G \to H)] \to [(F \to G) \to (F \to H)],$$

$$(\neg G \to \neg F) \to (F \to G).$$

2. *Axioms of Equality*. For any terms t, t_1, t_2, t_3;

$$t = t,$$

$$t_1 = t_2 \rightarrow t_2 = t_1,$$

$$t_1 = t_2 \rightarrow (t_2 = t_3 \rightarrow t_1 = t_3).$$

For every relation symbol r and every function symbol f ($n = \arg(r)$ and $m = \arg(f)$),

$$t_1 = t_2 \wedge \cdots \wedge t_n = s_n \rightarrow \big(r(t_1, \ldots, t_n) \rightarrow r(s_1, \ldots, s_n)\big),$$

$$t_1 = t_2 \wedge \ldots \wedge t_m = s_m \rightarrow \big(f(t_1, \ldots, t_m) = f(s_1, \ldots, s_m)\big).$$

3. *Substitution Axioms*. For any formula F, variable x and term t such that the substitution $F(t/x)$ is proper,

$$\forall x F \rightarrow F(t/x).$$

4. *Axioms of Distributivity of a Quantifier*. For any formulas F, G and any variable x,

$$\forall x(F \rightarrow G) \rightarrow (\forall x F \rightarrow \forall x G).$$

5. *Adding a Redundant Quantifier*. For any formula F and any variable $x \notin V(F)$,

$$F \rightarrow \forall x\, F$$

A direct verification shows that the formulas occurring in groups 1 to 5 are universally true (in the case of group 3 we apply the rules in Lemma 7.1). Moreover, any closure of a universally true formula is also a universally true formula. Hence we obtain the following theorem.

8.2. Theorem. *Every axiom of logic is a universally true formula.*

8.3. The Definition of the Theorems of a Theory

Let T be an arbitrary set of formulas of a language L. The set T^* of the *theorems* of the theory T is defined by induction:

$$T^* = \bigcup\{T_n\colon\ n \in N\},$$

where

$$T_0 = T \cup \mathrm{LOG}$$

(LOG denotes the set of the axioms of logic) and

$$T_{n+1} = T_n \cup \{G: \exists F[F \in T_n \quad \text{and} \quad F \to G \in T_n]\}.$$

In other words, T_{n+1} is formed from T_n by adding all the conclusions obtained by an application of modus ponens to formulas from T_n.

Directly from this definition it follows that T^* is the least set of formulas containing $T \cup \text{LOG}$ and closed under the rule modus ponens.

Instead of $F \in T^*$ we usually write $T \vdash F$ and say that F is a theorem of T or F is provable in T or T proves F. In particular, $T^* \vdash F$ implies $F \in T^*$.

We have already used the word "theory" many times. Here let us make it clear that by a *theory* we mean an arbitrary set $T \subseteq \text{Fm}$.

We shall use the following induction principle for theorems (which follows immediately from the definition in Section 8.3): to show that a given property holds for all the theorems of a theory T, we check that it holds for the axioms $F \in T$ and for the axioms of logic $F \in \text{LOG}$, and next that it is preserved when passing from theorems of the form $F \to G$ and F to G. Here is an example.

8.4. Theorem. *If $T \vdash F$, then the formula F is true in every model of the theory T.*

Proof. Let $\langle \mathbb{A}, p \rangle \in \text{Mod}(T)$. Thus, $\mathbb{A} \models F[p]$ for every formula $F \in T$, and also for $F \in \text{LOG}$, since the latter is universally true. If now we have

$$\mathbb{A} \models F[p] \quad \text{and} \quad \mathbb{A} \models (F \to G)[p],$$

then we also have $\mathbb{A} \models G[p]$ directly from the definition of truth. Hence $\mathbb{A} \models F[p]$ for every theorem of theory T. □

Remarks

The definition of a theorem of T given in Section 8.3 requires justification. Theorem 8.4 is a part of it. It says that any theorem of T, according to this definition, behaves as it should—it is true in every model of T. But there is still another problem, presumably much more difficult: does the definition capture all potential theorems? Fortunately the word "potential" can be here defined precisely with the help of the truth relation—every true formula in every model of T is a potential theorem and nothing else. Hence, in order to justify the definition of a theorem we have to prove the implication converse to 8.4, that is,

If F is true in every model of the theory T, then $T \vdash F$.

This will follow from the completeness theorem proved in Chapter 11.

8.5. Theorem. *If $T \vdash F$, then there is a finite sequence of formulas $\langle F_0, \ldots, F_n \rangle$ such that $F_n = F$ and every term F_i, $i = 0, \ldots, n$, is either an axiom of T or of* LOG *or follows by modus ponens from some two earlier terms.*

Proof. The proof goes by induction on the theorems of T. If $F \in T \cup \mathrm{LOG}$, then the one-element sequence $\langle F \rangle$ satisfies the required conditions. If we have such sequences $\langle F_0, \ldots, F_n, F \rangle$ and $\langle H_0, \ldots, H_m, F \to G \rangle$ for the theorems F and $F \to G$, then the sequence $\langle F_0, \ldots, F_n, F, H_0, \ldots, H_m, F \to G, G \rangle$ is as demanded for G, which finishes the proof. □

Every sequence $\langle F_0, \ldots, F_n \rangle$ satisfying the conditions stated in the theorem 8.5 is called a *logical* (or *formal*) *proof* in T (of the formula F_n). Hence, if $\langle F_0, \ldots, F_n \rangle$ is a proof in T, then $T \vdash F_i$ for each $i = 0, \ldots, n$. In fact, each initial segment of a proof is a proof as well.

8.6. Corollary. *If $T \vdash F$, then there is a finite subset $T_0 \subseteq T$ such that $T_0 \vdash F$.*

Proof. F has a proof $\langle F_0, \ldots, F_n \rangle$ in T and we take as T_0 the set of those $H \in T$ that occur in this proof. □

If $T = \emptyset$, then instead of $\emptyset \vdash F$, we write simply $\vdash F$. In this case the formula F has a proof in which the only axioms are the axioms of logic and the other formulas follow from them by modus ponens. Thus, if $\vdash F$ holds, then F is universally true. The formulas F for which we have $\vdash F$ are called the *theorems of logic* (*tautologies*).

8.7. *Example*. We have $\vdash F \to F$, for any formula F. Let us show the proof of $F \to F$ in LOG. The formulas

$$H_1\colon F \to [(F \to F) \to F],$$

$$H_2\colon \{F \to [(F \to F) \to F]\} \to \{[F \to (F \to F)] \to (F \to F)\}$$

are in LOG. Applying the modus ponens we get

$$H_3\colon\ [F \to (F \to F)] \to (F \to F).$$

Also the following formula is an axiom of logic:

$$H_4\colon\ F \to (F \to F)$$

and by modus ponens again we derive $F \to F$. Thus, the sequence

$$\langle H_1, H_2, H_3, H_4, F \to F \rangle$$

is a proof of the formula $F \to F$ in logic.

EXERCISES

A Boolean algebra $\mathbb{A}$ is called *atomic* if for every element $a > \mathbb{O}$ there exists an atom less than or equal to a.

8.1. Every finite Boolean algebra is atomic.

8.2. Show that in a finite Boolean algebra every element greater than $\mathbb{O}$ is a sum of a number of atoms. Hence, a finite Boolean algebra is isomorphic with the power-set algebra $P(X)$, where X is the set of all the atoms.

8.3. Generalization of Exercise 8.2: if an algebra $\mathbb{A}$ is atomic and complete, then $\mathbb{A}$ is isomorphic with $P(X)$, where X is the set of the atoms of the algebra $\mathbb{A}$ (see Exercise 2.8).

8.4. In a finite Boolean algebra every filter is principle.

8.5. In an infinite Boolean algebra there always are nonprinciple ultrafilters. (*Hint*: if the algebra has any atoms, then the family of their complements is centered).

8.6. Let $S = S(\mathbb{A})$ be the set of ultrafilters of a given Boolean algebra $\mathbb{A}$. For any $a \in A$ define $U_a = \{p \in S(\mathbb{A}): \ a \in p\}$.

Show that the family $B = \{U_a: \ a \in A\}$ is a field of sets and that the mapping $h(a) = U_a$, for $a \in A$, is an isomorphism of $\mathbb{A}$ onto B.

8.7. Strengthening of Exercise 8.6: we introduce the Stone topology on the set $S = S(\mathbb{A})$, treating the sets U_a as the basic open sets. Notice that the sets U_a are open–closed in this topology. Prove that the topological space S is a compact Hausdorff space and that every open–closed set has the form U_a (Stone's theorem: cf. [S6]). The topological space $S = S(\mathbb{A})$ is called the *Stone space of the algebra* $\mathbb{A}$.

8.8. Check that isomorphic algebras have homeomorphic Stone spaces.

8.9. To a finite algebra $P(X)$, where X has n elements, there corresponds a discrete Stone space S with n elements.

8.10. An ultrafilter p of an algebra $\mathbb{A}$ is principal if and only if p is an isolated point in the space $S(\mathbb{A})$.

8.11. The Stone space for the algebra $FC(X)$ (see Exercise 2.6) is a one-point compactification of the discrete space X.

9

THEOREMS OF THE LOGICAL CALCULUS

In this chapter we investigate more deeply the relation of provability. In particular, we introduce and investigate the notion of the consistency of a theory. Also, we define the logical equivalence (more generally, the equivalence in T), which permits us to look on the set Fm of formulas as on a Boolean algebra.

Let us simplify the notation a bit: instead of $T \cup \{F_1, \ldots, F_n\} \vdash F$ we shall often write $T, F_1, \ldots, F_n \vdash F$. In particular, for $T = \emptyset$ we usually write $F_1, \ldots, F_n \vdash F$.

9.1. Theorem (On deduction). *For every theory T and formula F (in a given language L) we have*

$$(T, F \vdash G) \quad \text{if and only if} \quad (T \vdash F \to G)$$

for an arbitrary formula G of the language L.

Proof. The proof of the implication to the right goes by induction on the theorems G of $T \cup \{F\}$. Assume $T, F \vdash G$. If G is an axiom and $G \in T \cup \mathrm{LOG}$, then we have $T \vdash G$, and since $G \to (F \to G)$ is an axiom of logic, we get $T \vdash F \to G$ by modus ponens. If $G = F$, then we have $\vdash F \to G$ in view of Example 8.7, which implies immediately $T \vdash F \to G$. Now, assume the conclusion for the theorems H and $H \to G$ of the theory $T \cup \{F\}$. Thus, we have

$$T \vdash (F \to H) \quad \text{and} \quad T \vdash (F \to (H \to G)).$$

Since the formula $[F \to (H \to G)] \to [(F \to H) \to (F \to G)]$ is an axiom of logic, hence, by applying the modus ponens twice, we obtain $T \vdash (F \to G)$.

The converse implication is obtained very easily; the condition $T \vdash (F \to G)$ implies obviously $T, F \vdash (F \to G)$, and since clearly $T, F \vdash F$, hence also $T, F \vdash G$, which completes the proof. □

9.2. Theorem (On indirect proofs). *If* $T, \neg G \vdash \neg F$, *then* $T \vdash (F \to G)$.

Proof. From 9.1, the deduction theorem, the condition $T, \neg G \vdash \neg F$ implies $T \vdash (\neg G \to \neg F)$. The formula $(\neg G \to \neg F) \to (F \to G)$ is an axiom of logic, and so we have also $T \vdash (F \to G)$. □

A set of formulas (a theory) T is called *inconsistent* if there is a formula F for which $T \vdash F$ and $T \vdash \neg F$ hold simultaneously. Otherwise, we say that T is *consistent*. If T has a model, then T is consistent in view of Theorem 8.4.

Note the following (from Corollary 8.6): *If every finite subset of a theory* T *is consistent, then* T *is also consistent.*

Since, if $T \vdash F$ and $T \vdash \neg F$ for some F, then for some finite subsets $T_0, T_1 \subseteq T$ we would have $T_0 \vdash F$ and $T_1 \vdash \neg F$. But then the finite subset $T_0 \cup T_1$ of T would be inconsistent.

9.3. Theorem. *If* T *is an inconsistent theory, then* $T \vdash G$ *holds for every formula* G.

Proof. The formula $\neg F \to (F \to G)$ (the so-called Duns Scotus law) is (for any F, G) a theorem of logic; we have clearly $\neg F, \neg G \vdash \neg F$, which implies $\neg F \vdash (F \to G)$, by Theorem 9.2, whence the Duns Scotus law follows from the deduction theorem. Now, if T is inconsistent, that is, $T \vdash F$ and $T \vdash \neg F$ for some formula F, then, by applying the modus ponens twice, we infer $T \vdash G$ for any formula G. □

Note also the following fact

9.4. *If* T *is consistent then so is the set* $\{F\colon\ T \vdash F\}$ *of the theorems of* T.

This follows immediately from the definition of the relation $\vdash$.

Let us derive now a few rules of the propositional calculus.

9.5. *If* $\vdash \big(F \to (F \to G)\big)$, *then also* $\vdash (F \to G)$.

Proof. From the deduction theorem it follows that the condition $\vdash \big(F \to (F \to G)\big)$ implies $F, F \vdash G$, that is, $F \vdash G$, whence $\vdash (F \to G)$, from the deduction theorem. □

9.6. *We have* $\{F' \to F, G \to G'\} \vdash \big((F \to G) \to (F' \to G')\big)$.

Proof. We have (by modus ponens)

$$F' \to F,\ G \to G',\ F \to G,\ F' \vdash G',$$

whence 9.6 follows by the deduction theorem. □

The formulas F, F' are called *logically equivalent* (write $F \equiv_l F'$), if both implications $F \to F'$ and $F' \to F$ are theorems of logic;

$$\vdash (F \to F') \quad \text{and} \quad \vdash (F' \to F).$$

Thus, from 9.6 it follows that

If $F \equiv_l F'$ and $G \equiv_l G'$, then $(F \to G) \equiv_l (F' \to G')$.

9.7. *$F \equiv_l \neg\neg F$, for any formula F.*

Proof. We have

$$\vdash (\neg\neg F \to (\neg F \to \neg\neg\neg F))$$

by the Duns Scotus law, whence

$$\neg\neg F \vdash (\neg F \to \neg\neg\neg F).$$

Since the formula

$$(\neg F \to \neg\neg\neg F) \to (\neg\neg F \to F)$$

is an axiom of logic, we obtain $\neg\neg F \vdash (\neg\neg F \to F)$, whence $\vdash (\neg\neg F \to F)$, by 9.5 and by the deduction theorem. To prove $\vdash (F \to \neg\neg F)$ notice that the following holds:

$$\vdash (\neg\neg\neg F \to \neg F),$$

by the part of the theorem already proved. On the other hand, we have the axiom of logic

$$(\neg\neg\neg F \to \neg F) \to (F \to \neg\neg F),$$

whence $\vdash (F \to \neg\neg F)$ by the modus ponens. $\square$

9.8. *$(F \to G) \equiv_l (\neg G \to \neg F)$ for arbitrary F, G.*

Proof. The implication $(\neg G \to \neg F) \to (F \to G)$ is an axiom. From 9.6 we infer

$$\{\neg\neg F \to F,\ G \to \neg\neg G,\ F \to G\} \vdash (\neg\neg F \to \neg\neg G),$$

whence

$$(F \to G) \vdash (\neg\neg F \to \neg\neg G),$$

on the basis of 9.7. The formula

$$(\neg\neg F \to \neg\neg G) \to (\neg G \to \neg F)$$

is an axiom. We obtain thus $(F \to G) \vdash (\neg G \to \neg F)$, whence $\vdash ((F \to G) \to (\neg G \to \neg F))$, which completes the proof. □

From 9.8 we immediately infer that

$$\textit{if } F \equiv_l G, \quad \textit{then} \quad \neg F \equiv_l \neg G.$$

9.9. *The following formulas are theorems of logic:*

(a) $\neg(F \to G) \to F$,

(b) $\neg(F \to G) \to \neg G$,

(c) $F \to [G \to \neg(F \to \neg G)]$.

Proof. From the Duns Scotus law $\vdash (\neg F \to (F \to G))$ using 9.8 we obtain

$$\vdash (\neg F \to G) \to \neg\neg F).$$

Now, we replace $\neg\neg F$ by F, on the basis of 9.6 and 9.7, which yields a).

The formula in (b) is, in view of 9.8, logically equivalent with the axiom $G \to (F \to G)$.

Proof of (c): We have $F, (F \to \neg G) \vdash \neg G$, whence

$$F \vdash ((F \to \neg G) \to \neg G.$$

Applying 9.8 we infer

$$F \vdash (\neg\neg G \to \neg(F \to \neg G)).$$

Now, we can replace $\neg\neg G$ by G so that

$$F \vdash (G \to \neg(F \to \neg G)),$$

which immediately gives (c). □

Since $\neg(F \to \neg G)$ is denoted by $F \wedge G$ (the conjunction), the formulas in 9.9 can be written as

(a) $(F \wedge G) \vdash F$.

(b) $(F \wedge G) \vdash G$.

(c) $F, G \vdash (F \wedge G)$.

Hence we obtain the following equivalence:

$$F \equiv_l G \quad \text{if and only if} \quad \vdash (F \equiv G).$$

9.10. *For arbitrary F, G we have*

$$\vdash (\neg F \to G) \to [(\neg F \to \neg G) \to F]).$$

Proof. By 9.9 (c), we have

$$\neg F, G \vdash (\neg(\neg F \to \neg G))$$

On the other hand, we have $\{\neg F \to G, \neg F\} \vdash G$, whence

$$\neg F \to G, \neg F \vdash \neg(\neg F \to \neg G),$$

so by Theorem 9.2,

$$(\neg F \to G) \vdash ((\neg F \to \neg G) \to F),$$

which finishes the proof. □

9.11. Theorem (Reductio ad absurdum). *If the formula $\neg F$ is inconsistent with T (i.e., the set $T \cup \{\neg F\}$ is inconsistent), then $T \vdash F$.*

Proof. There is a formula G for which

$$T, \neg F \vdash G \quad \text{and} \quad T, \neg F \vdash \neg G,$$

that is

$$T \vdash (\neg F \to G) \quad \text{and} \quad T \vdash (\neg F \to \neg G).$$

Applying 9.10 we obtain $T \vdash F$, which completes the proof. □

Obviously, the converse of Theorem 9.11 is also true.

We say that a formula F is *decidable* on the basis of the theory T if we have either $T \vdash F$ or $T \vdash \neg F$. Otherwise, if neither $T \vdash F$ nor $T \vdash \neg F$, we say that the formula F is *undecidable* or *independent* of T. Thus, the independence of F of T is equivalent with the consistency of two theories: T enlarged by F and T enlarged by $\neg F$.

9.12. Lindenbaum's Theorem. *Every consistent set of formulas T has a consistent extension $T^* \supseteq T$ such that for every formula F we have either $T^* \vdash F$ or $T^* \vdash \neg F$. Thus, every formula is decidable on the basis of T^*.*

Proof. The family R of all consistent extensions of the set T is nonempty (since $T \in R$) and partially ordered by inclusion. If $R_0 \subseteq R$ is a linearly ordered subfamily, then $\bigcup R_0 \in R$, that is, $\bigcup R_0$ is a consistent set. If a finite subset $\{F_1, \ldots, F_n\} \subseteq \bigcup R_0$, then for some $S_1, \ldots, S_n \in R_0$ we have $F_1 \in S_1, \ldots, F_n \in S_n$. Then $F_1, \ldots, F_n \in S$, where S is the largest of the sets $S_1, \ldots, S_n$. Hence the set $\{F_1, \ldots, F_n\}$ is consistent as a subset of a consistent set S. Thus, the set $\bigcup R_0$ is consistent, since all its finite subsets are. By the Kuratowski–Zorn lemma there exists a maximal consistent extension T^* of the set T, that is, a maximal element of the family R. Let F be any formula. If $T^* \nvdash F$ then $T \cup \{\neg F\}$ is a consistent set by Theorem 9.11, whence $\neg F \in T^*$, by maximality. In particular, $T^* \vdash \neg F$. □

Let us note that for any maximal consistent set S the condition $S \vdash F$ is equivalent to $F \in S$ (because $\{F\colon S \vdash F\}$ is consistent; see 9.4).

9.13. The Lindenbaum Algebra

We can treat the logical connectives as certain operations on the set Fm of formulas of the given language L.

Consider the algebraic system

$$\mathbb{F}\mathrm{m} = \langle \mathrm{Fm}, \vee, \wedge, \neg, \neg(H_0 \to H_0), H_0 \to H_0 \rangle,$$

in which the signs $\vee$, $\wedge$, and $\neg$ denote the operations of forming disjunction, conjunction, and negation, respectively, and H_0 is a fixed sentence. This system is not a Boolean algebra since, for example, $(F \vee G) \neq (G \vee F)$, unless $F = G$. Let f be an arbitrary n-ary operation with respect to Fm determined by a term of the language of Boolean algebras (see Section 5.7). The formula $f(F_1, \ldots, F_n)$ is for obvious reasons called a *Boolean combination* of the formulas $F_1, \ldots, F_n$. The rules of the propositional calculus in 9.6 to 9.9 prove (by induction on n) that the formula

9.14 $$(F_1 \equiv G_1) \wedge \ldots \wedge (F_n \equiv G_n) \to \big(f(F_1, \ldots, F_n) \equiv f(G_1, \ldots, G_n)\big)$$

is a theorem of logic.

For $T \subseteq \mathrm{Fm}$ we say that the formulas F, G are equivalent in T (we write $F \equiv_T G$), if both $T \vdash (F \to G)$ and $T \vdash (G \to F)$ hold [or equivalently $T \vdash (F \equiv G)$]. From 9.14 it follows that $f(F_1, \ldots, F_n) \equiv_T f(G_1, \ldots, G_n)$, provided $F_i \equiv_T G_i$, $i = 1, \ldots, n$. At the same time, for arbitrary F, G, H we have

$$\vdash (F \equiv F),$$

$$\vdash \big((F \equiv G) \to (G \equiv F)\big),$$

$$\vdash \big(F \equiv G \wedge G \equiv H) \to F \equiv H),$$

whence it follows that the relation "$\equiv_T$" is a congruence in $\mathbb{F}\mathrm{m}$. The factor algebra $\mathbb{F}(T) = \mathbb{F}\mathrm{m}/\equiv_T$ is called the Lindenbaum algebra of the theory T. The algebra $\mathbb{F}(\emptyset) = \mathbb{F}\mathrm{m}/\equiv_l$ is denoted by the symbol $\mathbb{F}$. It is easy to verify that

9.15. *If T is a consistent set, then $\mathbb{F}(T)$ is a Boolean algebra.*

The unit of the algebra $\mathbb{F}(T)$ is the equivalence class

$$[H_0 \to H_0] = \{F\colon\ T \vdash (F \ \equiv\ (H_0 \to H_0))\} = \{F\colon\ T \vdash F\},$$

that is, this is the class of all the theorems of the theory T. Since, if $T \vdash ((H_0 \to H_0) \to F)$, then also $T \vdash F$, because evidently $T \vdash (H_0 \to H_0)$. Conversely, from $T \vdash F$ it follows that $T \vdash (F \to (H_0 \to H_0))$, by the deduction theorem and also that $T \vdash ((H_0 \to H_0) \to F)$, since, more generally, we have $T \vdash (G \to F)$ for any formula G. Thus, the zero of the algebra consists of the negations of the theorems:

$$\mathbb{O} = \{\neg F\colon\ T \vdash F\}.$$

The assumption that T is consistent implies $\mathbb{O} \neq \mathbb{1}$.

Let us check, for instance, the axiom of commutativity:

$$[F] + [G] = [G] + [F].$$

It is sufficient to prove

$$\vdash ((F \vee G) \ \equiv\ (G \vee F)).$$

But $F \vee G$ is the formula $\neg F \to G$, which is logically equivalent with $\neg G \to F$, that is, with $G \vee F$, by 9.7 and 9.8. In a similar way, the other axioms can be verified.

EXERCISES

9.1. Prove the rule 9.14.

9.2. Show that the Boolean ordering in the algebra $\mathbb{F}(T)$ satisfies the condition

$$[F] \leq [G] \quad \text{if and only if} \quad T \vdash (F \to G).$$

(*Hint*: $\vdash ((F \to G) \to [(\neg F \to G) \to G])$ or use Exercise 4.2.)

9.3. Show that a formula F is consistent with T if and only if $[F] > \mathbb{O}$ in $\mathbb{F}(T)$ and that a set S of formulas is consistent if the set $[S] = \{[F]\colon F \in S\}$ is centered in $\mathbb{F}$ (Exercise 3.9).

9.4. If $T \subseteq \text{Fm}$ is a consistent set, then $\mathcal{F}_T = \{[F]: T \vdash F\}$ is a filter in the algebra $\mathbb{F}$ and every filter in $\mathbb{F}$ is of this form. In addition, show that

(a) $\mathcal{F}_T$ is principle if and only if the theory T has a finite axiomatization;

(b) $\mathcal{F}_T$ is an ultrafilter if and only if there are no formulas independent of T.

9.5. The factor algebra $\mathbb{F}/\mathcal{F}_T$ is isomorphic with $\mathbb{F}(T)$ (see Exercise 9.4).

9.6. Let Y be a set of variables of the language L. The algebra $\mathbb{F}_Y(T)$ is defined in the same way as $\mathbb{F}(T)$, except that we restrict ourselves only to those formulas F, for which $V(F) \subseteq Y$. Show that the function $h([F]_Y) = [F]$, assigning to the elements $[F]_Y \in \mathbb{F}_Y(T)$ the class $[F] \in \mathbb{F}(T)$, is an embedding.

9.7. We may treat the algebra $\mathbb{F}_Y(T)$, in view of Exercise 9.6, as a subalgebra of the algebra $\mathbb{F}(T)$. Show that $\mathbb{F}(T)$ is a directed sum of the subalgebras $\mathbb{F}_Y(T)$, where Y runs over the finite subsets of the set of variables.

9.8. Every open formula (i.e., a formula without quantifiers) is logically equivalent with a disjunction of conjunctions $F_1 \wedge \cdots \wedge F_n$, where each F_i is an atomic formula or a negation of an atomic formula (see Exercise 7.11).

9.9. Let $\mathbb{A}$ be a Boolean algebra generated by a set $E \subseteq A$. Show that if a given function h: $E \longrightarrow B$ (where $\mathbb{B}$ is an arbitrary Boolean algebra) satisfies the condition: *if* $\varepsilon_1 a_1 \ldots \varepsilon_n a_n = \mathbb{O}$, *then* $\varepsilon_1 h(a_1) \ldots \varepsilon_n h(a_n) = \mathbb{O}$ *for any* $a_1, \ldots,$ $a_n \in E$ *and* $\varepsilon = \pm 1$, then h can be extended to a homomorphism of the algebra $\mathbb{A}$ into $\mathbb{B}$ (Sikorski's extension theorem [S2]). (*Hint*: Apply Exercise 7.11).

9.10. Prove that the algebra $B(X)$ of the open closed subsets of the Cantor cube $X(S)$ (Exercise 2.9) is a free algebra with card S many free generators (with respect to the class of all Boolean algebras). (*Hint*: Use the Sikorski extension theorem, Exercise 9.9.).

10

GENERALIZATION RULE AND ELIMINATION OF CONSTANTS

We continue to formulate laws of the logical calculus. We discuss here such quantifier rules as generalization and change of bound variables. Also, we convince ourselves that addition of new constants to the language does not destroy the consistency of a theory.

10.1. Theorem (The generalization rule). *Assume that $x \notin V(F)$ for every $F \in T$. Then the condition $T \vdash F$ implies $T \vdash \forall xF$.*

Proof. The proof goes by induction with respect to the theorems F of the theory T. If $F \in \mathrm{LOG}$, then also $\forall xF \in \mathrm{LOG}$, thus $T \vdash \forall xF$. If $F \in T$, then by the assumption $x \notin V(F)$. Thus, $F \to \forall xF$ is an axiom of logic and the sequence $\langle F, F \to \forall xF, \forall xF \rangle$ is a proof of the formula $\forall xF$ in T. Assume now that F is inferred by the modus ponens from H and $H \to F$ and that the conclusion holds for the latter formulas. Thus, there are proofs $d_1 = \langle H_1, \ldots, H_n \rangle$ of the formula $\forall xH$ and $d_2 = \langle G_1, \ldots, G_m \rangle$ of the formula $\forall x(H \to F)$. It follows that the sequence

$$\langle H_1, \ldots, H_n, G_1, \ldots, G_m$$

$$\forall x(H \to F) \to (\forall x\, H \to \forall xF), \forall xH \to \forall xF, \forall xF \rangle$$

is a proof of the formula $\forall xF$ in T. $\square$

10.2. Corollary. *If T is a set of sentences and $T \vdash F$, then also $T \vdash \forall x_1, \ldots, x_n\, F$ for any sequence of variables $x_1, \ldots, x_n$.*

In particular, for $T = \emptyset$ we obtain the following rule:

If F is a theorem of logic, then any closure $\forall x_1, \ldots, x_n\, F$ is also a theorem of logic.

10.3. Change of Variables. *If a variable y does not occur in F (neither free nor bound), then the formulas $\forall xF$ and $\forall yF(y/x)$ are logically equivalent.*

Proof. The formula $\forall xF \to F(y/x)$ is an axiom of logic. Thus, we have $\forall xF \vdash F(y/x)$ and if we apply the generalization rule stated in Theorem 10.1, then we obtain $\forall xF \vdash \forall yF(y/x)$, and hence $\vdash (\forall xF \to \forall yF(y/x))$. Conversely, substituting x for y in $F(y/x)$ we come back to F:

$$F(y/x)(x/y) \text{ is } F.$$

Therefore the formula $\forall yF(y/x) \to F$ is an axiom of logic, whence $\forall yF(y/x) \vdash F$, and thus $\forall yF(y/x) \vdash \forall xF$, by Theorem 10.1, since $x \notin V(\forall yF(y/x))$. Thus, we have $\vdash (\forall yF(y/x) \to \forall xF)$, which completes the proof. □

10.4. Corollary. *Assume that we are given a term t and a variable z. For every formula F there exists a formula G obtained from F by a change of bound variables such that $F \equiv_l G$ and the substitution $G(t/z)$ is proper.*

Proof. First, let us notice that adjoining a quantifier preserves the logical equivalence

$$\textit{if } F \equiv_l G \textit{ then also } \forall xF \equiv_l \forall x\, G.$$

This follows directly from the distributivity axiom

$$\forall x(F \to G) \to (\forall xF \to \forall x\, G),$$

since from $\vdash F \to G$, we infer $\vdash \forall x(F \to G)$, by Corollary 10.2, whence $\vdash (\forall xF \to \forall xG)$, and similarly for the converse implication.

Now to obtain G we apply induction on F. Only the quantifier case $\forall xF$ is relevant (in the connective cases we use only the fact that the connectives preserve the logical equivalence). So assume that we have a $G \equiv_l F$ obtained by a change of variables such that $G(t/z)$ is proper. If $x \in V(t)$, choose a y occurring neither in G nor in t. By Corollary 10.2, $\forall yG(y/x) \equiv_l \forall xG$ and $\forall xG \equiv_l \forall xF$ as just proved. Clearly, t can be properly substituted for z in $\forall yG(y/x)$. □

10.5. Theorem. *Assume that a constant c does not occur in any formula of the set T. Under this assumption if $T \vdash F$, then $T \vdash \forall yF(y/c)$ for some variable y, and moreover there is a proof from T of the formula $\forall yF(y/c)$ in which the constant c does not occur.*

Proof. Let $d = \langle F_0, \dots, F_n \rangle$ be a proof of the formula F in T. Choose a variable y not occurring in this proof (i.e., in any formula F_i, $i = 0, \dots, n$). It is easy to see that the sequence $d(y/c) = \langle F_0(y/c), \dots, F_n(y/c) \rangle$ is a proof in T of the formula $F(y/c)$. Since, if $F_i \in \text{LOG}$, then also $F_i(y/c) \in \text{LOG}$, by a direct verification. If $F_i \in T$, then by the assumption $F_i(y/c)$ is identical with F_i. Finally, if F_l is $F_k \to F_i$ for some $k, l \leq i$ (i.e., F_i is derived by modus ponens from F_k and F_l), then $F_l(y/c)$ is the formula $F_k(y/c) \to F_i(y/c)$, and thus $F_i(y/c)$ is derived from $F_k(y/c)$ and $F_l(y/c)$ by modus ponens. Hence, we have $T_0 \vdash F_i(y/c)$, where $T_0 = \{F_i(y/c)\colon i = 0, \dots, n\}$. Since the proof $d(y/c)$ does not contain the constant c, we have $T_0 \vdash F(y/c)$ in the language L' arising from the original language L by rejecting the constant c. Applying the generalization rule in Theorem 10.1, we infer in this case $T_0 \vdash \forall y F(y/c)$ in L', and hence also $T \vdash \forall y F(y/c)$ in L. In this way the proof of Theorem 10.5 has been completed. □

10.6. Corollary (Elimination of constants). *Assume that the constant c does not occur in any formula of the set T. Then, provided c does not occur in F, the condition $T \vdash F(c/x)$ implies $T \vdash \forall x F$.*

Proof. By Theorem 10.5 we obtain $T \vdash \forall y F(c/x)(y/c)$ for some variable y that does not occur in $F(c/x)$. But then $F(c/x)(y/c)$ is identical with $F(y/x)$. Thus, we have $T \vdash \forall y F(y/x)$ and hence $T \vdash \forall x F$, since $\forall x F \equiv_l \forall y F(y/x)$ by Section 10.3. □

Let the language L^* be an expansion of language L by new constants (see Section 7.3). In the language L^* we have more formulas and therefore, more sequences of formulas and hence more proofs than in L.

10.7. Corollary. *Let $T \subseteq Fm(L)$ and $F \in Fm(L)$. If $T \vdash F$ in L^*, then also $T \vdash F$ in L.*

Proof. Let $d = \langle F_i\colon i = 1, \dots, n \rangle$ be a proof of the formula F in the language L^* and let $c_1, \dots, c_m$ be all the new constants which occur in formulas F_i of this proof. Thus, d is a proof in the language $L\{c_1, \dots, c_m\}$—the expansion of the language L by the constants $c_1, \dots, c_m$. The constants $c_1, \dots, c_m$ do not occur in the formulas from the set T. Applying Theorem 10.5, we obtain

$$T \vdash \forall y_1, \dots, y_m F(y_1/c_1, \dots, y_m/c_m) \quad \text{in } L.$$

The constants $c_1, \dots, c_m$ do not occur in F either, and thus the last formula is identical with $\forall y_1, \dots, y_m F$. However, since y_1 does not occur in F,

$$\forall y_1, \dots, y_m F \to \forall y_2, \dots, y_m F$$

is an axiom of logic. So we obtain $T \vdash \forall y_2, \dots, y_m\, F$. Repeating this m times we get $T \vdash F$. □

10.8. Corollary. *If T is a consistent set in a language L, then T is also consistent in any expansion L^* of language L by new constants.*

Proof. If T were inconsistent in L^*, then, in view of Theorem 9.3, we would have $T \vdash F$ in L^* for every formula $F \in \mathrm{Fm}(L)$. Applying Corollary 10.7, we would obtain $T \vdash F$ in L, for every formula $F \in \mathrm{Fm}(L)$, and T would be inconsistent in L. □

EXERCISES

10.1. If $x \notin V(F)$, then $F \equiv_l \forall xF$ and $F \equiv_l \exists xF$.

10.2. If $F \equiv_l G$, then also $\exists xF \equiv_l \exists xG$.

10.3. Show that if a variable y does not occur in F, then $\exists xF \equiv_l \exists yF(y/x)$.

10.4. Show the commutativity laws for the quantifiers

$$\forall x, yF \equiv_l \forall y, xF \quad \text{and} \quad \exists x, y\, F \equiv_l \exists y, xF.$$

10.5. Show that for every formula F there is an open formula G and a sequence of mutually different variables $y_1, \ldots, y_n$ such that $F \equiv_l \mathbb{Q}_1 y_1 \ldots \mathbb{Q}_n y_n G$, where $\mathbb{Q}_1, \ldots, \mathbb{Q}_n$ denote the quantifiers $\forall$ of $\exists$ (the normal form theorem).

10.6. Check that $\vdash (\forall xF \to \exists xF)$ and $\vdash \exists x(x = x)$.

10.7. Show that $\vdash \big(F(t/x) \to \exists xF\big)$.

10.8. Prove $\forall x(F \to G) \vdash (\exists xF \to \exists xG)$.

10.9. Let $L\{c\}$ be an expansion of the language L by a new constant c. Show that for $F, G \in \mathrm{Fm}(L)$, if $x \notin V(G)$, then the condition $T, \exists xF \vdash G$ in L is equivalent with the condition $T, F(c/x) \vdash G$ in $L\{c\}$.

10.10. Let a term $t = t(x_1, \ldots, x_n)$ be given. Show that for $y \notin V(t)$ we have

$$\vdash \forall x_1, \ldots, x_n \exists! y\big(t(x_1, \ldots, x_n) = y\big).$$

10.11. If $F \equiv_l G$, then also $F(t/x) \equiv_l G(t/x)$.

10.12. Assume that we are given a formula F and a variable x. Changing, if necessary, the variables we can regard always $F(t/x)$ as a result of a proper substitution of a term t for x in F. Let T be a consistent set of sentences. Show that in the Lindenbaum algebra $\mathbb{F}(T)$ the following rules hold:

$$[\forall xF] = \inf\{[F(t/x)]\colon t \in \mathrm{Tm}\},$$

$$[\exists xF] = \sup\{[F(t/x)]\colon t \in \mathrm{Tm}\}.$$

The above rules hold also if t runs over variables only.

11

THE COMPLETENESS OF THE LOGICAL CALCULUS

Here we prove the central theorem of logic—the completeness theorem. The proof will use almost all the results proved up to now. Recall that the completeness theorem is not only of interest on its own, but, first of all, it justifies the definition of provability (cf. the remark after Theorem 8.4).

The completeness theorem was proved by Gödel in 1930 [G2]. The proof given below is due to Henkin [H1].

11.1. Theorem. *Every consistent set of formulas of a given language L has a model.*

The proper completeness theorem; *if a formula F is true in every model of the theory T, then $T \vdash F$,* follows easily from Theorem 11.1. To prove this suppose that $T \nvdash F$. Then, in view of Theorem 9.10, the set $T \cup \{\neg F\}$ would be consistent and thus, by Theorem 11.1, it would have a model $\langle \mathbb{A}, p \rangle$. But then $\langle \mathbb{A}, p \rangle$ would be a model of the theory T in which the formula F would be false, which contradicts the assumption.

Proof of Theorem 11.1. Let T be a consistent set of formulas of a given language L. First we shall expand the set T to a consistent set S that additionally has the following Henkin property: *for every formula F and variable x there is a constant $c = c(F, x)$ such that the formula*

11.2 $$H(F, x)\colon\ \exists x F \to F(c/x)$$

belongs to S.

To construct the set S one has, in general, to expand language L by new constants (since, e.g., L could have no constants at all). Thus, for every formula F and variable x we fix a new constant $c = c(F, x)$ and we put

$$L^* = L\{c(F, x)\colon\ F \in \mathrm{Fm}(L) \wedge x \in X\},$$

[the expansion of language L by the new constants $c(F,x)$], and

$$T^* = T \cup \{H(F,x)\colon\ F \in \mathrm{Fm}(L) \wedge x \in X\}.$$

The set T^* is consistent, because otherwise some finite subset of the set T^* would be inconsistent, and therefore, some subset of the form

$$T_m = T \cup \{H_1, \ldots, H_m\},$$

where $H_i = H(F_i, x_i)$, $i = 1, \ldots, m$, are Henkin formulas, would be inconsistent, too. Thus, it suffices to show that every set of the form T_m is consistent. For $m = 0$ we have $T_m = T_0 = T$, but T is consistent in L, by the assumption, so it is also in L^*, by corollary 10.8. Assume inductively that it is so for T_m (i.e., for the extension of the set T by m arbitrary Henkin formulas). Then $T_{m+1} = T_m \cup \{H\}$, where formula H has the form $\exists x F \to F(c/x)$, for some $F \in \mathrm{Fm}(L)$ and a new constant c . If the set T_{m+1} were inconsistent, then, by Theorem 9.11, we would have

$$T_m \vdash \neg(\exists x F \to F(c/x)),$$

and hence also [see 9.9, (c)]

$$T_m \vdash \neg\forall x \neg F \quad \text{and} \quad T_m \vdash \neg F(c/x).$$

Since the constant c does not occur either in T_m or in F we may apply Corollary 10.6 on the elimination of constants to the relation $T_m \vdash (\neg F)(c/x)$ so that we obtain finally $T_m \vdash \neg\forall x \neg F$ and $T_m \vdash \forall x \neg F$, that is, the set T_m would be inconsistent, contradicting the assumption.

We have shown that the set T^* is consistent. However, the Henkin condition need not be satisfied, since condition $H(F,x) \in T^*$ holds only for the formulas F of the old language L. Therefore, we define by induction a sequence of languages L_n and a sequence of sets of formulas $S_n \subseteq \mathrm{Fm}(L_n)$ as follows:

$$L_0 = L \quad \text{and} \quad S_0 = T; \qquad L_{n+1} = L_n^* \quad \text{and} \quad S_{n+1} = S_n^*.$$

Thus, we have $L_{n+1} = L_n\{C_n\}$ for some set of constants C_n. Let $C = \bigcup\{C_n\colon\ n \in N\}$. Then the language $L\{C\}$ is a common expansion of all the languages L_n, for $n \in N$. We set $S = \bigcup\{S_n\colon\ n \in N\}$. Then $T \subseteq S$ and S satisfies the Henkin condition. This follows from the fact that every formula F of the language $L\{C\}$ contains a finite number of new constants $c_0, \ldots, c_n$, and thus $c_0, \ldots, c_n \in C_m$ for an $m \in N$, whence we infer $F \in \mathrm{Fm}(L_m)$ and consequently $H(F,x) \in S_{m+1}$. Finally, S is a consistent set because each of the sets S_n is consistent, since this property—as we proved before—is preserved under the operation $S_{n+1} = S_n^*$. On the other hand, every finite subset of the set $S = \bigcup\{S_n\colon\ n \in N\}$ is contained in some S_n, and thus it is also consistent, which implies the consistency of the whole set S (see Corollary 10.8).

Thus, we have completed the first part of the proof—a given consistent set T has been extended in some expansion $L\{C\}$ of the language L by new constants to a consistent set S satisfying the Henkin condition.

Now, let us construct a model $\langle \mathbb{A}, p \rangle$ for S. The reduct of the model $\langle \mathbb{A}, p \rangle$ to the type of language L will be then a model for T (see Lemma 7.5).

According to 9.12, Lindenbaum's Theorem, we may additionally assume (enlarging S if necessary),

11.3 $$\text{either } F \in S \text{ or } \neg F \in S,$$

for every formula F of the language $L\{C\}$.

Note that condition $S \vdash F$ implies $F \in S$.

First consider the following structure:

$$\mathbb{B} = \langle B;\ \{r^B\}, \{f^B\}, \{c^B\} \rangle,$$

where $B = \mathrm{Tm}(L\{C\})$ is the set of all the terms of the language $L\{C\}$, and the relations r^B, operations f^B, and distinguished elements c^B are defined as follows:

$$r^B(t_1, \ldots, t_n) \text{ if and only if } r(t_1, \ldots, t_n) \in S, \text{ for every relation symbol } r;$$

$$f^B(t_1, \ldots, t_m) = f(t_1, \ldots, t_m), \text{ for every function symbol } f;$$

$$c^B = c, \text{ for every constant of the language } L\{C\}.$$

From the equality axioms (Section 8.1, paragraph 2) it follows immediately that the relation

$$(t =_S s) \quad \text{if and only if} \quad (\text{the formula } t = s \text{ belongs to } S)$$

is a congruence in $\mathbb{B}$. Let $\mathbb{A} = \mathbb{B}/{=_S}$ be the factor system. Thus, we can write

11.4 $$\mathbb{A} = \langle A;\ \{r^A\}, \{f^A\}, \{c^A\} \rangle,$$

where $A = \{[t]\colon\ t \in \mathrm{Tm}(L\{C\})\}$ is the set of equivalence classes of the relation "$=_S$" and we have the equivalence

$$r^A([t_1], \ldots, [t_n]) \quad \text{if and only if} \quad r(t_1, \ldots, t_n) \in S$$

for any relation symbol r, since the conditions

$$t_1' =_S t_1, \ldots, t_n' =_S t_n \text{ and } r^B(t_1', \ldots, t_n') \text{ imply } r^B(t_1, \ldots, t_n),$$

on the basis of the equality axioms. Moreover

$$f^A([t_1], \ldots, [t_m]) = [f(t_1, \ldots, t_m)],$$

for every function symbol f and

$$c^A = [c], \quad \text{for every constant } c.$$

Let p be an assignment in $\mathbb{A}$ defined for all the variables as follows: $p(x) = [x]$. It is easy to see that $t[p] = [t]$, for $t \in \mathrm{Tm}(L\{C\})$. We apply induction. We have $x[p] = p(x) = [x]$, for any variable x and for any constant c, $c[p] = c^A = [c]$, and

$$f(t_1, \ldots, t_m)[p] = f^A(t_1[p], \ldots, t_m[p]) = f^A([t_1], \ldots, [t_m]) = [f(t_1, \ldots, t_m)].$$

Now, by induction on F we prove the following equivalence:

11.5 $$\mathbb{A} \models F[p] \text{ if and only if } F \in S.$$

For atomic formulas we have

$$\mathbb{A} \models (t = s)[p] \quad \text{iff} \quad t[p] = s[p] \quad \text{iff} \quad [t] = [s] \quad \text{iff} \quad t =_S s \quad \text{iff} \quad (t = s) \in S;$$

$$\begin{aligned} \mathbb{A} \models r(t_1, \ldots, t_n)[p] \quad & \text{iff} \quad r^A(t_1[p], \ldots, t_n[p]) \\ & \text{iff} \quad r^A([t_1], \ldots, [t_n]) \quad \text{iff} \quad r(t_1, \ldots, t_n) \in S. \end{aligned}$$

Assume 11.5 for all $F \in \mathrm{Fm}_l$, for some l. We prove that it holds for all $H \in \mathrm{Fm}_{l+1}$. If H is $\neg F$, where $F \in \mathrm{Fm}_l$, then

$$\mathbb{A} \models \neg F[p] \quad \text{iff} \quad \mathbb{A} \not\models F[p] \quad \text{iff} \quad F \notin S \quad \text{iff} \quad \neg F \in S,$$

in view of 11.3.

If H is $F \to G$, where $F, G \in \mathrm{Fm}_l$, then

$$\begin{aligned} \mathbb{A} \models (F \to G)[p] \quad & \text{iff} \quad (\mathbb{A} \not\models F[p] \text{ or } \mathbb{A} \models G[p]) \\ & \text{iff} \quad (F \notin S \text{ or } G \in S) \quad \text{iff} \quad (F \to G) \in S. \end{aligned}$$

The last equivalence follows from 11.3 and the laws in 9.9.

Finally, let H be $\forall x F$, where $F \in \mathrm{Fm}_l$. By the assumption, 11.5 holds for any formula G which is obtained from F by changing the bound variables and substitution of terms (see Exercise 7.5). It follows that,

$$\begin{aligned} (\mathbb{A} \not\models \forall x\, F[p]) & \equiv \big(\exists t \in \mathbb{A} \not\models F\big[p([t]/x)\big]\big) \\ & \equiv \big(\exists t \in \mathbb{A} \not\models F\big[p(t[p]/x)\big]\big). \end{aligned}$$

Changing the bound variables in F we will obtain a formula $G \equiv_l F$ for which $G(t/x)$ is a proper substitution. Hence condition $\mathbb{A} \not\models \forall x F[p]$ implies $\mathbb{A} \not\models G(t/x)[p]$ for some t (see Lemma 7.1), whence $G(t/x) \notin S$. Thus $\forall x G \notin S$, since the formula $\forall x G \to G(t/x)$ is an axiom of logic. But $G \equiv_l F$ implies $\forall x G \equiv_l \forall x F$, whence $\forall x F \notin S$. Thus, we have shown if $\forall x F \in S$, then $\mathbb{A} \models \forall x F[p]$. Conversely, assume now that $\forall x F \notin S$. By 11.3 we get $\neg \forall x F \in S$, that is, $\exists x \neg F \in S$. We apply now the Henkin property: the formula

$\exists x \neg F \to \neg F(c/x)$ is in S for some constant c, whence $\neg F(c/x) \in S$, and thus $F(c/x) \notin S$. For the formula $F(c/x)$ the inductive assumption is valid. Thus we obtain $\mathbb{A} \not\models F(c/x)[p]$, which is equivalent with the condition $\mathbb{A} \not\models F[p([c]/x)]$, which obviously implies $\mathbb{A} \not\models \forall x F[p]$. Thus we have proved 11.5 for every formula F of the language $L\{C\}$. Hence, it follows that the pair $\langle \mathbb{A}, p\rangle$ is a model of the set S, which completes the proof of the theorem. □

The model 11.4 is called the *model on terms*, since the elements of the universe are equivalence classes of terms.

From Theorem 11.11, the completeness theorem, immediately follows a theorem frequently used in model theory.

11.6. Compactness Theorem (Malcev [M1]). *If every finite subset of a given set T has a model, then T also has a model.*

It follows immediately from the assumption that set T is consistent.

Now, let us estimate the cardinality (of the universe) of the obtained model 11.4 on terms. By the cardinality of the language we mean the joint cardinality of the set of all the symbols of the language. Assume that language L is countable. It is easy to check that then the sets $\mathrm{Tm}(L)$ and $\mathrm{Fm}(L)$ are also countable because we have card $\mathrm{Tm}_0 = \omega$, since Tm_0 is the union of the set of variables and the set of constants. Assume inductively card $\mathrm{Tm}_l = \omega$. Since

$$\mathrm{Tm}_{l+1} = \mathrm{Tm}_l \cup \bigcup_f \{f\} \times [\mathrm{Tm}_l]^{m(f)},$$

where f runs over the function symbols and $m(f)$ is the number of arguments of the symbol f, we infer that also card $\mathrm{Tm}_{l+1} = \omega$, since a countable union of countable sets is still a countable set. Hence also card $\mathrm{Tm} =$ card $\bigcup\{\mathrm{Tm}_l\colon\ l \in N\} = \omega$. Similarly we prove the equality card $\mathrm{Fm} = \omega$.

Hence it follows that the set of constants

$$C = \{c(F, x)\colon\ F \in \mathrm{Fm}(L) \wedge x \in X\}$$

is countable, and hence also card $L\{C\} = \omega$, whence card $\mathrm{Tm}(L\{C\}) = \omega$.

Thus, we obtain the following improvement of the Theorem 11.1:

11.7. *A consistent set of formulas T of a countable language L always has a finite or a countable model.*

A similar remark applies to the compactness theorem.

EXERCISES

11.1. Show that for any language L we have

$$\text{card Tm} = \text{card }(F \cup C \cup X),$$

where F, C, X denote, respectively, the set of function symbols, the set of constants, and the set of variables.

11.2. Show that for any language L we have card Fm = card $(R \cup \text{Tm})$, where R is the set of relation symbols. Thus we always have card $\text{Fm}(L) =$ card L, that is, the cardinality of L coincides with the cardinality of the set of formulas of L.

11.3. Let $L\{C\}$ be the language used in the proof of Theorem 11.1. Show the equality card $\text{Tm}(L\{C\}) =$ card L. Hence, it follows that a consistent set of formulas of the language L has a model of power less than or equal to card L.

11.4. Notice the following property of a model on terms: every class $[t]$ contains a constant c, whence $[t] = [c]$. Hence, that model can be referred to as a *model on constants*.

11.5. Let L be a countable language. Extend a consistent set T to a set S with the Henkin property in one step, that is, without iterating the operation "star."

11.6. Let T be a countable set of sentences in a countable language L. Construct a model for T without introducing new constants: instead of constants $c(F, x)$ use the variables of language L.

11.7. Let E be a subset A of a given Boolean algebra $\mathbb{A}$ with the bounds sup E and inf E. We say that an ultrafilter Q of $\mathbb{A}$ is E-complete if the following condition holds:

$$\textit{if } \sup E \in Q, \textit{ then } E \cap Q \neq \emptyset$$

or equivalently

$$\textit{if } E \subseteq Q, \textit{ then } \inf E \in Q.$$

Check that any principle ultrafilter is E-complete.

11.8. A subset D of an algebra $\mathbb{A}$ is called *dense*, if $\mathbb{O} \notin D$ and the following condition holds:

$$\forall a > \mathbb{O} \exists b \in D[b \leq a].$$

Check that for any set E such that $\mathbb{O} \notin E$, the set

$$d(E) = \{a > \mathbb{O}\colon \forall b \in E(a \cdot b = \mathbb{O})\} \cup \{a > \mathbb{O}\colon \exists b \in E(a \leq b)\}$$

is dense.

11.9. *The Rasiowa–Sikorski lemma* (see [RS]). Let $\{E_n\colon n \in N\}$ be a countable family of subsets (in a given Boolean algebra $\mathbb{A}$) having the sup's and inf's. Show that there is an ultrafilter Q which is E_n-complete for all $n \in N$, cf. Exercises 11.7 and 11.8. (*Hint*: Choose inductively a decreasing sequence $a_n \in d(E_n)$ and find an ultrafilter containing all the elements a_n).

11.10. Let T be a consistent set of sentences in a countable language L. In the Lindenbaum algebra $\mathbb{F}(T)$ choose an ultrafilter Q which is $E(F,x)$-complete, for all $F \in \mathrm{Fm}$ and $x \in X$ (cf. Exercise 11.9), where

$$E(F,x) = \{[F(y/x)]\colon\ y \in X\}.$$

Construct a model on variables for T using the relation defined as follows

$$x =_Q y \quad \text{if and only if} \quad [(x = y)] \in Q,$$

and proving an analogue of the condition in 11.5.

12

DEFINABILITY

Accepting a new definition in a theory T is technically nothing else than expanding the language of T be a new relation, operation, or constant symbol and adding a new defining axiom. But axioms of this sort do not strengthen the theory—each new theorem is equivalent to an old one obtained by an obvious elimination process.

Let T be a consistent set of sentences in the language L. If $\mathbb{A}$ is a relational system (of the same type as language L), then any formula $H(x_1, \ldots, x_n)$ determines a relation r_H^A on the universe A defined by the condition

$$r_H^A(a_1, \ldots, a_n) \quad \text{if and only if} \quad \mathbb{A} \models H[a_1/x_1, \ldots, a_n/x_n].$$

We say the relation r_H^A is *definable* in $\mathbb{A}$ by the formula H.

Suppose that a given formula $H(x_1, \ldots, x_n, y)$ has the property

12.1 $$\mathbb{A} \models \forall x_1, \ldots, x_n \exists! y H.$$

In this case the condition

$$f_H^A(a_1, \ldots, a_n) = b \quad \text{if and only if} \quad \mathbb{A} \models H[a_1/x_1, \ldots, a_n/x_n, b/y]$$

defines an operation $f_H^A \colon A^n \longrightarrow A$. We say that the operation f_H^A is *definable* in $\mathbb{A}$ by the formula H. If we have $T \vdash \forall x_1, \ldots, x_n \exists! y H$, then 12.1 holds for every model $\mathbb{A}$ of the theory T. We say in this case that H *defines* the operation f_H in T.

Now, suppose that $V_f(H) = \{x\}$ and the following condition holds:

$$\mathbb{A} \models \exists! x H.$$

Then there is exactly one element $c_H^A \in A$, for which $\mathbb{A} \models H[c_H^A/x]$. We say in this case that the element c_H^A is *definable* in $\mathbb{A}$ by the formula H. If $T \vdash \exists! x\ H$, then H *defines* the element c_H^A in every model $\mathbb{A}$ of the theory T. We also say in this case that the formula H *defines* a constant in T.

Example. The relations r^A, operations f^A, and distinguished elements c^A of a given structure $\mathbb{A}$ are definable in $\mathbb{A}$ by the formulas $r(x_1, \ldots, x_n)$, $f(x_1, \ldots, x_m) = y$, and $c = x$, respectively. The formulas of the form $t(x_1, \ldots, x_n) = y$, where $t \in \mathrm{Tm}$, define the operations f_t^A (see Section 5.7). All these objects are definable in logic, that is, in the case of $T = \emptyset$. The formula $x + y = 0$ defines, in a ring $\mathbb{A}$, the operation $f^A(a) = -a$, the operation of taking the inverse element with respect to the addition. The set theoretical formula $\forall z(z \in x \to z \in y)$ defines the inclusion relation in every transitive system $\mathbb{A}$, while the formula $\forall z(z \in y \equiv z = x)$ defines the operation $f(a) = \{a\}$, for $a \in A$.

Let D_r, D_f, D_c be the sets of formulas defining relations, operations, and constants in T, respectively. Thus, $D_r = \mathrm{Fm}$ and the following conditions hold:

$$T \vdash \forall x_1, \ldots, x_n \exists! y\ H \quad \text{for every } H \in D_f,$$

$$T \vdash \exists! x H \quad \text{for every } H \in D_c.$$

Let L^* be an expansion of language L by new relation symbols r_H for $H \in D_r$, function symbols f_H, for $H \in D_f$, and constants c_H, for $H \in D_c$. Let T^* be the set obtained from T by adding all the definitions, that is, sentences of the form

$$\forall x_1, \ldots, x_n [r_H(x_1, \ldots, x_n) \equiv H], \quad \text{for } H \in D_r,$$

12.2
$$\forall x_1, \ldots, x_n, y [f_H(x_1, \ldots, x_n) = y \equiv H], \quad \text{for } H \in D_f,$$

$$\forall x [x = c_H \equiv H], \quad \text{for } H \in D_c.$$

Now, every model $\mathbb{A} \in \mathrm{Mod}(T)$ determines the expansion

12.3
$$\mathbb{A}^* = \langle \mathbb{A};\ \{r_H^A\}_{H \in D_r}, \{f_H^A\}_{H \in D_f}, \{c_H^A\}_{H \in D_c} \rangle,$$

and clearly, we have $\mathbb{A}^* \in \mathrm{Mod}(T^*)$. Conversely, every model $\mathbb{A}^*$ of the theory T^* has form 12.3, where $\mathbb{A} \in \mathrm{Mod}(T)$. In particular, T^* is consistent.

12.4. Theorem (On elimination of definitions). *For each formula F of L^* there is a formula $\tilde{F}$ of L such that*

$$T^* \vdash F \equiv \tilde{F}.$$

Moreover, if F is a sentence, then the condition $T^* \vdash F$ implies $T \vdash \tilde{F}$.

Proof. First, we show how to eliminate new relation symbols. For an atomic F of the form $r_H(t_1, \ldots, t_n)$, where $H \in D_r$ and $t_1, \ldots, t_n$ are arbitrary terms, define $\varphi(F)$ as the formula $H(t_1/x_1, \ldots, t_n/x_n)$ (renaming, if necessary, the bound variables to make the indicated substitution proper). For other atomic F, let $\varphi(F)$ be F itself. The condition

$$T^* \vdash \forall x_1, \ldots, x_n \big(r_H(x_1, \ldots, x_n) \equiv H(x_1, \ldots, x_n)\big)$$

implies

$$T^* \vdash r_H(t_1, \ldots, t_n) \equiv H(t_1, \ldots, t_n),$$

by the axiom of substitution, so that $T^* \vdash F \equiv \varphi(F)$ for all atomic F.

We extend φ on the whole of $\mathrm{Fm}(L^*)$ by induction as follows

$$\varphi(\neg F) \ \text{ is } \ \neg\varphi(F)$$

$$\varphi(F \to G) \ \text{ is } \ \varphi(F) \to \varphi(G)$$

$$\varphi(\forall x\, F) \ \text{ is } \ \forall x\, \varphi(F).$$

Clearly, no new relation symbol occurs in $\varphi(F)$ and also $T^* \vdash F \equiv \varphi(F)$, by the rules of logic.

Now, we want to eliminate the new function symbols. First, let us define the number $n_f(t)$—the number of occurrences of f in t—by induction, as follows:

$$n_f(x) = n_f(c) = 0,$$

$$n_f(f'(t_1, \ldots, t_m)) = n_f(t_1) + \cdots + n_f(t_m), \quad \text{if } f \neq f'$$

and

$$n_f\big(f(t_1, \ldots, t_m)\big) = n_f(t_1) + \cdots + n_f(t_m) + 1.$$

Define also $n_f(F)$, for atomic F as follows:

$$n_f(t = s) = n_f(t) + n_f(s)$$

and

$$n_f\big(r(t_1, \ldots, t_n)\big) = n_f(t_1) + \cdots + n_f(t_n).$$

Finally, let

$$n(F) = \sum_f n_f(F),$$

where f runs over the new function symbols occurring in F.

Thus, $n(F)$ is the number of occurrences of all new function symbols in F. We apply induction on $n(F)$, for atomic F. If $n(F) = 0$ there is nothing to eliminate and we put $\psi(F) = F$. So let $n(F) > 0$ and assume that for each G with $n(G) < n(F)$ there is a $\psi(G) \equiv_{T^*} G$ without new function symbols. Since there occur new function symbols in F there must occur also a term of the form

$$s = f_H(t_1, \ldots, t_m),$$

where $H \in D_f$ and $t_1, \ldots, t_m \in \mathrm{Tm}(L)$.

Let F' be

$$\exists y \big(F(y/s) \wedge H(t_1/x_1, \ldots, t_m/x_m, y)\big),$$

where y is a variable not occurring in F and a suitable change of bound variables has been made in H. The substitution $t(y/s)$ is defined as follows:

$$t(y/s) = \begin{cases} t, & t \neq s, \\ y, & t = s, \end{cases}$$

for $t \in \mathrm{Tm}_k$, where $k = \min\{i\colon\ s \in \mathrm{Tm}_i\}$, and

$$f(t_1, \ldots, t_n)(y/s) = f(t_1(y/s), \ldots, t_n(y/s)).$$

For atomic F the substitution $F(y/s)$ is now obvious). Clearly, $n(F') < n(F)$.

To prove $T^* \vdash F' = F$ we let $\mathbb{B} \in \mathrm{Mod}(T^*)$ and let p be an arbitrary assignment in $\mathbb{B}$.

Using the substitution lemma we obtain

$$\begin{aligned} \mathbb{B} \models F'[p] \quad &\text{iff} \quad \exists b \in B\ \mathbb{B} \models F(y/s)[p(b/y)] \\ &\qquad \text{and} \quad \mathbb{B} \models H(t_1/x_1, \ldots, t_m/x_m)[p(b/y)] \\ &\text{iff} \quad b = f_H^B(t_1[p], \ldots, t_m[p]) \quad \text{and} \quad \mathbb{B} \models F(y/s)[p(b/y)] \\ &\text{iff} \quad \mathbb{B} \models F(y/s)[p(s^{\mathbb{B}}[p]/y)] \\ &\text{iff} \quad \mathbb{B} \models F(y/s)(s/y)[p] \quad \text{iff} \quad \mathbb{B} \models F[p]. \end{aligned}$$

Note that $t_i[p(b/y)] = t_i[p]$, for $i = 1, \ldots, m$, since $y \notin V_f(F)$.

Hence, $T^* \vdash F' \equiv F$, by the completeness theorem.

Now, by the inductive assumption, we have $\psi(F') \equiv_{T^*} F' \equiv_{T^*} F$ and $\psi(F') = \psi(F)$ has no new function symbols.

We extend ψ at nonatomic F as follows:

$$\psi(\neg F) \text{ is } \neg\psi(F), \quad \psi(F \to G) \text{ is } \psi(F) \to \psi(G) \quad \text{and} \quad \psi(\forall x F) \text{ is } \forall x \psi(F).$$

Thus, by an easy induction we obtain $T^* \vdash \psi(F) \equiv F$, for each $F \in \mathrm{Fm}(L^*)$, and $\psi(F)$ has no new function symbol.

Finally, we define the function χ eliminating the new constants. Let $c_{H_1}, \ldots, c_{H_n}$ be all the new constants occurring in F. Let $\chi(F)$ be the formula

$$\exists y_1, \ldots, y_n [F(y_1/c_{H_1}, \ldots, y_n/c_{H_n}) \wedge H_1(y_1/x) \wedge \cdots \wedge H_n(y_n/x)].$$

Clearly, $\chi(F)$ has no new constants.

For an arbitrary $\mathbb{B} \in \mathrm{Mod}(T^*)$ and an assignment p,

$$\mathbb{B} \models \chi(F)[p]$$

$$\text{iff} \quad \exists b_1, \dots, b_n \in B\ \mathbb{B} \models \big[F(y_1/c_1, \dots, y_n/c_n)[p(b_1/y_1, \dots, b_n/y_n)]$$

$$\wedge H_1(y_1/x)[p(b_1/y_1, \dots, b_n/y_n)]$$

$$\cdots \wedge H_n(y_n/x)[p(b_1/y_1, \dots, b_n/y_n)]\big]$$

$$\text{iff} \quad \exists b_1, \dots, b_n \in B\big[b_1 = c_{H_1}^B \wedge \cdots \wedge b_n = c_{H_n}^B$$

$$\wedge \mathbb{B} \models F(y_1/c_{H_1}, \dots, y_n/c_{H_n})[p(b_1/y_1, \dots, b_n/y_n)]\big]$$

$$\text{iff} \quad \mathbb{B} \models F(y_1/c_{H_1}, \dots, y_n/c_{H_n})[p(c_{H_1}^B/y_1, \dots, c_{H_n}^B/y_n)]$$

$$\text{iff} \quad \mathbb{B} \models F(y_1/c_{H_1}, \dots, y_n/c_{H_n})(c_{H_1}^B/y_1, \dots, c_{H_n}^B/y_n)[p]$$

$$\text{iff} \quad \mathbb{B} \models F[p].$$

Hence, $T^* \vdash \chi(F) \equiv F$, by the completeness theorem.

Now, we define $\tilde{F}$ as $\chi\psi\varphi(F)$. Thus, $\tilde{F} \in \mathrm{Fm}(L)$ and $T^* \vdash \tilde{F} \equiv F$. To prove the last assertion we let F be a sentence and assume $T^* \vdash F$ and let $\mathbb{A} \in \mathrm{Mod}(T)$. Then $\mathbb{A}^* \in \mathrm{Mod}(T^*)$, and hence $\mathbb{A}^* \models F$, whence $\mathbb{A}^* \models \tilde{F}$. However, $\tilde{F}$ is a sentence of language L, so it is true in the reduct, that is, $\mathbb{A} \models \tilde{F}$. By the completeness theorem, we infer $T \vdash \tilde{F}$, which completes the proof. □

Intuitively, theory T^* contains new objects which are definable by means of the primitive notions of theory T. Let us define by induction a sequence of languages L_n and a sequence of theories T_n as follows:

$$L_0 = L \quad \text{and} \quad T_0 = T; \qquad L_{n+1} = (L_n)^* \quad \text{and} \quad T_{n+1} = (T_n)^*.$$

In this way we introduce the definable notions by means of the primitive notions and the notions defined at earlier stages. A multiple application of Theorem 12.4, the elimination theorem, shows that also those more complex definitions can be reduced to the original language L. Thus, the formulas of language L^* and of its iterations can be, with no fear of misunderstanding, treated as formulas of language L. And so, for $F \in \mathrm{Fm}(L^*)$ condition $T \vdash F$ means $T \vdash \tilde{F}$, and similarly, $\mathbb{A} \models F[p]$ means $\mathbb{A} \models \tilde{F}[p]$, for $\mathbb{A} \in \mathrm{Mod}(T)$, and so on.

Example. The formula $F(x, y)$: $\forall z(z \in x \to z \in y)$ defines in set theory the inclusion relation. Let $\mathbb{A} = \langle A, \in^A \rangle$ be a transitive system. We usually write

$$\mathbb{A} \models (x \subseteq y)[a/x, b/y],$$

if $\mathbb{A} \models F[a/x, b/y]$, that is, if a is included in b.

12.5. Definability with Parameters

Let us distinguish some of the free variables $y_1, \ldots, y_m$ of the formula F. Let $V_f(F) = \{x_1, \ldots, x_n, y_1, \ldots, y_m\}$. Every sequence $b_1, \ldots, b_m$ of the elements of the universe of a system $\mathbb{A}$ determines the set

$$Z(F;\ b_1, \ldots, b_m)$$
$$= \{\langle a_1, \ldots, a_n\rangle\colon\ \mathbb{A} \models F[a_1/x_1, \ldots, a_n/x_n, b_1/y_1, \ldots, b_m/y_m]\}.$$

We say in this case that the set (relation) $Z(F;\ b_1, \ldots, b_m)$ is *definable* in $\mathbb{A}$ by the formula F with the parameters $b_1, \ldots, b_m$. If $b_1 = c_1^A, \ldots, b_m = c_m^A$ are certain distinguished elements, then $Z(F;\ b_1, \ldots, b_m)$ is definable in $\mathbb{A}$ by the formula $F(c_1/y_1, \ldots, c_m/y_m)$, that is, it is definable without parameters. More generally, if the parameters $b_1, \ldots, b_m$ are definable in $\mathbb{A}$ by the formulas $H_1(y_1), \ldots, H_m(y_m)$, then also the set $Z(F;\ b_1, \ldots, b_m)$ can be defined in $\mathbb{A}$ without parameters; namely, it is definable by the formula

$$\exists y_1, \ldots, y_m[F \wedge H_1 \wedge \cdots \wedge H_m].$$

However, in general, the class of sets which are parametrically definable in $\mathbb{A}$ is larger than the class of sets *parameter-free definable*.

Similarly, we define the parametric definability of operations and elements. For instance, if the formula $F(x_1, \ldots, x_n, y, z_1, \ldots, z_m)$ has the property

$$\mathbb{A} \models \forall x_1, \ldots, x_n\, \exists! y F[b_1/z_1, \ldots, b_m/z_m]$$

for some sequence of parameters $b_1, \ldots, b_m$, then the set $Z(F;\ b_1, \ldots, b_m)$ is an operation parametrically definable in $\mathbb{A}$. Obviously, every element $a \in A$ (i.e., a one-element set $\{a\}$) is parametrically definable in $\mathbb{A}$, namely $\{a\} = Z(x = y;\ a)$.

EXERCISES

12.1. Let $\mathrm{Def}_c(A)$ denote the family of those subsets of A^n which are parametrically definable in $\mathbb{A}$ by formulas of the form $H(x_1, \ldots, x_n)$. Check that

(a) $\mathrm{Def}_n(\mathbb{A})$ is a field of sets.

(b) $\mathrm{Def}_n(\mathbb{A})$ contains the field $FC(A^n)$ (cf. Exercise 2.6).

(c) If $n < m$ and $Z \in \mathrm{Def}_m(\mathbb{A})$, then the projection of Z onto A^n belongs to $\mathrm{Def}_n(A)$.

12.2. The map

$$h([F]_T) = \{\langle a_1, \ldots, a_n\rangle\colon\ \mathbb{A} \models F[a_1/y_1, \ldots, a_n/y_n]\}$$

is a homomorphism of the Lindenbaum algebra $\mathbb{F}_Y(T)$ onto the field $\mathrm{Def}_n(\mathbb{A})$, where $\mathbb{A} \in \mathrm{Mod}(T)$. Show that if T is a complete set of sentences (i.e., for any sentence F, we have either $T \vdash F$ or $T \vdash \neg F$), then for all $n \in N$ the fields $\mathrm{Def}_n(\mathbb{A})$, $\mathbb{A} \in \mathrm{Mod}(T)$, are isomorphic.

12.3. Let $f = f_H$ be an operation definable in a theory T by the formula $H = H(x_1, \ldots, x_n, y)$. Show that for an arbitrary formula $F \in \mathrm{Fm}$, the formulas

$$\exists y[H(x_1, \ldots, x_n, y) \wedge F] \quad \text{and} \quad \forall y[H(x_1, \ldots, x_n, y) \to F]$$

are equivalent with the substitution $F\big(f(x_1, \ldots, x_n)/y\big)$.

Exercises 12.14 to 12.13 are a continuation of Exercises 6.8 to 6.12.

The Zermelo–Fraenkel system of axioms is usually denoted by ZF and the Zermelo system by Z.

12.4. Show that the axiom of subsets is a theorem of the theory ZF. Thus, ZF is a nonweaker theory than Z.

12.5. Let $H_S(x, y)$ be the formula

$$\forall z[z \in y \ \equiv \ (z \in x \vee z = x)],$$

defining in Z the successor operation $S(x) = x \cup \{x\}$. Check that in every transitive system $\mathbb{A} = \langle A, \in^A \rangle$ we have

$$S(a) = b \ \equiv \ \mathbb{A} \models H_S[a, b] \quad \text{for } a, b \in A.$$

Thus, the notion of the successor remains unchanged when defined within a transitive structure $\mathbb{A}$.

12.6. The set ω of natural numbers is defined in set theory as the least set containing $\emptyset$(as an element) and closed under the successor operation S. Thus, the formula defining ω may look as follows:

$$H_\omega(x){:}\ \ W(x) \wedge \forall y[W(y/x) \to x \subseteq y],$$

where $W(x)$ is the formula

$$\emptyset \in x \wedge \forall y[y \in x \to S(y) \in x].$$

Check that for an arbitrary transitive system $\mathbb{A}$ we have

$$a = \omega \quad \text{if and only if} \quad \mathbb{A} \models H_\omega[a], \quad \text{for } a \in A$$

(cf. the remark in Exercise 12.5).

12.7. Let Inf denote the axiom of infinity. Show that for any transitive system $\mathbb{A}$ we have

$$\mathbb{A} \models \mathrm{Inf} \quad \text{if and only if} \quad \omega \in A.$$

12.8. Find a formula $H(x, y)$ such that, for any transitive system $\mathbb{A}$ the condition $\mathbb{A} \models H[a, b]$ is equivalent with the condition: "a is a family of nonempty mutually disjoint sets, and b is a selector of the family a."

12.9. Prove that for every set A there is a transitive set B such that $A \subseteq B$ and in $\mathbb{B} = \langle B, \in^B \rangle$ all the axioms of set theory are valid except possibly the power-set axiom, the axiom of subsets, and the replacement axiom.

12.10. The formula $H_P(x, y)$: $\forall z[z \in y \equiv z \subseteq x]$ defines in Z (and so also in ZF) the operation of the power-set $y = P(x)$. Check that for any transitive system $\mathbb{A}$ we have

$$\mathbb{A} \models H_P[a, b] \quad \text{if and only if} \quad b = P(a) \cap A, \qquad \text{for } a, b \in A.$$

Show that the validity in $\mathbb{A}$ of the power-set axiom is equivalent with the condition $P(a) \cap A \in A$, for every $a \in A$.

12.11. Show that the validity in a transitive system of axioms of the replacement axiom is equivalent with the condition "for every parametrically definable in $\mathbb{A}$ operation f_H^A and for every $a \in A$ the image $f_H^A[a]$ of the set a belongs to A."

12.12. Show that in the system $\mathbb{R}_\omega = \langle R_\omega, \in \rangle$ (see Section 6.3) all the axioms of ZF except the infinity axiom are valid.

12.13. Show that the system $\mathbb{R}_{\omega+\omega}$ is a model of the theory Z.

13

PEANO ARITHMETIC

Peano arithmetic is an axiomatic theory of nonnegative integers. It was Peano who axiomatized the properties of natural numbers in the late nineteenth century. This theory has been extensively studied in mathematical logic. The famous incompleteness theorems of Gödel are usually proved only for arithmetic. These and other problems connected with Peano arithmetic are considered in Chapters 18 to 22. This chapter has an introductory character.

Consider the algebraic system

$$\mathbb{N} = \langle N;\ +, \cdot, 0, 1\rangle,$$

where N is the set of natural numbers (nonnegative integers) and $+$, $\cdot$ are the arithmetical operations of addition and multiplication, respectively. The numbers 0 and 1 are chosen as the distinguished elements. The system $\mathbb{N}$ is the main or standard model of the theory of natural numbers which is called the *Peano arithmetic*. As the axioms of the arithmetic, we choose some characteristic sentences which are true in $\mathbb{N}$. The language L corresponding to the system $\mathbb{N}$ is the same as the language of ring theory; it contains two binary function symbols which are denoted by $+$ and $\cdot$ (similarly to the operations in $\mathbb{N}$) and two constants Λ_0, Λ_1 denoting zero and one.

The axioms of arithmetic are closures (with respect to the occurring variables) of the following formulas:

1. *The laws of commutativity, associativity, distributivity and the properties of identity elements*

$$x + y = y + x, \quad x \cdot y = y \cdot x,$$

$$(x + y) + z = x + (y + z), \quad (x \cdot y) \cdot z = x \cdot (y \cdot z),$$

$$(x + y) \cdot z = x \cdot z + y \cdot z,$$

$$x + \Lambda_0 = x, \quad x \cdot \Lambda_1 = x.$$

2. *The law of subtraction* $x + z = y + z \rightarrow x = y$.

3. *Every number except zero is a successor*:

$$x \neq 0 \equiv \exists y[x = y + \Lambda_1].$$

4. *The axioms of induction*. For every formula F and every variable x we have the following axiom:

$$\Big(F(\Lambda_0/x) \wedge \forall x[F \to F((x+\Lambda_1)/x)]\Big) \to \forall x F.$$

The set of the above axioms is denoted by the symbol PA.

The usual induction principle, "if a set $X \subseteq N$ contains the number 0 and satisfies the condition $\forall n[n \in X \to n+1 \in X]$, then $X = N$," has been restricted to those sets X which are parametrically definable,

$$X = \{n \in N\colon\ \mathbb{N} \models F[n/x, b_1/y_1, \ldots, b_m/y_m]\},$$

for a formula $F(x, y_1, \ldots, y_m)$, and some $b_1, \ldots, b_m \in N$. This is the strongest possible form of induction expressible in language L. In fact, the variables of the language run over the universe, not over its subsets. Thus, the properties of the subsets can be expressed only by means of formulas defining them, and so we have to restrict ourselves to definable subsets.

Of course, we have $\mathbb{N} \models F$, for every sentence $F \in$ PA. The model $\mathbb{N}$ is called the *standard* model of PA.

Since PA is a set of sentences, by the logical substitution axiom and the generalization rule, we have PA $\vdash F$ if and only if PA $\vdash \forall x_1, \ldots, x_n F$, for every formula F and any sequence $x_1, \ldots, x_n$ of variables.

Thus, the properties expressed by the formulas of the form $\forall x_1, \ldots, x_n F$ can be expressed in shorter form by rejecting the initial universal quantifiers.

13.1. Lemma. *The following formulas are theorems of* PA:

$$\text{(a)}\ \ x \cdot \Lambda_0 = \Lambda_0,$$

$$\text{(b)}\ \ x + y = \Lambda_0 \to x = \Lambda_0.$$

Proof. By the completeness theorem the condition PA $\vdash F$ is equivalent with the validity of the formula F in every model $\mathbb{M}$ under an arbitrary assignment in $\mathbb{M}$. To simplify the notation, if

$$\mathbb{M} = \langle M;\ +^M, \cdot^M, \Lambda_0^M, \Lambda_1^M\rangle$$

is an arbitrary model, then we shall omit the superscripts of the operations and also we shall denote Λ_0^M by 0 and Λ_1^M by 1.

Let $a \in M$. By the algebraic axioms in paragraph 1 we have

$$a + 0 = a = a \cdot 1 = a(0+1) = a \cdot 0 + a \cdot 1 = a \cdot 0 + a,$$

and subtracting a from both sides, that is, applying the axiom in paragraph 2, we get $a \cdot 0 = 0$, which means $\mathbb{M} \models (x \cdot \Lambda_0 = \Lambda_0)[a]$.

Similarly, if $a, b \in M$ and $a \neq 0$, then $a = a_0 + 1$ for some a_0 (by the axiom in paragraph 3) and then $a + b = a_0 + b + 1$, that is, the element $a + b$ is a successor, whence $a + b \neq 0$ (again by axiom 3), which proves point (b). □

13.2. The Ordering

The usual ordering $\leq$ of the set N can be characterized by the equivalence

$$n \leq m \quad \text{if and only if} \quad \exists k \in N[n + k = m] \qquad \text{for } n, m \in N.$$

Moreover, we have also

$$n < m \quad \text{if and only if} \quad \exists k[n + k + 1 = m].$$

We introduce a new definable relation symbol $\leq$ and a new defining axiom

$$\forall x, y[x \leq y \equiv \exists z(x + z = y)].$$

Additionally, let $x < y$ be an abbreviation of the formula $x \leq y \wedge x \neq y$.

13.3. Theorem. *The formula $x \leq y$ defines a linear ordering of the universe, that is, all the formulas*

$$x \leq x, \quad (x \leq y \wedge y \leq x) \rightarrow x = y, \quad (x \leq y \wedge y \leq z) \rightarrow x \leq z$$

and

$$x \leq y \vee y \leq x$$

are theorems of PA.

Proof. Let $\mathbb{M}$ be an arbitrary model of PA. For an arbitrary $a \in M$, we have $a + 0 = a$, which gives $a \leq^M a$, that is, $\mathbb{M} \models (x \leq x)[a]$.

Assume that $a \leq^M b$ and $b \leq^M a$. Thus, for some $u, w \in M$ we have $a + u = b$ and $b + w = a$, whence

$$a + u + w = a = a + 0.$$

From the subtraction axiom in paragraph 2 we get $u + w = 0$, whence $u = 0$, by point (b) of Lemma 13.1, and thus $a = b$. Assume now that $a \leq^M b$ and $b \leq^M c$. We have $a + u = b$ and $b + w = c$, for some $u, w \in M$, whence $a + u + w = c$, that is, $a \leq^M c$.

It remains to prove that the ordering $\leq^M$ is linear, that is,

$$a \leq^M b \vee b \leq^M a \quad \text{for } a, b \in M.$$

We apply induction. Let $F(x, y)$ be the formula $x \leq y \vee y \leq x$ and let $a \in M$. We have to show that the parametrically definable set

$$Y = \{b \in M\colon\ \mathbb{M} \models F[a/x, b/y]\}$$

coincides with M. Since $0 + a = a$, we have $0 \leq^M a$, and thus $0 \in Y$. Assume that $b \in Y$, that is, $a \leq^M b \vee b \leq^M a$. If $a \leq^M b$, then also $a \leq^M b + 1$. On the other hand, if $b \leq^M a$, that is, $b + u = a$, for some $u \in M$, then either $u = 0$ and then $b = a$, whence $a \leq^M b + 1$, or else $u \neq 0$ and then $u = w + 1$, for some w (by the axiom in paragraph 3). Hence $b + 1 + w = a$, that is, $b + 1 \leq^M a$. In either case we obtain $a \leq^M b + 1$ or $b + 1 \leq^M a$. By the induction principle we have then $Y = M$, which completes the proof. □

Let us list now a few simple properties of the ordering $\leq$.

13.4. $\mathrm{PA} \vdash \forall x(\Lambda_0 \leq x)$, *that is, zero is the least element.*

This follows immediately from the axiom $\forall x(x + \Lambda_0 = x)$.

13.5. $\mathrm{PA} \vdash (\Lambda_0 < \Lambda_1)$, *that is one is greater than zero.*

Of course we have $0 \leq^M 1$ and the equality $0 = 1$ contradicts the axiom in paragraph 3.

13.6. $\mathrm{PA} \vdash \forall x, y((x < y + \Lambda_1) \equiv (x \leq y))$, *that is, the ordering is discrete: there is no element between b and b + 1.*

Let us observe that in any model $\mathbb{M}$ the condition $a <^M b + 1$ means $a + u = b + 1$ for some $u \in M$. We have $u \neq 0$, hence $u = w + 1$, for some $w \in M$. Thus, $a + w = b$, that is, $a \leq^M b$.

Let F be a formula of the form $\forall x[x \leq y \to G]$. In this case we say that the quantifier $\forall x$ is *bounded* in F. Similarly, for the existential quantifier, the formula F has the form $\exists x[x \leq y \wedge G]$. We call the quantifiers bounded also in the case where instead of $\leq$ there occurs $<$. The above formulas with a bounded quantifier are written more briefly as $\forall x \leq yG$ and $\exists x \leq yG$. Also, a formula F is called bounded if all its quantifiers are bounded. It can be proved (see Exercise 19.11) that a proper substitution of an arbitrary term t for y in a bounded formula yields in effect a bounded formula. In other words, quantifiers of the form $\forall x \leq t$ or $\exists x \leq t$ are equivalent to bounded quantifiers.

We have

13.7 $$\mathrm{PA} \vdash \forall x[\forall y < x(F(y/x) \to F)] \to \forall xF$$

for every formula F and every variable y (this is called the scheme of the complete induction).

In the proof we apply the induction scheme for the formula $G(x)$: $\forall y \leq x\, F(y/x)$. Let $\mathbb{M}$ be an arbitrary model of PA. Assume that we have

13.8 $$\mathbb{M} \models \forall x[\forall y < x(F(y/x) \to F)]$$

and let $Z = \{a \in M\colon\ \mathbb{M} \models G[a]\}$. If in F there are free variables $z_1, \ldots, z_m$ other than x, then they occur also in G and we substitute for $z_1, \ldots, z_m$ arbitrary parameters $b_1, \ldots, b_m \in M$. From 13.8 it follows immediately that $\mathbb{M} \models F[0]$, whence $\mathbb{M} \models G[0]$, and hence $0 \in Z$. Assume now that $a \in Z$. We have thus $\mathbb{M} \models F[b]$, for every $b \leq^M a$, and also for $b = a +^M 1$, by the assumption. Therefore, $\mathbb{M} \models G[a+1]$, and thus $a + 1 \in Z$. Hence $Z = M$, whence it follows immediately $\mathbb{M} \models \forall x F$. This completes the proof of 13.7.

We have

13.9 $$\mathrm{PA} \vdash (\exists x\, F \to \exists x[F \wedge \forall y < x \neg F(y/x)]),$$

for any formula F and variable x. This is called the *minimum scheme*. Condition 13.9 means that in any model every nonempty parametrically definable set has a least element. The proof can be obtained immediately by changing the implication in 13.9 to the contrapositive one and applying 13.7 to the formula $\neg F$.

We have

13.10 $$\mathrm{PA} \vdash (x \leq y \to \exists! z(x + z = y)).$$

If in a given model $\mathbb{M}$ we have $a + u = b$ and $a + w = b$ then $a + u = a + w$, whence $u = w$. Thus, we may introduce a new operation $y - x$ (*the bounded difference*) by means of the definition

$$(y - x = z) \ \equiv\ ((x \leq y \wedge x + z = y) \vee (y < x \wedge z = 0)).$$

Hence, we have the following:

$$\mathrm{PA} \vdash (x \leq y \to x + (y - x) = y).$$

Similarly, we introduce the absolute value

$$(|x - y| = z) \ \equiv\ ((x \leq y \wedge x + z = y) \vee (y \leq x \wedge y + z = x)).$$

13.11. Numerals

We define the terms Λ_n, for $n > 1$, by induction: $\Lambda_{n+1} = \Lambda_n + \Lambda_1$. Hence

$$\Lambda_n = \Lambda_1 + \cdots + \Lambda_1 \ n\text{-times}, \qquad \text{for } n \geq 1.$$

In the standard model $\mathbb{N}$ the values of the terms Λ_n, $n \in N$, are consecutive natural numbers: $\Lambda_n^N = n$, for every $n \in N$.

Now we prove the following properties:

13.12

$$\mathrm{PA} \vdash (\Lambda_n + \Lambda_m = \Lambda_{n+m}),$$

$$\mathrm{PA} \vdash (\Lambda_n \cdot \Lambda_m = \Lambda_{n \cdot m}),$$

$$\mathrm{PA} \vdash (x \leq \Lambda_n \rightarrow x = \Lambda_0 \vee \cdots \vee x = \Lambda_n).$$

Proof. Let $\mathbb{M}$ be an arbitrary model. We have

$$\Lambda_n^M + \Lambda_0^M = \Lambda_n^M.$$

Assume that $\Lambda_n^M + \Lambda_m^M = \Lambda_{n+m}^M$. Then

$$\Lambda_n^M + \Lambda_{m+1}^M = \Lambda_n^M + \Lambda_m^M + \Lambda_1^M = \Lambda_{n+m}^M + \Lambda_1^M = \Lambda_{n+m+1}^M,$$

which proves the first property. Similarly, we have

$$\Lambda_n^M \cdot \Lambda_0^M = \Lambda_0^M$$

by Lemma 13.1, and assuming $\Lambda_n^M \cdot \Lambda_m^M = \Lambda_{n \cdot m}^M$ we infer

$$\Lambda_n^M \cdot \Lambda_{m+1}^M = \Lambda_n^M \cdot (\Lambda_m^M + \Lambda_1^M) = \Lambda_{n \cdot m}^M + \Lambda_n^M = \Lambda_{n \cdot m+n}^M = \Lambda_{n(m+1)}^M,$$

which gives the second rule. The third one for $n = 0$ takes the form

$$\mathrm{PA} \vdash (x \leq \Lambda_0 \rightarrow x = \Lambda_0),$$

and thus it holds true, by 13.4. Assuming its validity for n we have

$$(\mathrm{PA}, x \leq \Lambda_{n+1}) \vdash (x \leq \Lambda_n \vee x = \Lambda_{n+1})$$

by 13.6, and hence

$$(\mathrm{PA}, x \leq \Lambda_{n+1}) \vdash (x = \Lambda_0 \vee \cdots \vee x = \Lambda_n \vee x = \Lambda_{n+1}),$$

which completes the proof. □

13.13. Nonstandard Models

Let $\mathbb{M}$ be an arbitrary model of PA. Consider the function h: $N \longrightarrow M$ defined by the equality $h(n) = \Lambda_n^M$, for $n \in N$.

13.14. *The function h is an embedding of the standard model* $\mathbb{N}$ *into* $\mathbb{M}$.

We have

$$h(0) = \Lambda_0 = 0^M, \quad h(1) = \Lambda_1^M = 1^M,$$

and, using 13.12, we get

$$h(n+m) = \Lambda_{n+m}^M = \Lambda_n^M + \Lambda_m^M = h(n) + h(m).$$

Similarly, we obtain

$$h(n \cdot m) = h(n) \cdot h(m),$$

also by 13.12. Moreover, the function h is one-to-one since if $n \neq m$ and, for example, $n < m$, then $\Lambda_m = \Lambda_n + \Lambda_{m-n}$, whence $\Lambda_n^M <^M \Lambda_m^M$, since $m - n > 0$.

13.15. *The image $h[N]$ is an initial segment of the set M with respect to the ordering $\leq^M$.*

This follows immediately from the third rule in 13.12.

13.16. *The function h is the unique embedding of $\mathbb{N}$ into $\mathbb{M}$.*
Clearly, if g: $\mathbb{N} \longrightarrow \mathbb{M}$ is an embedding, then

$$g(0) = 0^M = \Lambda_0^M \quad \text{and} \quad g(1) = 1^M = \Lambda_1^M.$$

Assuming inductively $g(n) = h(n)$ we obtain

$$g(n+1) = g(n) + g(1) = h(n) + \Lambda_1^M = \Lambda_n^M + \Lambda_1^M = \Lambda_{n+1}^M = h(n+1)$$

and therefore $g = h$.

Thus, we see that a given model $\mathbb{M}$ is nonstandard (i.e., nonisomorphic to $\mathbb{N}$), if the unique embedding h: $\mathbb{N} \longrightarrow \mathbb{M}$ is not "onto," that is, there are elements $a \in M \setminus h[N]$. The latter condition is, by 13.15, equivalent with $a^M > h[N]$, that is, $\Lambda_n^M <^M a$, for every $n \in N$. Of course, the part $h[N]$ of the universe of $\mathbb{M}$ can be identified with N. Then, the condition guaranteeing the nonstandardness of $\mathbb{M}$ can be expressed thus: there are elements $a \in M$ such that $n <^M a$, for every $n \in N$. Such elements a (if any) are called *nonstandard* or *infinite*.

13.17. Theorem (Skolem, [S4]). *There exist countable nonstandard models of* PA *which are equivalent with the model* $\mathbb{N}$.

Proof. Let T consist of all the sentences true in $\mathbb{N}$ and of the formulas $x > \Lambda_n$, for $n \in N$. Every finite subset T_0 of the set T has a model because in T_0 there are finitely many sentences true in $\mathbb{N}$ and finitely many formulas of the form $x > \Lambda_{n_1}, \ldots, x > \Lambda_{n_k}$. If $a = \max\{n_1, \ldots, n_k\} + 1$, then the pair $\langle \mathbb{N}, a \rangle$ is a model of T_0. By Theorem 11.6, the compactness theorem (see also 11.7), the set T has a countable model $\langle \mathbb{M}, a \rangle$. Of course, $\mathbb{M}$ is a model of PA and it is nonstandard, since we have $\mathbb{M} \models (\Lambda_n < x)[a]$, for every $n \in N$, that is, a is an infinite element.

Let F be an arbitrary sentence. If $\mathbb{N} \models F$, then $F \in T$, hence $\mathbb{M} \models F$. If $\mathbb{N} \not\models F$, then $\mathbb{N} \models \neg F$, hence $\neg F \in T$, whence $\mathbb{M} \models \neg F$, that is, $\mathbb{M} \not\models F$. We have shown the equivalence

$$\mathbb{N} \models F \quad \text{if and only if} \quad \mathbb{M} \models F, \quad \text{for every sentence } F,$$

that is, we have shown that the models $\mathbb{M}$ and $\mathbb{N}$ are equivalent. □

Remark. The initial segment N of a given nonstandard model $\mathbb{M}$ is not a parametrically definable set in $\mathbb{M}$. Because, in the opposite case, since $0 \in N$ and N is closed under successor, we would have $N = M$, contradicting the fact that $\mathbb{M}$ is nonstandard.

EXERCISES

13.1. *The maximum principle*. Show that for any formula F the formula

$$\exists xF \wedge \exists z \forall x[F \rightarrow x \leq z] \rightarrow \exists x\big[F \wedge \forall y(x < y \rightarrow \neg F(y/x))\big]$$

is a theorem of PA.

13.2. *The Dirichlet Principle*. Show that the formula

$$\exists z \forall x \exists! y(y < z \wedge F) \rightarrow \exists x_1, x_2, y\big[x_1 \neq x_2 \wedge F(x_1/x) \wedge F(x_2/x)\big],$$

where F is any formula such that $z \notin V_f(F)$ is a theorem of PA.

13.3. Show the following equivalences

$$\forall x \leq t \exists y\, F \equiv_{\text{PA}} \exists v \forall x \leq t \exists y \leq vF,$$
$$\exists x \leq t \forall yF \equiv_{\text{PA}} \forall v \exists x \leq t \forall y \leq vF$$

for any F such that $v \notin V(F)$. Thus, every arithmetical formula is equivalent in PA with a formula of the form $\mathbb{Q}_1 x_1, \ldots, \mathbb{Q}_m x_m F$, where $\mathbb{Q}_i$ is the quantifier $\forall$ or $\exists$, and all the quantifiers in F are bounded.

13.4. Let PA* arise from PA by the replacement of the induction axioms by the minimum scheme. Show that the theories PA and PA* are equivalent, that is,

$$(\text{PA} \vdash F) \equiv (\text{PA}^* \vdash F), \quad \text{for any formula } F.$$

13.5. We define a new relation symbol (*divisibility*) by means of the axiom

$$\forall x, y[x|y \equiv \exists z(x \cdot z = y)].$$

Show that $\text{PA} \vdash (x|y \wedge y \neq 0 \rightarrow x \leq y)$.

13.6. Define the operation $\mathrm{GCD}(x, y) = z$, of the *greatest common divisor.* Prove that

$$\mathrm{PA} \vdash \big(\mathrm{GCD}(x, y) = z \to \exists u, v(z = |ux - vy|)\big).$$

13.7. The theorem on divisibility with remainder. Show that

$$\mathrm{PA} \vdash \big(y \neq \Lambda_0 \to \exists! q, r(x = q \cdot y + r \wedge r < y)\big).$$

13.8. Show that in every ordered ring with subtraction a finite system of equations (i.e., a formula of the form $t_1 = s_1 \wedge \cdots \wedge t_n = s_n$, where t_i, s_i are arbitrary terms) is equivalent with one equation, that is, with a formula of the form $t = 0$. Show the same for a disjunction of systems of equations. (*Hint*: $a_1^2 + \cdots + a_n^2 = 0 \;\equiv\; a_1 = \cdots = a_n = 0$ for arbitrary elements $a_1, \ldots, a_n$ of the ring).

13.9. For an arbitrary model $\mathbb{M}$ of PA, construct an ordered ring $Z(\mathbb{M})$ for which M is the nonnegative part; more precisely, $\mathbb{M}$ is the subsystem of nonnegative elements of the reduct of $Z(\mathbb{M})$ to the type of the language of arithmetic.

13.10. Prove that every open (quantifier free) formula of PA is equivalent in PA with a formula of the form $\exists x_1, \ldots, x_n[t = s]$. Thus, any arithmetical formula is equivalent in PA with a formula of the form $\mathbb{Q}_1 x_1 \cdots \mathbb{Q}_n x_n[t = s]$, where $\mathbb{Q}_i$ is the quantifier $\forall$ or $\exists$. (*Hint*: The formula $t \neq s$ is equivalent with the formula $\exists x(t + x + 1 = s \vee s + x + 1 = t)$, and hence, every open formula is equivalent with a formula of the form $\exists x_1, \ldots, x_n\; F$, where F is a disjunction of systems of equations; apply the preceding exercise.)

13.11. Let $\mathbb{Z}_+[x]$ consist of the polynomials of the ring $\mathbb{Z}[x]$, for which the leading coefficient (the coefficient of the largest power of x) is positive, and of the zero polynomial. Show that all the axioms of PA except that of induction are true in $\mathbb{Z}_+[x]$ (with the usual ring operations).

13.12. Let a be an infinite element of a nonstandard model $\mathbb{M}$. Check that the set $M_0 = \{w(a)\colon\; w \in \mathbb{Z}_+[x]\}$ is closed under the operations of the model $\mathbb{M}$. Show that the subsystem determined by M_0 is isomorphic with the $\mathbb{Z}_+[x]$ from the preceding exercise.

13.13. *The Overspill Principle.* Let $\mathbb{M}$ be a nonstandard model. Show that if $\mathbb{M} \models F[n]$, for arbitrarily large $n \in N$, then also $\mathbb{M} \models F[a]$ for an infinite element $a \in M$.

13.14. Check that any proper initial segment of a given nonstandard model $\mathbb{M}$ without a greatest element is not parametrically definable in $\mathbb{M}$.

13.15. Let $\mathbb{M}$ be a countable nonstandard model. Show that the ordering $\langle M, \leq^M \rangle$ is isomorphic with the ordering $\langle N \cup (Q \times Z), \leq \rangle$, where the relation $\leq$ orders N in the usual way and after the elements of the set N

follow the pairs $\langle q, z \rangle$ (q is a rational and z is an integer) ordered lexicographically (i.e., $\langle q', z' \rangle \leq \langle q'', z'' \rangle$ if $q' < q''$ or $q' = q''$ and $z' < z''$).

13.16. Find a subsystem $\mathbb{M}_0 \subseteq \mathbb{M}$ of a given nonstandard countable model $\mathbb{M}$ such that in $\mathbb{M}_0$ all the axioms of PA except the induction are true and the orderings $\langle M, \leq^M \rangle$ and $\langle M_0, \leq^{M_0} \rangle$ are isomorphic.

13.17. *The Pairing Function.* Let $J(m,n) = ((m+n) \cdot (m+n+1)/2) + m$ for $m, n \in N$. Show that the function J maps $N \times N$ onto N in a one-to-one way.

13.18. Let $K(l) = m$, if there is an n such that $l = J(m,n)$, and let $L(l) = n$ if there is an m such that $l = J(m,n)$. Then, we have $J\big(K(l), L(l)\big) = l$, for every $l \in N$. Show that $K(l) \leq l$ and $L(l) \leq l$, for $l \in N$. Show that for a fixed n the function $J_n(m) = J(m,n)$ is increasing (and, symmetrically for a fixed m).

13.19. Show that the functions J, K, L from Exercises 13.17 and 13.18 are definable in PA.

13.20. *Digital Representation at a Base.* Show that for every m, the formula F_m,

$$z > \Lambda_1 \to \forall x < z^{m+1} \exists x_0, \cdots, x_m < z[x = x_0 + x_1 \cdot z + \cdots + x_m \cdot z^m],$$

is provable in PA. Moreover, the digits $x_0, \ldots, x_m$ are uniquely determined.

13.21. Check that the bounded formula $\varphi(x,y)$,

$$\exists x_0, \ldots, x_m < z \exists y_0, \ldots, y_m < z\{x = x_0 + x_1 \cdot z + \cdots + x_m \cdot z^m$$

$$\wedge\, y = y_0 + y_1 \cdot z + \cdots + y_m \cdot z^m \wedge \bigvee_{0 \leq i \leq m} [(x_i < y_i) \wedge \bigwedge_{i < j \leq m} (x_j = y_j)]\},$$

define the antilexicographical ordering $<_l$ for numbers less than z^{m+1} (cf. Exercise 13.20). Show that the formula $x < y \equiv x <_l y$ is provable in PA, that is, the antilexicographical ordering provably coincides with the ordinary one.

14

SKOLEM-LÖWENHEIM THEOREMS

Any theory T having an infinite model has models in every power greater than or equal to card L. In particular, there are always nonisomorphic infinite models of T. We give a proof that uses the technique of Skolem functions, commonly used in model theory.

We recall that by the power of a relational system $\mathbb{A}$ we mean the power of its universe, that is, the cardinal number card A and the power of a language L is the power of all the basic symbols jointly. It is easy to check that card $L = \text{card Fm}(L)$, that is, the power of the language is equal to the power of the set of all formulas (see Exercise 11.2).

14.1. Theorem. *If a consistent set of sentences T of the language L has an infinite model, then it has models of arbitrarily large cardinality.*

Proof. Let $C = \{c_i\colon\ i \in I\}$, where $c_i \neq c_j$ for $i \neq j$, be a set of new constants. We construct a new set of sentences T^* in the language $L\{C\}$ by adding to T all the sentences of the form $\neg(c_i = c_j)$, for $i \neq j \in I$. We claim that every finite subset $S \subseteq T^*$ has a model, since, we may assume that S has the form

$$S = \{F_1, \ldots, F_n\} \cup \{c_i \neq c_j\colon\ i,j \in I_0 \quad \text{and} \quad i \neq j\},$$

where $F_1, \ldots, F_n \in T$ and I_0 is a finite set. Let $\mathbb{A}$ be an infinite model of the set T. In the universe A we find elements a_i for $i \in I_0$ such that $a_i \neq a_j$ for $i \neq j \in I_0$. Clearly, the expanded system $\langle \mathbb{A}, \{c_i^A\colon\ i \in I_0\}\rangle$, in which $c_i^A = a_i$, for $i \in I_0$, is a model of the set S. Hence, in view of the compactness theorem , the set T^* has a model $\langle \mathbb{B}, \{c_i^B\colon\ i \in I\}\rangle$. Obviously, $\mathbb{B} \in \text{Mod}(T)$ and $c_i^B \neq c_j^B$ for $i \neq j$, whence card $B \geq$ card C, which completes the proof. □

Remark. Assume additionally that card $C \geq$ card L. Then, as can easily be seen, card $L\{C\} =$ card C. On the other hand, the set T^* has a model of power less than or equal to card $L\{C\} =$ card C (see Exercise 11.13), and hence it has a model of power exactly card C. Therefore, Theorem 14.1 can be strengthened as follows: *if a set T of sentences of the language L has an infinite model, then it has a model of every power greater than or equal to card L.* This is the *Skolem–Löwenheim* theorem in a weaker version, see Skolem [S3] and Löwenheim [L2].

We say that a subsystem $\mathbb{A} \subseteq \mathbb{B}$ is an *elementary subsystem* (an *L-subsystem*; see [TV]), if for every formula $F(x_1, \ldots, x_n)$ of the language L we have

14.2 $\quad (\mathbb{A} \models F[a_1, \ldots, a_n])$ if and only if $(\mathbb{B} \models F[a_1, \ldots, a_n])$

for arbitrary elements $a_1, \ldots, a_n \in A$. We write in this case $\mathbb{A} \leq_L \mathbb{B}$ or $\mathbb{A} \leq \mathbb{B}$ if the language is clear. More generally, an embedding $h\colon \mathbb{A} \longrightarrow \mathbb{B}$ is called an *elementary embedding* (an *L-embedding*) if the equivalence

14.3 $\quad (\mathbb{A} \models F[a_1, \ldots, a_n])$ if and only if $\big(\mathbb{B} \models F[h(a_1), \ldots, h(a_n)]\big)$

holds for any formula $F(x_1, \ldots, x_n)$ of language L and any sequence of elements $a_1, \ldots, a_n \in A$.

From the above definitions it follows that if $h\colon \mathbb{A} \longrightarrow \mathbb{B}$ is an elementary embedding, then $h[\mathbb{A}] \leq_L \mathbb{B}$ holds.

If $\mathbb{A} \leq \mathbb{B}$, then 14.2 holds for all sentences F of language L:

$$(\mathbb{A} \models F) \quad \text{if and only if} \quad (\mathbb{B} \models F).$$

Hence, condition $\mathbb{A} \leq \mathbb{B}$ implies condition $\mathbb{A} \equiv_L \mathbb{B}$, that is, the equivalence of the systems $\mathbb{A}$ and $\mathbb{B}$ in L. In particular, if $\mathbb{A} \in \mathrm{Mod}(T)$ and $\mathbb{A} \leq \mathbb{B}$, then $\mathbb{B} \in \mathrm{Mod}(T)$ as well.

14.4. Theorem. *Every infinite system $\mathbb{A}$ has an elementary extension of an arbitrarily large power.*

Proof. We expand language L by new constants c_a corresponding to all elements $a \in A$. We obtain the language $L(\mathbb{A})$, called sometimes the *language of the model* $\mathbb{A}$. Let $T = \mathrm{Th}(\mathbb{A})$ be the set of all the sentences of language $L(\mathbb{A})$ that are true in $\langle \mathbb{A}, \{c_a^A\colon\ a \in A\}\rangle$, where $c_a^A = a$, for all $a \in A$. Thus, T has an infinite model. By Theorem 14.1, T has a model $\langle \mathbb{B}, \{c_a^B\colon\ a \in A\}\rangle$, whose power is not smaller than a given cardinal. We shall show that the mapping $h\colon \mathbb{A} \longrightarrow \mathbb{B}$, defined by $h(a) = c_a^A$, for $a \in A$, is an L-embedding. Applying Lemma 7.1 on substitution, we obtain

$$(\mathbb{A} \models F[a_1/x_1, \ldots, a_n/x_n])$$

$$\text{iff} \quad (\langle \mathbb{A}, \{c_a^A\colon\ a \in A\}\rangle \models F(c_{a_1}/x_1, \ldots, c_{a_n}/x_n)) \quad \text{iff} \quad F(c_{a_1}, \ldots, c_{a_n}) \in T.$$

Hence, condition $\mathbb{A} \models F[a_1, \ldots, a_n]$ implies condition

$$\langle \mathbb{B}, \{c_a^B\colon\ a \in A\}\rangle \models F(c_{a_1}, \ldots, c_{a_n}),$$

and thus, also $\mathbb{B} \models F[c^B_{a_1}, \ldots, c^B_{a_n}]$, that is $\mathbb{B} \models F[h(a_1), \ldots, h(a_n)]$ for any formula F. Conversely, $\mathbb{A} \not\models F[a_1, \ldots, a_n]$ means $\mathbb{A} \models \neg F[a_1, \ldots, a_n]$, which implies

$$\mathbb{B} \models \neg F[h(a_1), \ldots, h(a_n)],$$

and so

$$\mathbb{B} \not\models F[h(a_1), \ldots, h(a_n)].$$

Hence, 14.2 holds for any formula $F(x_1, \ldots, x_n)$ and any sequence $a_1, \ldots, a_n \in A$. Taking as F the formula $x \neq y$ we see that h is one-to-one. Similarly, taking as F the atomic formulas of the form $f(x_1, \ldots, x_m) = y$, $r(x_1, \ldots, x_n)$ or $x = c_a$ we find out that h is an embedding, which completes the proof. □

The next theorem is called the *Tarski–Vaught criterion*, see [TV].

14.5. Theorem. *If* $\mathbb{A} \subseteq \mathbb{B}$ *and the following condition is fulfilled: for every formula* F*, variable* x*, and assignment* p *in* $\mathbb{A}$*, if* $\mathbb{B} \models \exists x F[p]$*, then there exists an* $a \in A$ *such that* $\mathbb{B} \models F[p(a/x)]$*, then* $\mathbb{A}$ *is an L-subsystem of* $\mathbb{B}$*,* $\mathbb{A} \leq \mathbb{B}$.

Proof. We check 14.2 by induction on F. From the assumption we have $\mathbb{A} \subseteq \mathbb{B}$, whence 14.2 holds for atomic formulas (see Exercise 5.6). Assuming 14.2 for F and negating both sides we obtain 14.2 for $\neg F$. The case of the implication is equally easy, so, having assumed 14.2 for F, we pass to the case of formula $\forall x F$. We have

$$(*) \quad \begin{aligned} (\mathbb{A} \not\models \forall x\, F[a_1, \ldots, a_n]) \quad &\text{iff} \quad (\exists a \in A\ \mathbb{A} \not\models F[a/x, a_1, \ldots, a_n]) \\ &\text{iff} \quad (\exists a \in A\ \mathbb{B} \not\models F[a/x, a_1, \ldots, a_n]). \end{aligned}$$

The last condition obviously implies

$$\mathbb{B} \not\models \forall x\, F[a_1, \ldots, a_n].$$

Conversely, if $\mathbb{B} \not\models \forall x\, F[a_1, \ldots, a_n]$, then we have $\mathbb{B} \models \exists x \neg F[a_1, \ldots, a_n]$. Then, by the assumption of the theorem, there exists an $a \in A$ such that $\mathbb{B} \models \neg F[a/x, a_1, \ldots, a_n]$ holds, and so, $\mathbb{B} \not\models F[a/x, a_1, \ldots, a_n]$, that is, the last condition of $(*)$ holds. □

14.6. Skolem Functions

Now we define the so-called Skolem functions. Assume that we are given a relational system $\mathbb{A}$. Every formula F and a variable $x \in V(F)$ determines a family of parametrically definable sets

$$Z(F, x; a_1, \ldots, a_n) = \{a \in A\colon\ \mathbb{A} \models F[a/x, a_1/y_1, \ldots, a_n/y_n]\},$$

where $y_1, \ldots, y_n$ are all the free variables of the formula F different from x. Let g be a choice function for nonempty subsets of the set A. Then $g(Z) \in Z$, provided $Z \subseteq A$ and $Z \neq \emptyset$. Fix an element $b \in A$ and put additionally $g(\emptyset) = b$.

Define

$$f_{F,x}(a_1, \ldots, a_n) = g\big(Z(F, x; a_1, \ldots, a_n)\big)$$

for any formula F and variable x. If $V(F) = \{x\}$, then $f_{F,x}$ is a single element of the set A. Directly from the definition the property follows:

14.7. *If* $\mathbb{A} \models \exists x\, F[a_1, \ldots, a_n]$, *then* $\mathbb{A} \models F[a/x, a_1, \ldots, a_n]$, *where* $a = f_{F,x}(a_1, \ldots, a_n)$.

Intuitively speaking, the function $f_{F,x}$ provides an example for every existential property. Every function f for which 14.7 holds is called a *Skolem function* for F, x.

Let $E \subseteq A$ be an arbitrary subset of the universe and take a set $\mathcal{F}$ consisting of the functions $f_{F,x}$—the Skolem functions for all the formulas F and variables x. Define by induction an increasing sequence $E_0 \subseteq E_1 \subseteq \cdots$ as follows: $E_0 = E$, and E_{n+1} is obtained from E_n by adding all the values $f(a_1, \ldots, a_m)$ for $a_1, \ldots, a_m \in E_n$ and $f \in \mathcal{F}$, $m = \arg(f)$. Hence the set $E^* = \bigcup\{E_n\colon\ n \in N\}$ is the least set containing E and closed under all the Skolem functions from $\mathcal{F}$. E^* is called the *Skolem hull* of the set E in the given system $\mathbb{A}$.

14.8. *The Skolem hull E^* of a set E contains all distinguished elements of the system $\mathbb{A}$ and is closed under all the operations f^A of the system $\mathbb{A}$.*

For the proof let c^A be a distinguished element and let f be a Skolem function for the formula F: $x = c$. We have $\mathbb{A} \models \exists x F$ and hence $\mathbb{A} \models (x = c)[f]$ by 14.7, whence $f = c^A$. Similarly, we find that Skolem functions for formulas of the form $f(x_1, \ldots, x_n) = x$ coincide with the operations f^A.

From 14.8 it follows that the Skolem hull E^* of a set E is the universe of some subsystem $\mathbb{S}(\mathbb{A}, E) \subseteq \mathbb{A}$.

14.9. *We always have $\mathbb{S}(\mathbb{A}, E) \leq_L \mathbb{A}$; that is, any Skolem hull is an L-submodel.*

This follows immediately from the Tarski–Vaught criterion 14.5 and 14.7.

Let us estimate the power of the system $\mathbb{S}(\mathbb{A}, E)$, that is, the power of the set E^*. Every Skolem function is determined by a pair $\langle F, x\rangle$; the set of such functions is at most of power card $\mathrm{Fm}(L) = \text{card } L$. Hence, when passing from E_n to E_{n+1} at most $\max\{\text{card } L, \text{card } E_n\}$-many elements are added, whence $\text{card } E \leq \text{card } E^* \leq \max\{\text{card } E, \text{card } L\}$. Thus, if $\text{card } E \geq \text{card } L$, then $\text{card } E^* = \text{card } E$.

Summing up the results of this chapter we obtain the following strengthening, due to Tarski, of the Skolem–Löwenheim theorem.

14.10. Theorem. *Let T be a set of sentences of language L and let $\mathbb{A}$ be an infinite model of set T. Then:*

(a) *for every set $E \subseteq A$ such that* $\text{card } E \geq \text{card } L$ *there is a model $\mathbb{B} \leq_L \mathbb{A}$ containing E and of the same power as E;*

(b) *for every cardinal number $\kappa \geq max\{\text{card } A, \text{card } L\}$ there is a model $\mathbb{B}$ of power κ such that $\mathbb{A} \leq_L \mathbb{B}$.*

Proof. To prove (a) we take $\mathbb{S}(\mathbb{A}, E)$ as $\mathbb{B}$. To prove (b) notice that by Theorem 14.4 we have a model $\mathbb{B}$ of power greater than or equal to κ, such that $\mathbb{A} \leq_L \mathbb{B}$. If $\mathbb{B}$ has the power greater than κ, then there is a subset $E \subseteq B$ of power κ such that $A \subseteq E$. Then $\mathbb{A} \subseteq \mathbb{S}(\mathbb{B}, E) \leq_L \mathbb{B}$ and $\mathbb{S}(\mathbb{B}, E)$ has power κ. We have $\mathbb{A} \leq_L \mathbb{S}(\mathbb{B}, E)$. In fact, for any formula F and assignment p in $\mathbb{A}$ we have

$$\mathbb{A} \models F[p] \quad \text{iff} \quad \mathbb{B} \models F[p] \quad \text{iff} \quad \mathbb{S}(\mathbb{A}, E) \models F[p],$$

since $\mathbb{A} \leq_L \mathbb{B}$ and $\mathbb{S}(\mathbb{B}, E) \leq_L \mathbb{B}$.

14.11. Corollary. *If T has an infinite model, then it has a model of any power greater than or equal to* $\text{card } L$.

14.12. Example. Let K, L be algebraically closed fields. We show that the condition $K \subseteq L$ implies $K \leq_L L$.

In the proof we shall use a well-known Steinitz theorem (proved in [S5]):

Steinitz Theorem. *For any subfield $k \subseteq K$ where K is algebraically closed there is a set $X \subseteq K$ such that $K = \overline{k(X)}$ (i.e., K is the algebraic closure of the subfield generated by $k \cup X$), and moreover, $k(X)$ is isomorphic with the field of rational functions of variables from a set equipotent with X and with coefficients from k.*

Proof. From this theorem it follows easily that uncountable algebraically closed fields K_1, K_2 of the same power (and characterisitic) are isomorphic. For the proof we let K be their common countable subfield. We have $K_1 = \overline{K(X)}$, $K_2 = \overline{K(Y)}$. Since the algebraic closure of an infinite field has the same power as the field, the sets X and Y are uncountable and of the same power. Thus, there exists an isomorphism g: $K(X) \longrightarrow K(Y)$, and moreover, $g|K = \text{id}$. The isomorphism g can be extended, in the well known way, onto the algebraic closures. Assume now that $k \subseteq K$, where k, K are algebraically closed. By the Skolem-Löwenheim theorem there are uncountable algebraically closed fields K_1, K_2 of the same power that $k \leq_L K_1$ and $K \leq_L K_2$. Thus, by the remark just stated there is an isomorphism g: $K_1 \longrightarrow K_2$ such that, $g|k = \text{id}$.

Now we have

$$k \models F[a_1, \ldots, a_n] \quad \text{iff} \quad K_1 \models F[a_1, \ldots, a_n] \quad \text{iff} \quad K_2 \models F[g(a_1), \ldots, g(a_n)]$$

$$\text{iff} \quad K_2 \models F[a_1, \ldots, a_n] \quad \text{iff} \quad K \models F[a_1, \ldots, a_n],$$

whence $k \leq_L K$. □

EXERCISES

14.1. If a set of sentences T has finite models with an arbitrarily large number of elements, then T has an infinite model as well.

14.2. If a given sentence F is true in some fields of arbitrarily large positive characteristic, then F is true in some field of characteristic zero.

14.3. Show that the theory of fields of characteristic zero is not finitely axiomatizable.

14.4. Show that the class of all the groups containing an element of an infinite order is not finitely axiomatizable.

14.5. The ordered field $\mathbb{K} = \langle K, \leq, +, \cdot, 0, 1 \rangle$ is called *archimedean* if the following condition is fulfilled:

$$\forall a > 0 \, \exists n \in N [n \times 1 \geq a],$$

where $n \times 1 = 1 + \cdots + 1$ is the sum of n ones. Show that every archimedean field is L-equivalent with some non-archimedean field.

14.6. Check that the subsystem of the ordering $\langle N, \leq \rangle$ consisting of positive numbers is isomorphic and hence also equivalent with $\langle N, \leq \rangle$, but it is not an elementary subsystem of $\langle N, \leq \rangle$.

14.7. Show that the embedding h: $\mathbb{A} \longrightarrow \mathbb{B}$ is an L-embedding if and only if the systems $\langle \mathbb{A}, \{a\}_{a \in A} \rangle$ and $\langle \mathbb{B}, \{h(a)\}_{a \in A} \rangle$ are equivalent in $L(\mathbb{A})$.

14.8. Check that if card $\mathbb{A} \geq$ card L, then there exists a $\mathbb{B}$ such that card $\mathbb{B} =$ card $\mathbb{A}$ and $\mathbb{A} \underset{\neq}{\leq}_L \mathbb{B}$, that is, $\mathbb{B}$ is a proper L-extension of $\mathbb{A}$.

14.9. Let $\{\mathbb{A}_i\colon\ i \in I\}$ be a family of systems, where the set of indices I is directed by the relation "$\leq$." Check that if $\mathbb{A}_i \leq_L \mathbb{A}_j$ for $i \leq j$, then

$$\mathbb{A}_i \leq_L \bigcup \{\mathbb{A}_i,\ i \in I\} \quad \text{for every } i \in I.$$

14.10. *The Direct Limit.* Let $\langle \{\mathbb{A}_i,\ i \in I\};\ \{g_{ij}\colon\ i \leq j\} \rangle$ be a directed system with a directed family of embeddings, that is, g_{ij}: $\mathbb{A}_i \longrightarrow \mathbb{A}_j$ and moreover, g_{ij} is the identity on $\mathbb{A}_i$ and $g_{jk} \circ g_{ij} = g_{ik}$, whenever $i \leq j \leq k$. Moreover, we assume that the universes of the systems $\mathbb{A}_i$ are mutually disjoint: $A_i \cap A_j = \emptyset$, for $i \neq j$.

(a) Check that the relation $a \simeq b \;\equiv\; \exists k \geq i,j[g_{ik}(a) = g_{jk}(b)]$, where i,j are such that $a \in A_i$ and $b \in A_j$, is an equivalence on the set $\bigcup\{A_i\colon\; i \in I\}$.

(b) On the set A of equivalence classes, define relations, operations, and distinguished elements in such a way that the system obtained,

$$\mathbb{A} = \sum \langle \{A_i\}, \{g_{ij}\} \rangle$$

satisfies this condition: the functions $g_i\colon\; \mathbb{A}_i \longrightarrow \mathbb{A}$, defined by $g_i(a) = [a]$, for all $a \in A_i$, are embeddings, and $\mathbb{A}$ is the least structure with this property (i.e., if a $\mathbb{B}$ has this property then $\mathbb{A}$ can be embedded into $\mathbb{B}$).

(c) Check that if all the g_{ij}s are L-embeddings, then also the g_is are L-embeddings.

14.11. Let $\{\mathbb{A}_i\colon\; i \in I\}$ be a directed system. If $B_i = A_i \times \{i\}$, then the map $h_i(a) = \langle a, i \rangle$ determines a system $\mathbb{B}$ isomorphic with $\mathbb{A}_i$, and moreover the universes B_i are mutually disjoint. Let $g_{ij}(\langle a, i \rangle) = \langle a, j \rangle$ for $i \leq j$. Show that the directed union $\bigcup\{\mathbb{A}_i\colon\; i \in I\}$ is isomorphic with the direct limit $\sum \langle \{\mathbb{B}_i\}; \{g_{ij}\} \rangle$.

14.12. Assume that $T \vdash \forall y_1, \ldots, y_n \exists! y\; H$, that is, the formula H defines in T a new function symbol f_H. Check that in any model $\mathbb{A}$, f_H^A is a Skolem function for H, y.

14.13. Show that in the arithmetic PA for every formula F and variable x there exists a definable function symbol $f_{F,x}$ such that for every model $\mathbb{M}$, $f_{F,x}^M$ is a Skolem function for F, x.

14.14. Let $\mathbb{M} \models$ PA. Let Def($\mathbb{M}$) denote the subsystem of $\mathbb{M}$ whose universe consists of all the elements of M which are definable in $\mathbb{M}$. Show that Def($\mathbb{M}$) $\leq_L \mathbb{M}$.

14.15. For a relational system $\mathbb{B}$ let Aut($\mathbb{B}$) denote the set of automorphisms of $\mathbb{B}$. Let $\mathbb{A}$, $\mathbb{B}$ be given. If we have

$$\forall a_1, \ldots, a_n \in A\, \forall b \in B\, \exists g \in \mathrm{Aut}(\mathbb{B})[g(b) \in A \wedge g(a_1) = a_1 \wedge \cdots \wedge g(a_n) = a_n],$$

then $\mathbb{A} \leq_L \mathbb{B}$.

14.16. If X, Y are infinite and are, respectively, sets of generators of free algebras $\mathbb{A}[X]$ and $\mathbb{A}[Y]$ and $X \subseteq Y$, then $\mathbb{A}[X] \leq_L \mathbb{A}[Y]$.

15

ULTRAPRODUCTS

Here we shall get familiar with one of the most interesting model–theoretic operations the ultraproduct. On the one side, the ultraproduct is a method of constructing new models and on the other it is a powerful tool in proofs of many theorems. For example, using ultraproducts we obtain another proof of the compactness theorem and also a simple algebraic characterization of axiomatizable classes of structures. Let us note that the ultraproduct occurred for the first time in a work of Skolem [S4].

Filters in a power-set algebra $P(I)$ are called filters on the set I. Thus, a family $p \subseteq P(I)$ is an *ultrafilter on* I (i.e., a maximal filter), if $\emptyset \notin p$ and the following conditions are fulfilled:

1. *if* $U \in p$ *and* $U \subseteq W$, *then* $W \in p$,
2. *if* $U, W \in p$, *then* $U \cap W \in p$,
3. *either* $U \in p$ *or* $I \setminus U \in p$,

for any sets $U, W \subseteq I$. Condition 3 is equivalent with the maximality of a filter (see Exericise 3.10). Moreover, every ultrafilter p on I satisfies the condition

$$(U \cup W \in p) \quad \text{if and only if} \quad (U \in p \vee W \in p)$$

for any $U, W \subseteq I$, since

$$(U \cup W \notin p) \quad \text{iff} \quad (I \setminus (U \cup W) \in p) \quad \text{iff} \quad ((I \setminus U) \cap (I \setminus W) \in p)$$
$$\text{iff} \quad (I \setminus U) \in p \wedge (I \setminus W) \in p) \quad \text{iff} \quad (U \notin p \wedge W \notin p).$$

Let there be given an indexed family $\{\mathbb{A}_i\colon\ i \in I\}$ of relational systems of a common type τ. First consider the product $\mathbb{A} = \prod\{\mathbb{A}_i\colon\ i \in I\}$. The elements of the universe A are the functions defined on I such that $a(i) \in A_i$, for $i \in I$. For any ultrafilter p on I the relation $=_p$ defined by the condition

$$a =_p b \quad \text{if and only if} \quad (\{i \in I\colon\ a(i) = b(i)\} \in p) \quad \text{for, } a, b \in A$$

is a congruence. Obviously, this relation is reflexive and symmetric. We show the transitivity. Let us denote

$$U(a,b) = \{i \in I\colon\ a(i) = b(i)\}.$$

If now $a =_p b$ and $b =_p c$, then $U(a,b), U(b,c) \in p$, and hence also $U(a,b) \cap U(b,c) \in p$. Obviously, we have $U(a,b) \cap U(b,c) \subseteq U(a,c)$, whence also $U(a,c) \in p$, that is, $a =_p c$. Now, let f^A be an arbitrary operation of the product $\mathbb{A}$. If $a_1 =_p b_1, \ldots, a_m =_p b_m$, then the set $U = U(a_1, b_1) \cap \cdots \cap U(a_m, b_m)$ belongs to p and $a_k|U = b_k|U$ for $k = 1, \ldots, m$. Hence, for $i \in U$ we have

$$f^{A_i}(a_1(i), \ldots, a_m(i)) = f^{A_i}(b_1(i), \ldots, b_m(i)),$$

that is,

$$f^A(a_1, \ldots, a_m)(i) = f^A(b_1, \ldots, b_m)(i) \quad \text{for each} \quad i \in U,$$

whence immediately $f^A(a_1, \ldots, a_m) =_p f^A(b_1, \ldots, b_m)$.

The congruence $=_p$ has also the following additional property: *If for an arbitrary relation r^A of the product $\mathbb{A}$ we denote*

$$U(r, a_1, \ldots, a_n) = \{i \in I\colon\ r^{A_i}(a_1(i), \ldots, a_n(i))\},$$

then we have

15.1 $$U(r, a_1, \ldots, a_n) \in p \ \equiv\ U(r, b_1, \ldots, b_n) \in p,$$

provided $a_1 =_p b_1, \ldots, a_n =_p b_n$.

To see this, it is sufficient to notice the relationship

$$U(r, a_1, \ldots, a_n) \cap U(a_1, b_1) \cap \ldots \cap U(a_n, b_n) \subseteq U(r, b_1, \ldots, b_n).$$

The *ultraproduct* $\mathbb{A}_p = \prod_i \mathbb{A}_i/p$ is defined as follows: the universe $\mathbb{A}_p = \{[a]_p\colon\ a \in A\}$ consists of the equivalence classes of the elements $a \in A$ with respect to the congruence $=_p$ and

$$r^{A_p}([a_1]_p, \ldots, [a_n]_p) \ \equiv\ \{i \in I\colon\ r^{A_i}(a_1(i), \ldots, a_n(i))\} \in p,$$

$$f^{A_p}([a_1]_p, \ldots, [a_n]_p) = [f^A(a_1, \ldots, a_m)]_p,$$

$$c^{A_p} = [c^A]_p,$$

for any relation r^A, operation f^A, and distinguished element c^A of the product $\mathbb{A}$. Thus the relations are defined by means of a condition that is weaker than in a usual factor system. Property 15.1 guarantees that the relation r^{A_p} is well defined.

The following theorem establishes the connection between the truth of formulas in $\mathbb{A}_p$ with the truth in the systems $\mathbb{A}_i$.

15.2. Theorem (Łoś, [L1]). *If $\mathbb{A}_p = \prod_i \mathbb{A}_i/p$ is an ultraproduct of the relational systems $\mathbb{A}_i$, $i \in I$, then we have*

15.3
$$\mathbb{A}_p \models F\big[[a_1]_p, \ldots, [a_k]_p\big] \quad \text{if and only if} \quad \{i \in I\colon \mathbb{A}_i \models F[a_1(i), \ldots, a_k(i)]\} \in p,$$

for any formula $F(x_1, \ldots, x_k)$ and elements $a_1, \ldots, a_k \in A$.

Proof. First, we check that

$$(*) \qquad t\big[[a_1]_p, \ldots, [a_k]_p\big] = \big[t[a_1, \ldots, a_k]\big]_p$$

for any term t. Applying induction we have $x\big[[a]_p\big] = [a]_p = \big[x[a]\big]_p$ and $c^{A_p} = [c^A]_p$, by the definition. Next,

$$\begin{aligned}
&f(t_1, \ldots, t_m)\big[[a_1]_p, \ldots, [a_k]_p\big] \\
&\quad = f^{A_p}\big(t_1\big[[a_1]_p, \ldots, [a_k]_p\big], \ldots, t_m\big[[a_1]_p, \ldots, [a_k]_p\big]\big) \\
&\quad = f^{A_p}\big(\big[t_1[a_1, \ldots, a_k]\big]_p, \ldots, \big[t_m[a_1, \ldots, a_k]\big]_p\big) \\
&\quad = \big[f^A(t_1[a_1, \ldots, a_k], \ldots, t_m[a_1, \ldots, a_k])\big]_p \\
&\quad = \big[f(t_1, \ldots, t_m)[a_1, \ldots, a_k]\big]_p,
\end{aligned}$$

which proves $(*)$.

Now we shall prove the equivalence 15.3 by induction on F. We introduce the notation

$$U(F;\ a_1, \ldots, a_k) = \{i \in I\colon\ \mathbb{A}_i \models F[a_1(i), \ldots, a_k(i)]\}.$$

Let F be an atomic formula $t = s$. Using $(*)$ we obtain

$$\begin{aligned}
&\big(\mathbb{A}_p \models F\big[[a_1]_p, \ldots, [a_k]_p\big]\big) \\
&\qquad \text{iff} \quad t\big[[a_1]_p, \ldots, [a_k]_p\big] = s\big[[a_1]_p, \ldots, [a_k]_p\big] \\
&\qquad \text{iff} \quad t[a_1, \ldots, a_k] =_p s[a_1, \ldots, a_k] \\
&\qquad \text{iff} \quad \{i \in I\colon\ t[a_1, \ldots, a_k](i) = s[a_1, \ldots, a_k](i)\} \in p.
\end{aligned}$$

On the other hand,

$$t[a_1, \ldots, a_k](i) = t[a_1(i), \ldots, a_k(i)] \quad \text{for } i \in I.$$

Clearly, it is so for variables and constants and, by the definition of the operations of the product, we get

$$f(t_1,\ldots,t_m)[a_1,\ldots,a_k](i) = f^A(t_1[a_1,\ldots,a_k],\ldots,t_m[a_1,\ldots,a_k])(i)$$
$$= f^{A_i}(t_1[a_1,\ldots,a_k](i),\ldots,t_m[a_1,\ldots,a_k](i))$$
$$= f^{A_i}(t_1[a_1(i),\ldots,a_k(i)],\ldots,t_m[a_1(i),\ldots,a_k(i)])$$
$$= f(t_1,\ldots,t_m)[a_1(i),\ldots,a_k(i)].$$

Since, obviously

$$t[a_1(i),\ldots,a_k(i)] = s[a_1(i),\ldots,a_k(i)] \quad \text{iff} \quad \big(\mathbb{A}_i \models (t = s)[a_1(i),\ldots,a_k(i)]\big),$$

so we obtain finally

$$\big(\mathbb{A}_p \models (t = s)\big[[a_1]_p,\ldots,[a_k]_p\big]\big) \quad \text{iff} \quad U(t = s;\ a_1,\ldots,a_k) \in p.$$

Let now F be an atomic formula $r(t_1,\ldots,t_n)$. Using $(*)$ we have

$$\big(\mathbb{A}_p \models F\big[[a_1]_p,\ldots,[a_k]_p\big]\big)$$
$$\text{iff} \quad r^{A_p}\big(t_1\big[[a_1]_p,\ldots,[a_k]_p\big],\ldots,t_n\big[[a_1]_p,\ldots,[a_k]_p\big]\big)$$
$$\text{iff} \quad r^{A_p}\big(\big[t_1[a_1,\ldots,a_k]\big]_p,\ldots,\big[t_n[a_1,\ldots,a_k]\big]_p\big)$$
$$\text{iff} \quad \{i \in I\colon r^{A_i}\big(t_1[a_1,\ldots,a_k](i),\ldots,t_n[a_1,\ldots,a_k](i)\} \in p$$
$$\text{iff} \quad \{i \in I\colon r^{A_i}\big(t_1[a_1(i),\ldots,a_k(i)],\ldots,t_n[a_1(i),\ldots,a_k(i)]\} \in p$$
$$\text{iff} \quad U\big(r(t_1,\ldots,t_n);\ a_1,\ldots,a_k\big) \in p.$$

Thus 15.3 holds for atomic formulas. Assume now 15.3 for a given formula F. Negating both sides we obtain

$$\big(\mathbb{A}_p \models \neg F\big[[a_1]_p,\ldots,[a_k]_p\big]\big) \quad \text{iff} \quad U(F;\ a_1,\ldots,a_k) \notin p$$
$$\text{iff} \quad \big(I \setminus U(F;\ a_1,\ldots,a_k) \in p\big) \quad \text{iff} \quad U(\neg F;\ a_1,\ldots,a_k) \in p,$$

since obviously

$$U(\neg F;\ a_1,\ldots,a_k) = I \setminus U(F;\ a_1,\ldots,a_k).$$

Similarly, we have

$$U(F \to G;\ a_1,\ldots,a_k) = \big(I \setminus U(F;\ a_1,\ldots,a_k)\big) \cup U(G;\ a_1,\ldots,a_k).$$

Hence, assuming 15.3 for the formulas F and G, we obtain

$$\begin{aligned}
&\big(\mathbb{A}_p \models (F \to G)\big[[a_1]_p, \ldots, [a_k]_p\big]\big) \\
&\qquad \text{iff} \quad \big(I \setminus U(F;\ a_1, \ldots, a_k) \in p \ \text{ or } \ U(G;\ a_1, \ldots, a_k) \in p\big) \\
&\qquad \text{iff} \quad \big(I \setminus U(F;\ a_1, \ldots, a_k) \cup U(G;\ a_1, \ldots, a_k)\big) \in p \\
&\qquad \text{iff} \quad U(F \to G;\ a_1, \ldots, a_k) \in p.
\end{aligned}$$

Finally, consider the case of the quantifier. Assume 15.3 for formula F and notice the relationship

$$\forall a \in A[U(\forall x\ F;\ a_1, \ldots, a_k) \subseteq U(F;\ a, a_1, \ldots, a_k)].$$

Given $a \in A$, if $i \in U(\forall x\ F;\ a_1, \ldots, a_k)$, then we have $\mathbb{A}_i \models \forall x\ F[a_1(i), \ldots, a_k(i)]$, thus also $\mathbb{A}_i \models F[a(i), a_1(i), \ldots, a_k(i)]$, that is, $i \in U(F;\ a, a_1, \ldots, a_k)$. So, if

$$U(\forall xF;\ a_1, \ldots, a_k) \in p, \quad \text{then also} \quad (F;\ a, a_1, \ldots, a_k) \in p,$$

for every a, which yields

$$\mathbb{A}_p \models F\big[[a]_p, [a_1]_p, \ldots, [a_k]_p\big]$$

for any $a \in A$, whence

$$\mathbb{A}_p \models \forall x\ F\big[[a_1]_p, \ldots, [a_k]_p\big].$$

Suppose now that $U(\forall x\ F;\ a_1, \ldots, a_k) \notin p$. Hence

$$I \setminus U(\forall xF;\ a_1, \ldots, a_k) \in p, \quad \text{that is} \quad (\neg\forall xF;\ a_1, \ldots, a_k) \in p.$$

For every $i \in U(\neg\forall xF;\ a_1, \ldots, a_k)$ there is an element $a(i) \in A$ such that

$$\mathbb{A}_i \models \neg F(a(i), a_1(i), \ldots, a_k(i)].$$

For the remaining $i \in I$ we define the value $a(i) \in A$ in an arbitrary way. So we have

$$U(\neg\forall xF;\ a_1, \ldots, a_k) \subseteq I \setminus U(F;\ a, a_1, \ldots, a_k),$$

whence $U(F;\ a, a_1, \ldots, a_k) \notin p$ and consequently

$$\mathbb{A}_p \not\models F\big[[a]_p, [a_1]_p, \ldots, [a_k]_p\big].$$

Hence

$$\mathbb{A}_p \not\models \forall x\, F\big[[a_1]_p, \ldots, [a_k]_p\big],$$

which completes the proof. □

15.4. Corollary. *The sentence F is true in the ultraproduct* $\mathbb{A}_p = \prod_i \mathbb{A}_i/p$ *if and only if the set* $U(F) = \{i \in I\colon\ \mathbb{A}_i \models F\}$ *is in p.*

15.5. Corollary. *Let T be a consistent set of sentences. The ultraproduct of any family of models of T is also a model of the theory T.*

Since, if $\mathbb{A}_i \in \mathrm{Mod}(T)$ for $i \in I$, then for $F \in T$ we have $U(F) = \{i \in I\colon\ \mathbb{A}_i \models F\} = I$ and hence $U(F) \in p$, for any ultrafilter p on I.

15.6. *Proof of Compactness Theorem.* Applying ultraproducts we may prove Theorem 11.6, the compactness theorem, directly (i.e., without referring to the completeness theorem, as we did in Chapter 11). Assume that T is an infinite set of sentences of a given language L. Let I be the family of all finite subsets of T. Thus, every $i \in I$ has the form

$$i = \{F_1, \ldots, F_n\}, \quad \text{where} \quad F_1, \ldots, F_n \in T.$$

The assumption of the compactness theorem ensures that for every $i \in I$ there is a system $\mathbb{A}_i$ such that

$$\mathbb{A}_i \models F_1 \wedge \cdots \wedge F_n,$$

where $F_1, \ldots, F_n$ are all the elements of the set i. Let us denote

$$e(F) = \{i \in I\colon\ F \in i\}, \qquad \text{for an arbitrary formula F.}$$

It is easy to see that the family $E = \{e(F)\colon\ F \in T\}$ is centered. In fact, any intersection $e(F_1) \cap \cdots \cap e(F_n)$ contains the set $\{F_1, \ldots, F_n\}$. Thus, there exists an ultrafilter p on I such that $E \subseteq p$. Let us take the ultraproduct $\mathbb{A}_p = \prod_i \mathbb{A}_i/p$ and let us show that it is a model of T. Suppose that $F \in T$. We have then $e(F) \in E$, and hence $e(F) \in p$. On the other hand $e(F) \subseteq U(F)$, that is, $e(F) \subseteq \{i \in I\colon\ \mathbb{A}_i \models F\}$, since, if $i \in e(F)$, then $F \in i$, and hence $\mathbb{A}_i \models F$. Thus, we have $U(F) \in p$, which means $\mathbb{A}_p \models F$, by Corollary 15.4. □

The above proof is due to Frayne, Morel, and Scott; see [FMS] for other results and applications.

Now, we show another application of the ultraproduct. Let $\mathbb{K}$ be a class of some relational systems of a common type τ. If the class $\mathbb{K}$ is axiomatizable, that is we have $\mathbb{K} = \mathrm{Mod}(T)$ for a set of sentences T of a language $L = L(\tau)$, then $\mathbb{K}$ is closed under the equivalence in L; if $\mathbb{A} \equiv_L \mathbb{B}$ and $\mathbb{A} \in \mathbb{K}$, then $\mathbb{B} \in \mathbb{K}$. Moreover, class $\mathbb{K}$ is closed under ultraproducts (Corollary 15.5). It turns out that the above conditions are also sufficient.

15.7. Theorem (Frayne–Morel–Scott [FMS]). *If a class $\mathbb{K}$ is closed under L-equivalence and ultraproducts, then $\mathbb{K}$ is axiomatizable.*

Proof. Let T consist of those sentences of the language L that are true in all systems $\mathbb{A}$ from class $\mathbb{K}$. Of course, we have $\mathbb{K} \subseteq \mathrm{Mod}(T)$. It remains to prove the converse inclusion. So, let $\mathbb{B} \in \mathrm{Mod}(T)$. The set $\mathrm{Th}(\mathbb{B})$ consists of all the sentences true in $\mathbb{B}$:

$$\mathrm{Th}(\mathbb{B}) = \{F\colon\ \mathbb{B} \models F\}.$$

Let I be the family of all finite subsets of the set $\mathrm{Th}(\mathbb{B})$. Then, for every $i \in I$ there exists a system $\mathbb{A}_i \in \mathbb{K}$ such that $\mathbb{A}_i \models F$ for every $F \in I$. In other words, if $i = \{F_1, \dots, F_n\}$, then $\mathbb{A}_i \models F_1 \wedge \cdots \wedge F_n$ because if there were not such an $\mathbb{A}_i$, then the negation $\neg(F_1 \wedge \ldots \wedge F_n)$ would be true in every system $\mathbb{A} \in \mathbb{K}$, and hence $\neg(F_1 \wedge \cdots \wedge F_n) \in T$. Since $\mathbb{B} \in \mathrm{Mod}(T)$, then we would have

$$\mathbb{B} \models \neg F_1 \vee \cdots \vee \neg F_n,$$

whence $\mathbb{B} \models \neg F$ for some $F \in i$, which contradicts the definition of the set I.

Set

$$e(F) = \{i \in I\colon\ F \in i\}.$$

The family $E = \{e(F)\colon\ \mathbb{B} \models F\}$ is centered, since $\{F_1, \dots, F_n\} \in e(F_1) \cap \cdots \cap e(F_n)$. We extend E to an ultrafilter p on I. By the assumption, the ultraproduct $\mathbb{A}_p = \prod_i \mathbb{A}_i/p$ belongs to the class $\mathbb{K}$. We claim that $\mathbb{A}_p \equiv_L \mathbb{B}$. Since, if $\mathbb{B} \models F$, then $e(F) \in p$, and since $e(F) \subseteq \{i\colon\ \mathbb{A}_i \models F\}$, we have $\mathbb{A}_p \models F$. Conversely, if $\mathbb{B} \not\models F$, that is $\mathbb{B} \models \neg F$, then, as above, we obtain $\mathbb{A}_p \models \neg F$, i.e. $\mathbb{A}_p \not\models F$. Therefore,

$$\mathbb{A}_p \equiv \mathbb{B} \quad \text{and} \quad \mathbb{A}_p \in \mathbb{K},$$

and hence $\mathbb{B} \in \mathbb{K}$, which completes the proof. □

15.8. The Ultrapower

Assume that $\mathbb{A}_i = \mathbb{A}$, for all $i \in I$. In this case, the ultraproduct $\prod_i \mathbb{A}_i/p$ is called an *ultrapower* and is denoted by $\mathbb{A}^I/p$.

There is an elementary canonical embedding $h\colon\ \mathbb{A} \longrightarrow \mathbb{A}^I/p$ of a system $\mathbb{A}$ into its ultrapower. Namely, let $k_a\colon\ I \longrightarrow A$ denote the constant function, $k_a(i) = a$, for all $i \in I$.

Put $h(a) = [k_a]_p$ for $a \in A$. The function h maps $\mathbb{A}$ into the ultrapower $\mathbb{A}^I/p$ and it suffices to check that the condition

15.9 $$\mathbb{A} \models F[a_1, \ldots, a_n] \quad \text{if and only if} \quad \mathbb{A}^I/p \models F[h(a_1), \ldots, h(a_n)]$$

holds for every formula F and elements $a_1, \ldots, a_n \in A$. Applying the Łoś theorem we obtain

$$\begin{aligned} \mathbb{A}^I/p \models F[h(a_1), \ldots, h(a_n)] \quad &\text{iff} \quad \mathbb{A}^I/p \models F[[k_{a_1}]_p, \ldots, [k_{a_n}]_p] \\ &\text{iff} \quad \{i \in I\colon\ \mathbb{A} \models F[k_{a_1}(i), \ldots, k_{a_n}(i)]\} \in p. \end{aligned}$$

But $k_a(i) = a$, for all $i \in I$, and 15.9 follows immediately. □

15.10. Models of PA

Using the method of the ultraproduct, we may construct a nonstandard model of the arithmetic PA. First, let us note that there are nonprincipal ultrafilters on the set N; if $e_n = N \setminus \{n\}$, then the family $E = \{e_n\colon\ n \in N\}$ is centered since

$$e_{n_1} \cap \cdots \cap e_{n_k} = N \setminus \{n_1, \ldots, n_k\} \neq \emptyset.$$

Every ultrafilter $p \supseteq E$ consists exclusively of infinite sets because, if $\{n_1, \ldots, n_k\} \in p$, then since $\{n_1, \ldots, n_k\} = \{n_1\} \cup \ldots \cup \{n_k\}$, we have $\{n_i\} \in p$, for some i, which is impossible since also $e_{n_i} = N \setminus \{n_i\} \in p$. Consider the ultrapower N^N/p, where p is an arbitrary nonprincipal ultrafilter on N. There exists then (see Section 15.8), an elementary canonical embedding $h\colon \mathbb{N} \longrightarrow \mathbb{N}^N/p$. By 13.16 we have $h(n) = \Lambda_n(\mathbb{N}^N/p)$, hence the image $h[N]$ can be identified with N. It is easy to find infinite elements in $\mathbb{N}^N/p$. Let id: $N \longrightarrow N$ be the identity, $\text{id}(n) = n$ for $n \in N$. We show that $N < [\text{id}]_p$. Otherwise we would have $[\text{id}]_p = h(n) = [k_n]_p$ for some n (i.e. $[\text{id}]_p$ is a class of a constant function). Thus $\text{id} =_p k_n$, that is, $\{i \in N\colon\ \text{id}(i) = k_n(i)\} \in p$ and hence $\{i \in N\colon\ i = n\} = \{n\} \in p$, which is impossible, since the ultrafilter p is nonprincipal.

We have shown that the ultrapower $\mathbb{N}^N/p$ with respect to a nonprincipal ultrafilter p on N is always a nonstandard model of the arithmetic PA.

15.11. Another Application of Ultraproducts

We shall show that class $\mathbb{K}$ of fields of characteristic greater than zero is not axiomatizable. Note that, if it were $\mathbb{K} = \text{Mod}(T)$ for some set of sentences T, then $\mathbb{K}$ would be closed under the ultraproduct. Take the fields $\mathbb{Z}_i$ of characteristic i, where i runs over the prime numbers. We claim that the ultraproduct $\prod \mathbb{Z}_i/p$, where p is a nonprincipal ultrafilter on the set of the prime numbers, has characteristic zero. For the proof, suppose that $\prod \mathbb{Z}_i/p \models H_n$, where H_n is the sentence $1 + \cdots + 1 = 0$ (the sum of n ones). But then $\{i\colon\ \mathbb{Z}_i \models H_n\} \in p$, and thus infinitely many of the fields $\mathbb{Z}_i$ would have a constant characteristic, which gives a contradiction.

EXERCISES

15.1. If p is a principle ultrafilter on I, then $p = \{U \subseteq I\colon\ i_0 \in U\}$ for some $i_0 \in I$. Show that then the ultraproduct $\prod \mathbb{A}_i/p$ is isomorphic with $\mathbb{A}_{i_0}$.

15.2. If I is a finite set, then every ultrafilter p on I is principle.

15.3. Check that if $\mathbb{A}$ is a finite model, then the ultrapower $\mathbb{A}^I/p$ is isomorphic with $\mathbb{A}$.

15.4. Let $h_i\colon\ \mathbb{A}_i \longrightarrow \mathbb{B}_i$, for $i \in I$, be embeddings. Define the product embedding $h\colon \prod_i \mathbb{A}_i/p \longrightarrow \prod \mathbb{B}_i/p$ in an obvious way and show that h is an elementary embedding, provided that all the h_is are elementary.

15.5. In the ultrapower $\mathbb{N}^N/p$, where p is a nonprincipal ultrafilter, find an infinite prime (i.e., an element $a > N$ divisible only by 1 and a).

15.6. Let p be a nonprincipal ultrafilter on N. In the ultrapower $\mathbb{N}^N/p$ find an element divisible by every number $n \in N$.

15.7. The system $\mathbb{N}[x]$, consisting of the polynomials of variable x with natural coefficients with the usual operations, contains a subsystem isomorphic with $\mathbb{N}$. However, $\mathbb{N}[x]$ is not a model of the arithmetic PA. Check that the subset N of classes of constant functions is not parametrically definable in the ultrapower $\mathbb{N}[x]^N/p$.

15.8. Let the field $\mathbb{Q}_n$ be obtained from the field of rationals $\mathbb{Q}$ by adding all the algebraic numbers of degree less than or equal to n. Show that the ultraproduct $\prod_n \mathbb{Q}_n/p$, where p is a nonprincipal ultrafilter on N, is an algebraically closed field. Hence, the class of fields that are not algebraically closed is not axiomatizable.

15.9. Using the previous exercise, show that the class of algebraically closed fields is not finitely axiomatizable. A similar result holds for real closed fields.

15.10. Let $\mathbb{R}$ be the ordered field of reals. Show that for a nonprincipal ultrafilter p on N, the ultrapower $\mathbb{R}^N/p$ is a non-archimedean field.

15.11. Let $\mathbb{R}^*$ be a non-archimedean extension of the field of reals $\mathbb{R}$ (e.g., $\mathbb{R}^* = \mathbb{R}^N/p$, as in Exercise 15.10). Show that the subset $R_0^* = \{a \in R^*\colon\ \exists n \in N\ |a| \leq n\}$ of finite elements is a ring and that the set $J = \{a \in R_0^*\colon\ \forall n > 0\ |a| < 1/n\}$ is an ideal in $\mathbb{R}_0^*$.

Check that the factor ring $\mathbb{R}_0^*/J$ is isomorphic with $\mathbb{R}$.

15.12. (Sierpiński) Let $\{q_n\colon\ n \in N\}$ be an enumeration of the rationals. Thus, for every irrational a there is a set $Z_a = \{n_k\colon\ k \in N\}$ such that the sequence q_{n_k} is convergent to a. Check that the intersection $Z_a \cap Z_b$ is finite for $a \neq b$ (we say in this case that Z_a, Z_b are almost disjoint).

15.13. Using Exercise 15.12, show that the ultrapower $\mathbb{N}^N/p$ has the cardinality continuum for a nonprincipal p.

15.14. Show that if $\mathbb{A}$ is a countable model, then the ultrapower $\mathbb{A}^N/p$ is a proper extension of the system $h[\mathbb{A}]$, where h is the canonical embedding.

15.15. Let $\mathbb{N}^*$ consist of those equivalence classes $[f]_p$ of the ultrapower $\mathbb{N}^N/p$ that are classes of functions definable in $\mathbb{N}$. Identifying $\mathbb{N}$ with its canonical image, show that $\mathbb{N} \leq_L \mathbb{N}^* \leq_L \mathbb{N}^N/p$ and both elementary inclusions $\leq_L$ are proper for a nonprincipal p.

16

TYPES OF ELEMENTS

In this chapter we introduce another method of constructing models, realizing and omitting types (consistent sets of formulas). The theorem on omitting types is, in fact, a strengthening of Theorem 11.1, the completeness theorem. There are numerous applications of this method in logic. Here we shall discuss some of them, such as the existence and characterization of prime models, saturation, and categoricity.

Let T be a consistent set of sentences in a given language L. We recall that a formula F is said to be consistent with T if the set $T \cup \{F\}$ is consistent. More generally, if S is a consistent set of formulas, $S \subseteq \mathrm{Fm}(L)$, then S is consistent with T if $T \cup S$ is a consistent set. The sets S consistent with T are often called *types* of the theory T.

Let $Y = \{y_1, \ldots, y_n\}$ be a finite set of variables of a language L. If we write $S(y_1, \ldots, y_n)$, we mean that S consists of formulas F of the form $F(y_1, \ldots, y_n)$, that is, $V(F) \subseteq \{y_1, \ldots, y_n\}$. Let $\mathbb{F}_Y(T)$ be the Lindenbaum algebra of the formulas of the form $F(y_1, \ldots, y_n)$ for theory T (see Exercise 9.6). Then, the formula $F = F(y_1, \ldots, y_n)$ is consistent with T if $[F]_T > 0$, that is, if the class of the formulas equivalent with F in the theory T is a nonzero element of the algebra $\mathbb{F}_Y(T)$. Similarly, the set $S = S(y_1, \ldots, y_n)$ is consistent with T if and only if the family $S_T = \{[F]_T\colon\ F \in S\}$ is centered. In this case the filter $Q(S)$ generated by S_T consists of the classes of the theorems of the theory $T \cup S$:

$$Q(S) = \{[G]_T\colon\ (T \cup S) \vdash G\}.$$

Let $\mathbb{A} \in \mathrm{Mod}(T)$. We say that $\mathbb{A}$ realizes S if for some $a_1, \ldots, a_n \in A$ we have $\mathbb{A} \models F[a_1, \ldots, a_n]$, for every formula $F \in S$, that is, if $\langle \mathbb{A}, a_1, \ldots, a_n \rangle$ is a model of the set $T \cup S$. In this case we write in shorter form, $\mathbb{A} \models S[a_1, \ldots, a_n]$. If $\mathbb{A}$ does not realize a given S, we say that $\mathbb{A}$ *omits* S. In the sequel we assume about every type S that S has the form $S = S(y_1, \ldots, y_n)$, for some $y_1, \ldots, y_n$.

16.1. ***Example***. Let $Q(S)$ be a principal filter in $\mathbb{F}_Y(T)$: $Q(S) = \{[H]_T\colon [G]_T \leq [H]_T\}$ for some formula G. Then the realizing (or the omitting) of the set S is equivalent with the realizing (or the omitting) of the generator G. First, we have $T \vdash (G \to F)$ for every formula $F \in S$. Since $[G]_T > \mathbb{O}$, $T \cup \{G\}$ has a model $\langle \mathbb{A};\ a_1, \ldots, a_n \rangle$. We have $\mathbb{A} \models \exists y_1, \ldots, y_n\ G$, that is, $\mathbb{A}$ realizes G, and so S is also realized. If T is a complete theory, that is, either $T \vdash F$ or $T \vdash \neg F$ for every sentence F, then all models of T are equivalent, hence G and so also S are realized in every model of T.

16.2. ***Example***. Let $S(x) = \{x > \Lambda_n\colon\ n \in N\}$. Every nonstandard model $\mathbb{M}$ of PA realizes S. The standard model $\mathbb{N}$ omits S.

Let $S(x) = \{x > n \times 1\colon\ n \in N\}$, where $n \times 1$ is the sum of n ones; see Exercise 14.5. Every non-archimedean field realizes S and every archimedean field omits S.

Consider the following definition: a type $S(y_1, \ldots, y_n)$ of a theory T is called *nonprincipal* if every filter $Q \supseteq S_T$ in the Lindenbaum algebra $\mathbb{F}_Y(T)$ is non-principal. The last condition means that for every formula $G = G(y_1, \ldots, y_n)$ consistent with T, the relation $[G]_T \leq [F]_T$ cannot hold for all $F \in S$, and thus $T \nvdash (F \to G)$ for some formula $F \in S$, which, in turn, is equivalent with the consistency with T of the formula $G \wedge \neg F$. Therefore, a type S of a theory T is nonprincipal if and only if for every formula $G = G(y_1, \ldots, y_n)$ consistent with T there is a formula $F \in S$ such that the conjunction $G \wedge \neg F$ is consistent with T.

Ehrenfeucht and Vaught have shown the following theorem; see also Grzegorczyk, Mostowski, and Ryll-Nardzewski [GMR].

16.3. Theorem (On omitting types). *Let T be a consistent set of sentences in a countable language L. If S is a nonprincipal type of the theory T, then T has a countable model omitting S.*

Put a bit stronger, *T has a model omitting a given countable family of nonprincipal types S.*

Proof. The proof makes use of the Lindenbaum algebra $\mathbb{F}(T)$, similarly to the proof of the completeness theorem indicated in Exercises 11.7 to 11.10. The algebra $\mathbb{F}_Y(T)$ can be treated as a subalgebra of $\mathbb{F}(T)$. Then a given nonprincipal type S in $\mathbb{F}_Y(T)$ [i.e., $S_T \subseteq \mathbb{F}_Y(T)$] is also nonprincipal as a type in $\mathbb{F}(T)$. If S had a generator G in $\mathbb{F}(T)$, where G had some free variables $z_1, \ldots, z_m \notin Y$, then we would have $T \vdash (\exists z_1, \ldots, z_m\ G \to F)$ for every formula $F \in S$, hence $[\exists z_1, \ldots, z_m G]_T \leq [F]_T$ for every $F \in S$, contradicting the assumption.

Put

$$D_Y(S) = \{[H]_T \in \mathbb{F}(T)\colon\ [H]_T > \mathbb{O} \wedge \exists F \in S\ [H]_T \cdot [F]_T = \mathbb{O}\}.$$

The set $D_Y(S)$ is dense in $\mathbb{F}(T)$. Assume $[G]_T > \mathbb{O}$. Then there exists an $F \in S$ such that $[G \wedge \neg F]_T > \mathbb{O}$, since S is a nonprincipal type. But then $[G \wedge \neg F]_T \in D_Y(S)$ and $[G \wedge \neg F]_T \leq [G]_T$, which proves that $D_Y(S)$ is dense.

If $Y = \{y_1, \ldots, y_n\}$ and $Z = \{z_1, \ldots, z_n\}$, then the change of variables

$(z_1/y_1, \dots, z_n/y_n)$ determines an isomorphism of the algebras $\mathbb{F}_Y(T)$ and $\mathbb{F}_Z(T)$. Hence the type S in $\mathbb{F}_Y(T)$ is mapped to an isomorphic type S' in $\mathbb{F}_Z(T)$, where $S' = \{F(z_1/y_1, \dots, z_n/y_n)\colon\ F \in S\}$. Of course, S' is still a nonprincipal type, so it determines a dense set $D_Z(S')$, which will be denoted in the sequel by $D_Z(S)$. In this way a given nonprincipal type S determines a countable family of dense sets $D_Y(S)$, where Y runs over all the n-tuples of the variables. If there are countably many of such types S, then the family of the corresponding sets $D_Y(S)$ is also countable. By the Rasiowa–Sikorski theorem (Exercise 11.9), there is an ultrafilter p in the algebra $\mathbb{F}(T)$ intersecting all the sets $D_Y(S)$ which is complete w.r.t. the sets $E(F, x) = \{F(y/x)\colon\ y \in X\}$ (cf. Exercise 11.10). Then, the relation $=_F$ defined on the variables

$$x =_p y \quad \text{if and only if} \quad [x = y]_T \in p$$

determines a model $\mathbb{A}$ of the theory T with the universe $A = \{[x]_T\colon\ x \in X\}$. Moreover the following condition is satisfied:

16.4 $\mathbb{A} \models F\big[[z_1]_T/x_1, \dots, [z_n]_T/x_n\big]$ if and only if $F(z_1/x_1, \dots, z_n/x_n) \in p$

for any formula $F(x_1, \dots, x_n)$ of language L.

Let $S(y_1, \dots, y_n)$ be any of the types to be omitted. By the construction, for every sequence of variables $z_1, \dots, z_n$ there is a formula $F \in S$ such that $[\neg F(z_1/y_1, \dots, z_n/y_n]_T \in p$. Hence, by 16.4 it follows immediately that the model $\mathbb{A}$ omits S. In this way the proof has been completed. □

16.5. End Extensions

Let $\mathbb{M}_1$, $\mathbb{M}_2$ be models of PA. The model $\mathbb{M}_2$ is called an *end extension* of the model $\mathbb{M}_1$, $\mathbb{M}_1 \subseteq_e \mathbb{M}_2$, if $M_1 \subseteq M_2$ and every element $b \in M_2 \setminus M_1$ follows all the elements of M_1 in the ordering $\leq^{M_2}$ (in symbols, $M_1 <_{M_2} b$). For example, every nonstandard model $\mathbb{M}$ is an end extension of the standard model $\mathbb{N}$.

16.6. Theorem (McDowell and Specker [MS]). *Every countable model* $\mathbb{M}$ *of* PA *has a proper elementary end extension.*

Proof. We add new constants c_a for $a \in M$ to denote elements of M and a new constant c. Let

$$T = \mathrm{Th}(\mathbb{M}) \cup \{c > c_a\colon\ a \in M\}.$$

Clearly, if $\langle M_1;\ \{c_a^{M_1}\}, c^{M_1}\rangle$ is any model of a set T, then the mapping $h(a) = c_a^{M_1}$ for $a \in M_1$ is an elementary embedding of $\mathbb{M}$ into $\mathbb{M}_1$. Hence, identifying $\mathbb{M}$ with its image $h[\mathbb{M}]$ we obtain $\mathbb{M} \leq \mathbb{M}_1$. Therefore, it suffices to find a model $\mathbb{M}_1$ of the set T omitting all the types

$$S_a(x) = \{x < c_a \wedge x \neq c_b\colon\ b <^M a\}, \qquad \text{for } a \in M.$$

So we have to prove that $S_a(x)$ is a nonprincipal type. Notice the following property of a formula G containing the constant c and such that $V(G) \subseteq \{x\}$.

16.7. *The formula* $G(c)$ *is consistent with* T if and only if *the formula* $\forall y \exists z \exists x(y < z \wedge G(z/c))$ *is consistent with* Th($\mathbb{M}$).

Assume that G is consistent with T. Then $\mathbb{M}_1 \models G[b]$ for some $b \in M_1$, where $\mathbb{M}_1$ is some model of T. Thus $\mathbb{M}_1 \models \exists x G$. Since T contains the sentences $c > c_a$ for $a \in M$, we have $\mathbb{M}_1 \models \exists x \exists z (G(z/c) \wedge z > c_a)$, for any c_a. But $\mathbb{M} \leq \mathbb{M}_1$, and hence

$$\mathbb{M} \models \exists z \exists x (G(z/c) \wedge z > c_a)$$

for any c_a. Hence

$$\mathbb{M} \models \forall y \exists z \exists x (y < z \wedge G(z/c)),$$

which proves the consistency of the required formula with Th($\mathbb{M}$). The converse implication is proved similarly. By 16.7 it is already easy to prove that types $S_a(x)$ are nonprincipal. Let $G(x)$ be a formula consistent with T. G may contain some constants $c_{a_1}, \ldots, c_{a_n}$ and c. By 16.7, we obtain

$$\mathbb{M} \models \forall y \exists z \exists x (y < z \wedge G(z/c)).$$

Define

$$f(u) = \min\{w \in M\colon\ \mathbb{M} \models \exists z (y < z \wedge G(z/c))[u/y, w/x]\}.$$

Let $a \in M$. If, for arbitrarily large $u \in M$, we have $a \leq^M f(u)$, then the formula $G \wedge c_a \leq x$ is consistent with T by 16.7, and if $f(u) <^M a$ for almost all u (i.e., for all $u > u_0$, some u_0), then, in view of the Dirichlet principle (Exercise 13.3), there exists a $b <^M a$, such that $f(u) = b$, for arbitrarily large u. In this case we have

$$\mathbb{M} \models \forall y \exists z \exists x (y < z \wedge G(z/c) \wedge x = c_b);$$

therefore, again by 16.7, the formula $G \wedge x = c_b$ is consistent with T. Hence, for any formula $G = G(x)$, there is a formula $F \in S_a$ such that $G \wedge \neg F$ is consistent with T, and thus, S_a is nonprincipal, which completes the proof. □

16.8. Prime Models

A model $\mathbb{A}$ of a set of sentences T is called *prime* if $\mathbb{A}$ is elementarily embeddable into every model $\mathbb{B}$ of the theory T. In this case, we have $\mathbb{A} \equiv_L \mathbb{B}$, for every $\mathbb{B} \in \text{Mod}(T)$, that is, all the models of T are equivalent, which means that T is a complete set of sentences, cf. Exercise 12.2.

16.9. *Example* Let $T = \mathrm{Th}(\mathbb{N})$ be the set of all the sentences true in the standard model of PA. Then $\mathbb{N}$ is a prime model of the theory T. In other words, $\mathbb{N}$ is elementarily embeddable into every model $\mathbb{M}$ of PA such that $\mathbb{M} \equiv \mathbb{N}$; we always have a unique embedding $h\colon \mathbb{N} \longrightarrow \mathbb{M}$ defined as $h(n) = \Lambda_n^M$, for $n \in N$ (see Section 13.13). If $\mathbb{M} \equiv \mathbb{N}$, then we have

$$\mathbb{N} \models F[n_1/x_1, \ldots, n_k/x_k]$$

$$\text{iff} \quad \mathbb{N} \models F(\Lambda_{n_1}/x_1, \ldots, \Lambda_{n_k}/x_k)$$

$$\text{iff} \quad \mathbb{M} \models F(\Lambda_{n_1}/x_1, \ldots, \Lambda_{n_k}/x_k)$$

$$\text{iff} \quad \mathbb{M} \models F[h(n_1)/x_1, \ldots, h(n_k)/x_k].$$

The field of the algebraic numbers is a prime model since it is elementarily embeddable into every algebraically closed field of characteristic zero, (cf. Example 14.12).

16.10. Theorem (Vaught [V1]). *Let T be a complete set of sentences in a countable language L having infinite models. Then a model $\mathbb{A}$ of the theory T is prime if and only if $\mathbb{A}$ is countable and $\mathbb{A}$ omits all maximal nonprincipal types of theory T.*

Proof. Let $\mathbb{A}$ be a prime model and S be a nonprincipal type. By the omitting types theorem, Theorem 16.3, there is a countable model $\mathbb{B}$ omitting S. Since $\mathbb{A}$ is elementarily embeddable into $\mathbb{B}$, $\mathbb{A}$ omits S and is countable too. Conversely, assume that $\mathbb{A}$ is a countable model omitting all the maximal nonprincipal types. Enumerate the elements of the universe $A = \{a_n\colon\ n \in N\}$ and fix also an enumeration $\{x_n\colon\ n \in N\}$ of all variables. Let $\mathbb{B}$ be an arbitrary model of the theory T. Put

$$S_0(x_0) = \{F(x_0)\colon\ \mathbb{A} \models F[a_0]\}.$$

Then S_0 is a maximal type of the theory T realizable in $\mathbb{A}$, hence a principal. Let $G_0(x_0)$ be a generator of type S_0. We have $\mathbb{A} \models G_0[a_0]$, whence $\mathbb{A} \models \exists x_0 G_0$ and thus $\mathbb{B} \models \exists x_0 G_0$, since $\mathbb{B} \equiv \mathbb{A}$. Hence, there is an element $b_0 \in B$ such that $\mathbb{B} \models G_0[b_0]$. Now take the type

$$S_1(x_0, x_1) = \{F(x_0, x_1)\colon\ \mathbb{A} \models F[a_0, a_1]\}$$

and its generator $G_1(x_0, x_1)$. Thus we have $\mathbb{A} \models G_1[a_0, a_1]$, hence $\mathbb{A} \models \exists x_1\ G_1[a_0]$, that is, $\exists x_1 G_1$ belongs to S_0, whence $T \vdash (G_0 \to \exists x_1 G_1)$. Therefore, we have $\mathbb{B} \models \exists x_1\ G_1[b_0]$, since $\mathbb{B} \models G_0[b_0]$. Thus, there is a $b_1 \in B$ such that $\mathbb{B} \models G_1[b_0, b_1]$. In this way we may define by induction a sequence of types $S_n(x_0, \ldots, x_n)$ and of their generators $G_n(x_0, \ldots, x_n)$ and a sequence of elements

$\{b_n\colon\ n \in N\} \subseteq B$, so that for every n we have

$$\mathbb{A} \models G_n[a_0, \ldots, a_n] \quad \text{and} \quad \mathbb{B} \models G_n[b_0, \ldots, b_n].$$

Since G_n generates S_n, we have as well

$$\mathbb{A} \models S_n[a_0, \ldots, a_n] \quad \text{and} \quad \mathbb{B} \models S_n[b_0, \ldots, b_n].$$

Putting $h(a_n) = b_n$, for $n \in N$, we obtain by the maximality of S_n,

$$\mathbb{A} \models F[a_0, \ldots, a_n] \quad \text{if and only if} \quad \mathbb{B} \models F[h(a_0), \ldots, h(a_n)]$$

for any formula $F(x_0, \ldots, x_n)$ whence it follows that $h\colon \mathbb{A} \longrightarrow \mathbb{B}$ is an elementary embedding. □

The following characterization of a prime model as well as Theorem 16.14 on maximal models are also due to Vaught [V1]. Recall that the Boolean algebra is called atomic if under every nonzero element there lies an atom. Under the assumption of Theorem 16.10 we have the following theorem:

16.11. Theorem. *T has a prime model if and only if every algebra $\mathbb{F}_Y(T)$ is atomic (where $Y = \{y_1, \ldots, y_n\}$).*

Proof. Assume that $\mathbb{A}$ is a prime model and let $[F]_T$ be a nonzero element of the algebra $\mathbb{F}_Y(T)$. Hence, the formula F is consistent with T, and thus $\mathbb{B} \models F[b_1, \ldots, b_n]$ for some $\mathbb{B} \in \text{Mod}(T)$ and a sequence $b_1, \ldots, b_n \in B$. So we have $\mathbb{B} \models \exists y_1, \ldots, y_n F$, and hence $\mathbb{A} \models \exists y_1, \ldots, y_n F$, since $\mathbb{A} \equiv \mathbb{B}$. Thus, we have $\mathbb{A} \models F[a_1, \ldots, a_n]$ for some elements $a_1, \ldots, a_n \in A$. Let $S(y_1, \ldots, y_n)$ be the type of the sequence $a_1, \ldots, a_n$, that is,

$$S = \{H(y_1, \ldots, y_n)\colon\ \mathbb{A} \models H[a_1, \ldots, a_n]\}.$$

By Theorem 16.10, S has a generator G. Clearly, $[G]_T$ is a generator of the ultrafilter $S_T = \{[H]_T\colon\ H \in S\}$, and thus $[G]_T$ is an atom of the algebra $\mathbb{F}_Y(T)$ and $[G]_T \leq [F]_T$, since $F \in S$ (cf. Exercise 3.3). Conversely, assume that the algebras $\mathbb{F}_Y(T)$ are atomic. For every set $Y = \{y_1, \ldots, y_n\}$ of variables such that $\mathbb{F}_Y(T)$ is infinite we set

$$S_Y(y_1, \ldots, y_n) = \{\neg G\colon\ [G]_T \text{ is an atom in } \mathbb{F}_Y(T)\}.$$

Of course, every S_Y is a nonprincipal type, and because there are only countably many sets Y (applying the omitting types theorem), we obtain a countable model $\mathbb{A}$ omitting all the types S_Y. Suppose now that $S = S(y_1, \ldots, y_n)$ is a maximal type realized in $\mathbb{A}$ and let $\mathbb{A} \models S[a_1, \ldots, a_n]$. If S were nonprincipal we would have $S_Y \subseteq S$ and thus S_Y would be realized, contradicting the construction. Hence, S is a principal type; that is, $\mathbb{A}$ omits every nonprincipal type. By Theorem 16.10 it follows that $\mathbb{A}$ is a prime model. □

16.12. *Example*. Every type $S(y_1, \ldots, y_n)$ realized in the standard model $\mathbb{N}$ of PA is principal. If $\mathbb{N} \models S[n_1, \ldots, n_k]$, then the formula $y_1 = \Delta_{n_1} \wedge \cdots \wedge y_k = \Delta_{n_k}$ is a generator of type S.

16.13. Universal Models

We continue to consider a complete theory T in a countable language L having infinite models. A countable model $\mathbb{B}$ of the theory T is called *universal*, if every countable model $\mathbb{A}$ is elementarily embeddable into $\mathbb{B}$.

If T has such a universal model $\mathbb{B}$ then there are at most countably many maximal types $S(y_1, \ldots, y_n)$ [that is, every Lindenbaum algebra $\mathbb{F}_Y(T)$ has at most countably many ultrafilters]. Since a given type $S(y_1, \ldots, y_n)$ is realized in a countable model $\mathbb{A}$ of the theory T, and since $\mathbb{A}$ is elementarie embeddable into $\mathbb{B}$, S is realized in $\mathbb{B}$ too, that is, $\mathbb{B} \models S[b_1, \ldots, b_n]$, for some $b_1, \ldots, b_n \in B$. To different types there correspond different realizing sequences, since if $S_1 \neq S_2$, then there is a formula F such that $F \in S_1$ and $\neg F \in S_2$, which follows from maximality. Hence the number of types S is not larger than the number of finite sequences of elements of B, and thus there are countably many of them, since $\mathbb{B}$ is countable by the assumption.

16.14. Theorem (Vaught [V1]). *If every algebra $\mathbb{F}_Y(T)$ has at most countably many ultrafilters, then T has a universal model.*

Proof. Let $\mathbb{A}$ be an arbitrary countable model of the theory T. Let $L(\mathbb{A})$ be the language of the model $\mathbb{A}$ (i.e., $L(\mathbb{A})$ is built from L by adding new constants c_a for every $a \in A$), and let $T^* = \mathrm{Th}(\langle \mathbb{A};\ \{c_a^A\colon\ a \in A\}\rangle)$; This is the theory of $\mathbb{A}^* = \langle \mathbb{A};\ \{c_a^A\colon\ a \in A\}\rangle$, where $c_a^A = a$, for $a \in A$. For any elements $a_1, \ldots, a_n \in A$ we consider maximal types $S(x, c_{a_1}, \ldots, c_{a_n})$ of the theory $\mathrm{Th}(\langle \mathbb{A};\ a_1, \ldots, a_n\rangle)$. There are at most countably many such types since they are obtained from the types $S(x, y_1, \ldots, y_n)$ of the theory T by the substitution $S(x, c_{a_1}/y_1, \ldots, c_{a_n}/y_n)$. Let $S_n(z_n)$ be their enumeration. We claim that the set of formulas

$$T_0 = T^* \cup \bigcup \{S_n(z_n)\colon\ n \in N\}$$

of the countable language $L(\mathbb{A})$ is consistent. Let $F_i(z_i)$, $i = 1, \ldots, m$, be the conjunction of the formulas from $S_i(z_i)$ occurring in a given finite subset of the set T_0. Since $S_i(z_i) = S_i(z_i, c_{a_1^i}, \ldots, c_{a_n^i})$ is a type of theory $\mathrm{Th}(\langle \mathbb{A};\ a_1^i, \ldots, a_n^i\rangle)$, thus $\langle \mathbb{A};\ a_1^i, \ldots, a_n^i\rangle \models \exists z_i\ F_i$ and hence $\mathbb{A}^* \models \exists z_i\ F_i$ for $i = 1, \ldots, m$, which proves the claim.

Applying the compactness theorem, we find a countable model $\mathbb{B}_0^* \equiv \mathbb{A}^*$, in which all the types S_n are realized. Let $\mathbb{B}_0$ be the reduct of $\mathbb{B}_0^*$ to the type of model

$\mathbb{A}$. Then $\mathbb{A} \leq_L \mathbb{B}_0$. Define now by induction a sequence $\mathbb{B}_0 \leq_L \mathbb{B}_1 \leq_L \cdots$ of elementary extensions, $\mathbb{B}_{n+1}$ is obtained from $\mathbb{B}_n$ in the same way as $\mathbb{B}_0$ from $\mathbb{A}$. Clearly, the union $\mathbb{B} = \bigcup\{\mathbb{B}_n\colon\ n \in N\}$ is a countable model of the theory T, $\mathbb{A} \leq_L \mathbb{B}$, and $\mathbb{B}$ has the following property.

16.15. *For every finite number of elements $b_1, \ldots, b_n \in B$, every type $S(x)$ of the theory $Th(\langle \mathbb{B};\ b_1, \ldots, b_n\rangle)$ is realized in $\mathbb{B}$.*

Thus, we have proved that every countable model of theory T has a countable elementary extension $\mathbb{B}$, which is a finitely saturated model, that is which has property 16.15. We shall show that model $\mathbb{B}$ is universal. Let $\mathbb{A}$ be a given countable model of theory T. Let us enumerate the universe $A = \{a_n\colon\ n \in N\}$ and let us take type $S_0(x)$ of the element a_0;

$$S_0(x) = \{F(x)\colon\ \mathbb{A} \models F[a_0]\}.$$

This is a type of theory $T = \mathrm{Th}(\mathbb{A}) = \mathrm{Th}(\mathbb{B})$, since $\mathbb{B} \equiv_L \mathbb{A}$. Hence, by 16.15, there is a $b_0 \in B$ such that $\mathbb{B} \models S_0[b_0]$. Equivalently, we can write $\langle \mathbb{A};\ a_0\rangle \equiv \langle \mathbb{B};\ b_0\rangle$ in language L expanded by a constant c_{a_0}. Assume that we have already defined the elements $b_0, \ldots, b_n \in B$ so that we have

16.16 $$\langle \mathbb{A};\ a_0, \ldots, a_n\rangle \equiv \langle \mathbb{B};\ b_0, \ldots, b_n\rangle$$

in the language $L\{c_{a_0}, \ldots, c_{a_n}\}$. Let $S_{n+1}(x_0, \ldots, x_{n+1})$ be the type of the sequence $a_0, \ldots, a_{n+1}$, that is,

$$S_{n+1} = \{F(x_0, \ldots, x_{n+1})\colon\ \mathbb{A} \models F[a_0, \ldots, a_{n+1}]\}.$$

This means that $S_{n+1}(c_{a_0}, \ldots, c_{a_n}, x_{n+1})$ is a type of the theory

$$\mathrm{Th}(\langle \mathbb{A};\ a_0, \ldots, a_n\rangle) = \mathrm{Th}(\langle \mathbb{B};\ b_0, \ldots, b_n\rangle),$$

hence there is an element $b_{n+1} \in B$ such that $\mathbb{B} \models S_{n+1}[b_0, \ldots, b_{n+1}]$.

Thus, there is a sequence $\{b_n\colon\ n \in N\} \subseteq B$ such that 16.16 holds for every $n \in N$, whence $\langle \mathbb{A};\ \{a_n\}\rangle \equiv \langle \mathbb{B};\ \{b_n\}\rangle$, or, a bit more precisely,

$$\langle \mathbb{A};\ \{c_{a_n}^A\colon\ n \in N\}\rangle \equiv \langle \mathbb{B};\ \{c_{b_n}^B\colon\ n \in N\}\rangle$$

in language $L(\mathbb{A})$, where $c_{a_n}^A = a_n$ and $c_{b_n}^B = b_n$ for all $n \in N$. It follows immediately that the map $h(a_n) = b_n$, for $n \in N$, is an elementary embedding of $\mathbb{A}$ into $\mathbb{B}$, which completes the proof. □

16.17. *Example*. Let $\mathbb{K}$ be the class of algebraically closed fields of characteristic 0. According to the Steinitz theorem, every countable field $\mathbb{A} \in \mathbb{K}$ has (up to an isomorphism) the form $\mathbb{A} = \overline{\mathbb{Q}(X)}$, that is, it is the algebraic closure of the field

$\mathbb{Q}(X)$ of the rational functions with variables $x \in X$ and coefficients in the field $\mathbb{Q}$ of rational numbers, where the set of variables X is at most countable. If we denote

$$\mathbb{A}_n = \overline{\mathbb{Q}(x_1, \ldots, x_n)} \quad \text{and} \quad \mathbb{A}_\omega = \overline{\mathbb{Q}(\{x_n\colon\ n \in N\})},$$

we have obviously, $\mathbb{A}_0 \leq \mathbb{A}_1 \leq \cdots \leq \mathbb{A}_\omega$ (see Example 14.12). Hence the field of the algebraic numbers $\mathbb{A}_0$ is a prime model and the field $\mathbb{A}_\omega$ is a universal model. The same holds for any fixed positive characteristic under the condition that the field $\mathbb{Q}$ is replaced by a prime field of that characteristic.

16.18. ***Example***. The theory of the standard model of PA, $T = \text{Th}(\mathbb{N})$, has a prime model, but it has no universal one. In fact, we show that the algebra $\mathbb{F}_Y(T)$, $Y = \{y\}$ has continuum many ultrafilters. For any set $E \subseteq N$ consisting of prime numbers only, let us define

$$S_E(y) = \{\Lambda_m | y\colon\ m \in E,\ m \text{ prime}\} \cup \{\neg\Lambda_m | y\colon\ m \notin E,\ m \text{ prime}\},$$

where $a|b$ denotes the divisibility relation. Of course, S_E is a type of theory T, since any finite set of formulas $\Lambda_{m_1}|y, \ldots, \Lambda_{m_k}|y, \Lambda_{n_1} \nmid y, \ldots, \Lambda_{n_l} \nmid y$ has a model namely $\langle \mathbb{N};\ m_1 \cdot \cdots \cdot m_k \rangle$. If $E_1 \neq E_2$, then S_{E_1} contains a formula of the form $\Lambda_m|y$ and S_{E_2} its negation (or the other way round). In this way, different sets E determine different ultrafilters of the Lindenbaum algebra, which proves the claim.

16.19. Categoricity

Let κ be an infinite cardinal number. We say that a theory T is *categorical in the power* κ if all its models of power κ are isomorphic.

Example. The theory of algebraically closed fields of a given cardinality is categorical in every uncountable power (see Example 14.12). By the well-known Cantor theorem, it follows that the theory of dense orderings without end points is categorical in the countable power. A similar result holds also for the theory of atomless Boolean algebras.

Let T be a theory categorical in some power $\kappa \geq \text{card } L$ having only infinite models. Then T must be complete; if $\mathbb{A}, \mathbb{B} \in \text{Mod}(T)$, then, applying the Skolem–Löwenheim theorem 14.10, we find models $\mathbb{A}_1 \equiv \mathbb{A}$ and $\mathbb{B}_1 \equiv \mathbb{B}$ both of power κ, and hence isomorphic. In particular $\mathbb{A}_1 \equiv \mathbb{B}_1$, whence $\mathbb{A} \equiv \mathbb{B}$. Thus, all the models of theory T are equivalent, which implies the completeness of T. Now we shall prove an elegant characterization of categoricity in the countable cardinality.

16.20. Theorem (Ryll-Nardzewski [R-N]; see also Engeler [E2] and Svenonius [S8]). *Let T be a complete theory in a countable language L having infinite models.*

Then T is categorical in the countable power if and only if each algebra $\mathbb{F}_Y(T)$ is finite.

Proof. By categoricity, it follows immediately that each type S is realized in every countable model, hence S must be principal (since a nonprincipal type can be omitted). So, each algebra $\mathbb{F}_Y(T)$ has only principal ultrafilters, and therefore it is finite (see Exercise 8.5). Conversely, assume that all the Lindenbaum algebras $\mathbb{F}_Y(T)$ are finite. Hence all the types S of the theory T are principal and consequently, every countable model of the theory T is prime, by Theorem 16.10. Thus, if $\mathbb{A}$, $\mathbb{B}$ are countable models, then there is an embedding h: $\mathbb{A} \longrightarrow \mathbb{B}$. It suffices to modify the construction of the h given in the proof of Theorem 16.10 so that h be an embedding onto B (i.e., an isomorphism). To this end we apply the Cantor method: let $A = \{a_n\colon\ n \in N\}$ $B = \{b_n\colon\ n \in N\}$. If G_0 is a generator of the type S_0 of the element $u_0 = a_0$, then $\mathbb{B} \models \exists x_0 G_0$ and let w_0 be the earliest term of the sequence $\{b_n\}$ satisfying the formula G_0 in $\mathbb{B}$. Next, if w_1 is the least term of the sequence $\{b_n\}$ different from w_0, then we take the type of the pair w_0, w_1 as $S_1(x_0, x_1)$. If G_1 is a generator of the type S_1, then $\mathbb{A} \models \exists x_1\ G_1[u_0]$ and by u_1 we denote the earliest term of the sequence $\{a_n\}$ for which we have $\mathbb{A} \models G_1[u_0, u_1]$. In general, having defined $u_0, \ldots, u_n$ and $w_0, \ldots, w_n$ so that $\langle \mathbb{A};\ u_0, \ldots, u_n \rangle \equiv \langle \mathbb{B};\ w_0, \ldots, w_n \rangle$, we take, for an odd n, the earliest term of the sequence $\{a_n\}$ different from $u_0, \ldots, u_n$ as u_{n+1} and as G_{n+1}, a generator of the type S_{n+1} of the sequence $u_0, \ldots, u_{n+1}$, and we define w_{n+1} as the earliest term of the sequence $\{b_n\}$ for which the sequence $w_0, \ldots, w_{n+1}$ satisfies G_{n+1} in $\mathbb{B}$. For an even n we proceed similarly, beginning from the sequence $w_0, \ldots, w_n$. In this way we obtain two sequences $\{u_n\colon\ n \in N\}$, $\{w_n\colon\ n \in N\}$ for which we have

$$\langle \mathbb{A};\ \{u_n\colon\ n \in N\} \rangle \ \equiv\ \langle \mathbb{B};\ \{w_n\colon\ n \in N\} \rangle,$$

and, moreover, the elements u_n exhaust the set A, since any term a_n occurs as some u_j, for some $j \leq 2n$. Similarly, the elements w_n exhaust the set B. Hence the function $h(u_n) = w_n$, for $n \in N$, is an isomorphism of the models $\mathbb{A}$ and $\mathbb{B}$. $\square$

16.21. Countably Saturated Models

Types $S(x)$ of the theory $\mathrm{Th}(\mathbb{A}) = \{F\colon\ \mathbb{A} \models F\}$ are often called *types of the elements* of the model $\mathbb{A}$. Hence, a set of formulas $S(x)$ is a type of a model $\mathbb{A}$ if and only if for every finite number of formulas $F_0, \ldots, F_n \in S(x)$ we have $\mathbb{A} \models \exists x(F_0 \wedge \cdots \wedge F_n)$. Clearly, if $S(x)$ is a type of a model $\mathbb{A}$, then the set $\mathrm{Th}(\mathbb{A}) \cup S(x)$ is consistent, hence for $F_0, \ldots, F_n \in S$ there is a model $\mathbb{B} \in \mathrm{Mod}(\mathrm{Th}(\mathbb{A}))$ for which $\mathbb{B} \models \exists x(F_0 \wedge \cdots \wedge F_n)$. But $\mathbb{B} \equiv \mathbb{A}$, so also $\mathbb{A} \models \exists x(F_0 \wedge \ldots \wedge F_n)$. The converse implication follows immediately by the compactness theorem.

Generalizing the notion of the finite saturation (see 16.15), consider the following definition.

Definition. An infinite model $\mathbb{A}$ (of any power) is called *countably saturated* if for every countable subset $\{a_n\colon n \in N\} \subseteq A$ every type of the expanded model $\mathbb{A}^* = \langle \mathbb{A};\ \{a_n\colon n \in N\}\rangle$ is realized in $\mathbb{A}^*$.

16.22. *Remark.* A countably saturated model $\mathbb{A}$ has to be uncountable; if it were $A = \{a_n\colon n \in N\}$, then the type $S(x) = \{x \neq c_{a_n}\colon n \in N\}$ of the model $\mathbb{A}^* = \langle \mathbb{A}; \{a_n\colon n \in N\}\rangle$ would not be realized in $\mathbb{A}^*$.

16.23. Theorem (Keisler [K2]). *If L is a countable language, then any ultrapower $\prod_{n \in N} \mathbb{A}_n/p$ with respect to a nonprincipal ultrafilter p on N is countably saturated.*

(The systems $\mathbb{A}_n$ are here arbitrary—they do not need to be models of a common theory T. We assume only that card $\mathbb{A}_n \geq 2$, $n \in N$.)

Proof. First, we prove that every type $S(x)$ of the ultraproduct $\prod_m \mathbb{A}_m/p$ is realized in it. By the assumption, the set $\mathrm{Fm}(L)$, and thus the set S, is countable. Let $S = \{F_n\colon n \in N\}$. Since S is a type, we have for every $n \in N$

$$\prod_m \mathbb{A}_m/p \models \exists x(F_0 \wedge \cdots \wedge F_n).$$

Applying the Łoś theorem we obtain for every $n \in N$, the set

$$U_n = \{i \in N\colon\ \mathbb{A}_i \models \exists x(F_0 \wedge \cdots \wedge F_n)\}$$

belongs to p.

Let $N_k = \{i \in N\colon i \geq k\}$. The ultrafilter p, as nonprincipal, contains all the sets N_k. Put

$$W_n = U_n \cap N_n, \qquad \text{for } n \in N.$$

Clearly, the set $W = \bigcup\{W_n\colon n \in N\}$ belongs to p. Every element $i \in W$ belongs only to finitely many sets W_n, since $i \notin W_n$ for $n > i$.

Hence, the number $n_i = \max\{n\colon\ i \in W_n\}$ is well-defined for all $i \in W$. Now we define a function f on N as follows: if $i \in W$, then $i \in W_{n_i} \subseteq U_{n_i}$; hence we have

$$\mathbb{A}_i \models \exists x(F_0 \wedge \cdots \wedge F_{n_i}).$$

The value $f(i)$ is defined as an arbitrary element $a \in A_i$ satisfying $\mathbb{A}_i \models (F_0 \wedge \cdots \wedge F_{n_i})[a/x]$. If $i \notin W$, then we let $f(i)$ be an arbitrary element of the set A_i.

Let $n \in N$ be arbitrary. If $i \in W_n$, then $n_i \geq n$; thus we have $\mathbb{A}_i \models F_n[f(i)]$. So we have shown the inclusion

$$W_n \subseteq \{i \in N\colon\ \mathbb{A}_i \models F_n[f(i)]\},$$

for every $n \in N$. But we have $W_n \in p$, and thus, by the Łoś theorem, we infer

$$\prod_n \mathbb{A}_n/p \models F_n[[f]_p],$$

for every $n \in N$, that is, the element $[f]_p$ realizes the type $S(x)$.

Now it is already easy to show the countable saturation. Let $S^*(x)$ be a type of the expanded ultraproduct

$$\mathbb{A}^* = \left\langle \prod_n \mathbb{A}_n/p;\ \{[a_k]_p\colon\ k \in N\} \right\rangle.$$

We have $\mathbb{A}^* = \prod_n \mathbb{A}_n^*/p$, where $\mathbb{A}_n^* = \langle \mathbb{A}_n;\ \{a_k(n)\colon\ n \in N\}\rangle$, that is, the expanded ultraproduct is the ultraproduct of the expanded systems. The corresponding language L^* is obtained from L by adding a countable number of new constants, so it is still countable. Therefore, we can apply the part already proved of the proof of Theorem 16.23 to the type S^* of the ultraproduct $\mathbb{A}^* = \prod_n \mathbb{A}_n^*/p$. In this way the proof has been completed. □

Recall now the definition of the η_1-orderings of Hausdorff. Every linear dense ordering without end points $\langle A, \leq \rangle$ (called also an η_0-ordering) satisfies the following condition.

16.24. *For any finite sets $U, W \subseteq A$, if $U < W$ (i.e., $u < w$ for every $u \in U$ and every $w \in W$), then there is a $v \in A$ such that $U < v < W$.*

If in 16.24 we allow countable subsets $U, W \subseteq A$ as well, then we shall obtain so called *η_1-orderings* (or *orderings η_1*).

We admit the case $U = \emptyset$ or $W = \emptyset$.

16.25. Corollary. *If $\mathbb{A} = \langle A, \leq \rangle$ is a dense linear ordering, then the ultrapower $\mathbb{A}^N/p$, with respect to any nonprincipal ultrafilter p on N, is of type η_1.*

Proof. Since $\mathbb{A}^N/p \equiv \mathbb{A}$, the ultrapower is linearly densely ordered. Now, if U, W are countable subsets of the ultrapower such that $U < W$, then the set of formulas

$$S_{U,W}(x) = \{c_u < x \wedge x < c_w\colon\ u \in U \text{ and } w \in W\}$$

is a type of the ultrapower $\langle \mathbb{A}^N/p;\ \{u\}_U,\ \{w\}_W\rangle$, and hence is realized, which completes the proof. □

16.26. *Example* (Robinson [R2]). We shall show that if k and K are ordered real closed fields such that $k \subseteq K$, then $k \preceq K$.

Proof. The proof is similar to that for in the case of algebraically closed fields (see Example 14.12), with the difference that instead of the Steinitz theorem we shall use the following theorem. □

16.27. Theorem (Erdös Gillman Henriksen [EGH]). *If K_1, K_2 are real closed fields of power ω_1, whose orderings are η_1, then K_1 and K_2 are isomorphic.*

Moreover, the isomorphism g can be chosen in such a way that $g|K = \text{id}$ on a given common countable subfield K.

So let $k \subseteq K$ and consider the case where k, K are countable. We have the elementary canonical embeddings

$$j\colon\ k \longrightarrow k^N/p, \quad J\colon\ K \longrightarrow K^N/p,$$

where, as can be easily seen, $j(a) = J(a) \cap k^N$ for $a \in k$. Both ultrapowers k^N/p and K^N/p are real closed fields and their orderings are η_1 (by Corollary 16.12). Moreover, both are uncountable (see Remark 16.22). Assume the continuum hypothesis. Then the cardinality of both ultrapowers is ω_1, and hence there is an isomorphism $g\colon k^N/p \longrightarrow K^N/p$ such that $g|j[k] = \text{id}$ by Theorem 16.27. Thus, we have

$$k \models F[a_1, \ldots, a_n] \quad \text{iff} \quad k^N/p \models F[j(a_1), \ldots, j(a_n)]$$

$$\text{iff} \quad K^N/p \models F[gj(a_1), \ldots, gj(a_n)] \quad \text{iff} \quad K^N/p \models F[j(a_1), \ldots, j(a_n)]$$

$$\text{iff} \quad K^N/p \models F[J(a_1), \ldots, J(a_n)] \quad \text{iff} \quad K \models F[a_1, \ldots, a_n],$$

which means that $k \leq K$. If k or K is uncountable, then we apply the Skolem–Löwenheim theorem in the following manner: If $k \models F[a_1, \ldots, a_n]$, then we take a countable subfield $k_1 \leq k$ containing $a_1, \ldots, a_n$ and a countable subfield $K_1 \leq K$ containing k_1. Then, by the part of the theorem already proved, we have

$$k_1 \models F[a_1, \ldots, a_n] \quad \text{if and only if} \quad K_1 \models F[a_1, \ldots, a_n],$$

whence it follows immediately

$$k \models F[a_1, \ldots, a_n] \quad \text{if and only if} \quad K \models F[a_1, \ldots, a_n]. \quad \square$$

The continuum hypothesis can be eliminated from the proof. However, we shall not deal with this problem here, since in Chapter 23 we shall prove a stronger version of the theorem.

EXERCISES

16.1. Let $Y = \{y_1, \ldots, y_n\}$ and $Z = \{z_1, \ldots, z_m\}$. If $[G]_T$ is an atom in $\mathbb{F}_{Y \cup Z}(T)$, then $[\exists z_1, \ldots, z_m\, G]_T$ is an atom in $\mathbb{F}_Y(T)$.

16.2. Show that if $\mathbb{A}, \mathbb{B}$ are prime models of a theory T, then $\mathbb{A}$ and $\mathbb{B}$ are isomorphic (see Vaught [V1]).

16.3. Show that if a theory T has a countable universal model, then it also has a prime model.

16.4. Show that a complete theory T having infinite models has a model realizing all the types of theory T.

16.5. If $\mathbb{A}$, $\mathbb{B}$ are countable finitely saturated models, then $\mathbb{A}$ and $\mathbb{B}$ are isomorphic.

16.6. Show that if a theory T has a countable universal model, then it also has a finitely saturated countable model.

16.7. A Boolean algebra having at most a countable number of ultrafilters is atomic.

16.8. Show that an algebraically closed field K is finitely saturated if and only if the transcendence degree of K is infinite (i.e., if in the presentation $K = \overline{Q(X)}$ of K, X is an infinite set).

16.9. Let $\mathbb{A}$ be a prime model of a theory T. Show that for every sequence $a_0, \ldots, a_n \in A$, if a map f: $\{a_0, \ldots, a_n\} \longrightarrow A$ is such that

$$\langle \mathbb{A};\ a_0, \ldots, a_n \rangle \ \equiv\ \langle \mathbb{A};\ f(a_0), \ldots, f(a_n) \rangle,$$

then f can be extended to an automorphism of the model $\mathbb{A}$ (Vaught [V1]).

16.10. Show that a countable finitely saturated model has the property stated in Exercise 16.9 (models with this property are called *finitely homogeneous*) (Vaught [V1]).

16.11. Show that the theory of ordered fields has continuum many types.

16.12. The theory of real closed fields has a prime model, but it has no universal model.

16.13. Let T be the theory of the algebraically closed fields. Determine the atoms of the algebras $\mathbb{F}_Y(T)$.

16.14. Let $\mathbb{A}$ be a countable system and $\mathbb{B}$ be a finitely saturated one. Show that the existence of an embedding h: $\mathbb{A} \longrightarrow \mathbb{B}$ is equivalent to the condition

$$\text{if } \mathbb{A} \models F, \text{then } \mathbb{B} \models F,$$

for every existential sentence F (i.e., a sentence having the form $\exists x_1, \ldots, x_n\ G$, where G is an open formula).

16.15. Under the same assumptions as in Exercise 16.14 show that there is an elementary embedding h: $\mathbb{A} \longrightarrow \mathbb{B}$ if and only if $\mathbb{A} \ \equiv\ \mathbb{B}$.

16.16. Let $\mathbb{A}$ be a model of PA. Let $\mathbb{B} \subseteq \mathbb{A}$ be a subsystem of $\mathbb{A}$ such that B is an initial segment of A. Let F be a formula in which all the quantifiers are bounded (in this case we shall say that F is *bounded*). Then for any

$b_1, \ldots, b_n \in B$ we have

$$\mathbb{B} \models F[b_1, \ldots, b_n] \quad \text{if and only if} \quad \mathbb{A} \models F[b_1, \ldots, b_n].$$

16.17. Let $\mathbb{A}, \mathbb{B}$ be such as in Exercise 16.16 and let F be a formula of the form

$$\exists x_1, \ldots, x_k G(x_1, \ldots, x_k, y_1, \ldots, y_n),$$

where G is bounded, and let H be a formula of the form

$$\forall x_1, \ldots, x_k \; G(x_1, \ldots, x_k, y_1, \ldots, y_n),$$

where G is bounded. Then, for $b_1, \ldots, b_n \in B$ the following implications are true

$$\text{if } \mathbb{B} \models F[b_1, \ldots, b_n], \text{ then } \mathbb{A} \models F[b_1, \ldots, b_n],$$

$$\text{if } \mathbb{A} \models H[b_1, \ldots, b_n], \text{ then } \mathbb{B} \models H[b_1, \ldots, b_n].$$

16.18. Let $\mathbb{A}$ be a countably saturated model. Let S be a countable set of formulas in the language $L\{c_a: a \in A\}$, where the formulas from the set S can contain arbitrary free variables. Show that if every finite subset of the set S is realized in $\mathbb{A}$, then the whole set S is realized in $\mathbb{A}$.

16.19. Let K be a real closed field whose ordering is η_1. Let $k_1, k_2 \subseteq K$ be isomorphic countable subfields. Show that for every element $a \in K$ there is an element $b \in K$ such that any isomorphism $g: k_1 \longrightarrow k_2$ can be extended onto the real closures $\overline{k_1(a)}, \overline{k_2(b)}$.

16.20. Prove the Erdös–Gillman–Henriksen theorem, Theorem 16.27, for fields of power ω_1 and type η_1 (use Exercise 16.19).

17

SUPPLEMENTARY QUESTIONS

In this chapter we shall discuss several classical problems of different sorts that belong to the general exposition. Elimination of quantifiers is, in many cases, a property of particular importance. Here we show that it holds for algebraically closed fields. A more important case of real closed fields will be proved in Chapter 23. The Robinson consistency theorem, the Craig interpolation, and the Beth definability theorem will end this chapter and Part I.

17.1. Elimination of Quantifiers

Let T be a consistent theory in a language L. Assume that T consists exclusively of sentences. We say that T admits the *elimination of quantifiers*, if for every formula F of the language L there is an open formula (i.e., a quantifier-free formula) G such that $T \vdash (F \equiv G)$ holds and $V(F) = V(G)$.

Clearly, such a theory T is always *model complete*, that is, the condition $\mathbb{A} \subseteq \mathbb{B}$ implies $\mathbb{A} \leq \mathbb{B}$ for any $\mathbb{A}, \mathbb{B} \in \mathrm{Mod}(T)$.

Every formula F is logically equivalent with a formula $\mathbb{Q}_1 z_1 \cdots \mathbb{Q}_n z_n\ G$, where $\mathbb{Q}_1, \ldots, \mathbb{Q}_n$ are quantifiers $\forall$ or $\exists$ and G is an open formula with $V(G) = V(F)$ (cf. Exercise 10.5). Hence, to prove the theorem on the elimination of quantifiers for a given theory T, it suffices to eliminate a single quantifier, that is, to find, for a formula $\mathbb{Q}xF$, where F is open, an equivalent in T open formula G. Moreover, the elimination of the quantifier from formulas $\exists xF$ and $\exists x\neg F$ implies the elimination of the quantifier from the formulas $\forall xF$ and $\forall x\neg F$, respectively, and vice versa. Thus, it suffices to assume that $\mathbb{Q} = \forall$ or $\mathbb{Q} = \exists$.

17.2. Theorem (Tarski). *The theory of algebraically closed fields admits the elimination of quantifiers.*

Proof (Seidenberg [51]). The theory T consists of the usual axiom of field theory and of sentences F_n for $n \geq 1$, where F_n is

$$\forall y_0, \ldots, y_n \exists x[y_n \neq 0 \to y_0 + y_1 \cdot x + \cdots + y_n \cdot x^n = 0].$$

We treat the subtraction $x - y$ as a primitive operation. Thus, we may identify the terms $t(x_1, \ldots, x_n)$ of the language with polynomials $f(x_1, \ldots, x_n)$ of several variables and integer coefficients (see Example 5.8). The atomic formulas have the form $f(x_1, \ldots, x_n) = 0$, since the formula $f = g$ is equivalent with $f - g = 0$. Every open formula is equivalent with a disjunction $F_1 \vee \cdots \vee F_n$, where each F_i has the form $G_1 \wedge \cdots \wedge G_m$, where all the G_js are atomic formulas or negations of atomic formulas (see Exercise 9.8). Since the formula $\exists x(F_1 \vee \cdots \vee F_n)$ is equivalent with the disjunction $\exists xF_1 \vee \cdots \vee \exists xF_n$, it suffices to eliminate the quantifier from formulas of the form $\exists x(G_1 \wedge \ldots \wedge G_m)$, where each G_j is either atomic or a negation of an atomic formula. Thus, in our case the proof reduces to the elimination of the quantifier from the formulas

$$\exists x(f_1 = 0 \wedge \cdots \wedge f_n = 0 \wedge g_1 \neq 0 \wedge \cdots \wedge g_m \neq 0).$$

But a formula $g_1 \neq 0 \wedge \cdots \wedge g_m \neq 0$ is equivalent with the formula $g_1 \cdot \cdots \cdot g_m \neq 0$. Finally, it is sufficient to eliminate the quantifier from formulas F of the form

$$\exists x(f_1 = 0 \wedge \cdots \wedge f_k = 0 \wedge g \neq 0),$$

where $f_1, \ldots, f_k, g$ are polynomials of several variables with integer coefficients. We may assume that these polynomials have the form $a_0 + a_1x + \cdots + a_nx^n$, where $a_0, \ldots, a_n$ are polynomials of the variables $y_1, \ldots, y_m$.

Case 1. Formula F has the form $\exists x(g \neq 0)$. Since $K(b_0, \ldots, b_n)$ is infinite (for any field K), formula F is equivalent to $g \neq 0$, hence if $g(x) = b_0 + b_1x + \cdots + b_nx^n$, then F is equivalent with the formula $b_0 \neq 0 \vee \cdots \vee b_n \neq 0$.

Case 2. F has the form $\exists x(f = 0 \wedge g \neq 0)$. Then, $\neg F$ is equivalent with $\forall x(f(x) = 0 \rightarrow g(x) = 0)$, which, in turn, is in any algebraically closed field equivalent with the condition $f|g^n$, where n is the degree of f. Thus, it suffices to express the condition $f|g$, that is, the divisibility of polynomials, by an open formula.

Let

$$f(x) = a_0 + a_1x + \ldots + a_nx^n,$$

$$g(x) = b_0 + b_1x + \ldots + b_mx^m.$$

Let J be the pairing function (see Exercise 13.18). If $J(n, m) = 0$, that is $f = a_0$ and $g = b_0$, then in any algebraically closed field $f|g$ holds if and only if $(a_0 \neq 0 \vee b_0 = 0)$. Let $J(n, m) > 0$ and assume that, for every pair n_1, m_1 for which $J(n_1, m_1) < J(n, m)$, we already have an open formula expressing the divisibility of polynomials with the degrees n_1 and m_1. If $n > m$ and the open formula G expresses the divisibility $(f - a_nx^n)|g$, then

$$f|g \quad \text{if and only if} \quad ((a_n = 0 \wedge G) \vee (a_n \neq 0 \wedge b_0 = 0 \wedge \cdots \wedge b_m = 0)).$$

If $n \leq m$ and the open formulas G and H express, respectively, the divisibility $(f - a_n x^n)|g$ and $f|(a_n g - b_m x^{m-n} f)$, then

$$f|g \quad \text{if and only if} \quad \big((a_n = 0 \wedge G) \vee (a_n \neq 0 \wedge H)\big).$$

In either case the relation $f|g$ has been expressed by an open formula.

The general case, where F is of the form

$$\exists x[f_1(x) = 0 \wedge \cdots \wedge f_k(x) = 0 \wedge g(x) \neq 0],$$

is reduced to the preceding cases with the help of the following observation: a formula $\exists x[f(x) = 0 \wedge g(x) = 0 \wedge H]$, with an arbitrary H, is equivalent in T with the disjunction of formulas of the form

$$H_0 \wedge \exists x[\bar{f}(x) = 0 \wedge \bar{g}(x) = 0 \wedge H],$$

where H_0 is an open formula not containing x and $J(\deg(\bar{f}), \deg(\bar{g})) < J(\deg(f), \deg(g))$.

Let for example, $n = \deg(f) \geq \deg(g) = m$, where f and g are as above. Then

$$\begin{aligned}
&\exists x[f(x) = 0 \wedge g(x) = 0 \wedge H] \\
&\quad \equiv_T \big((b_m = 0 \wedge \exists x[f(x) = 0 \wedge g(x) - b_m x^m = 0 \wedge H]) \\
&\qquad \vee (b_m \neq 0 \wedge \exists x[b_m f(x) - a_n x^{n-m} g(x) = 0 \wedge g(x) = 0 \wedge H])\big).
\end{aligned}$$

Hence, after finitely many steps at least one of the polynomials will become a constant and Case 1 or Case 2 will apply. □

Also the theory of real closed fields admits the elimination of quantifiers. The proof will be given in Chapter 23.

17.3. Robinson's Theorem. Let $R(L)$, $F(L)$, and $C(L)$ denote, respectively, the set of relation symbols, the set of function symbols, and the set of constants of a given language L. If L_1 and L_2 are some languages, then the set of variables and the set of logical signs can be taken to be the same in both L_1, L_2. Hence, we may construct the language $L_1 \cup L_2$, called the union of L_1, L_2, in which

$$R(L_1 \cup L_2) = R(L_1) \cup R(L_2), \quad F(L_1 \cup L_2) = F(L_1) \cup F(L_2)$$

and

$$C(L_1 \cup L_2) = C(L_1) \cup C(L_2).$$

Similarly, we may define the common part $L_1 \cap L_2$, the difference $L_1 \setminus L_2$, and so on.

17.4. Robinson's Theorem (On Consistency; cf. [R1]). *Let T_1 and T_2 be consistent sets of sentences in languages L_1, L_2, respectively. If $T = T_1 \cap T_2$ is a complete set of sentences in the language $L = L_1 \cap L_2$, then the set $T_1 \cup T_2$ is consistent (in the language $L_1 \cup L_2$).*

Proof. Let $\mathbb{A}_0 \in \mathrm{Mod}(T_1)$ and $\mathbb{B}_0 \in \mathrm{Mod}(T_2)$. By the assumption we have $\mathbb{A}_0 \equiv \mathbb{B}_0$ in L, that is, the relationship $(\mathbb{A}_0 \models F)$ if and only if $(\mathbb{B}_0 \models F)$ holds for all sentences F of language L. Hence, it follows that the set

17.5 $$\mathrm{Th}(\langle \mathbb{A}_0^L;\ \{a:\ a \in A_0\}\rangle) \cup \mathrm{Th}(\langle \mathbb{B}_0;\ \{b:\ b \in B_0\}\rangle),$$

(A^L denotes the reduct to the type of L) is consistent. For the proof we let $F(c_{a_1}, \ldots, c_{a_n})$ be a conjunction of a finite number of sentences of the first theory and $G(c_{b_1}, \ldots, c_{b_n})$ be a conjunction of the second. Then, we have clearly

$$\mathbb{A}_0^L \models \exists y_1, \ldots, y_n F(y_1/c_{a_1}, \ldots, y_n/c_{a_n})$$

and

$$\mathbb{B}_0 \models \exists z_1, \ldots, z_m G(z_1/c_{b_1}, \ldots, z_m/c_{b_m}),$$

whence we obtain

$$\mathbb{B}_0 \models \exists y_1, \ldots, y_n \exists z_1, \ldots, z_m (F \wedge G),$$

because $\mathbb{A}_0 \equiv \mathbb{B}_0$, in L. The reduct $\mathbb{B}_1$ of a model of the set in 17.5 to the type of language L_2 has the following property: *there exists an L-embedding f_0: $\mathbb{A}_0 \longrightarrow \mathbb{B}_1$ and an L_2-embedding h: $\mathbb{B}_0 \longrightarrow \mathbb{B}_1$.*

Replacing in B_1 the elements $h(b)$ by b and identifying, in a suitable way, the relations, operations, and distinguished elements in $\mathbb{B}_1$, we may assume that h is the identity on B_0, that is, that $\mathbb{B}_0 \leq \mathbb{B}_1$. Now we have

$$\langle \mathbb{A}_0;\ \{a:\ a \in A_0\}\rangle \equiv \langle \mathbb{B}_1;\ \{f_0(a):\ a \in A_0\}\rangle$$

in $L(\mathbb{A}_0)$.

Repeating the above reasoning, we obtain a model $\mathbb{A}_1$ such that $\mathbb{A}_0 \leq \mathbb{A}_1$ and an $L(\mathbb{A}_0)$-embedding

$$g_1:\ \langle \mathbb{B}_1;\ \{f_0(a):\ a \in A_0\}\rangle \longrightarrow \langle \mathbb{A}_1;\ \{a:\ a \in A_0\}\rangle,$$

whence g_1 is an L-embedding, g_1: $\mathbb{B}_1 \longrightarrow \mathbb{A}_1$ and $g_1 f_0(a) = a$, for every $a \in A_0$. Now we have

$$\langle \mathbb{B}_1;\ \{b:\ b \in B_1\}\rangle \equiv \langle \mathbb{A}_1;\ \{g_1(b):\ b \in B_1\}\rangle$$

in $L(\mathbb{B}_1)$.

Repeating the construction again, we obtain a $\mathbb{B}_2 \geq \mathbb{B}_1$ and an $L(\mathbb{A}_1)$-embedding f_1: $\mathbb{A}_1 \longrightarrow \mathbb{B}_2$ such that $f_1 g_1(b) = b$, for $b \in B_1$ and

$$f_1(a) = f_1(g_1 f_0(a)) = (f_1 g_1)(f_0(a)) = f_0(a), \quad \text{for } a \in A_0,$$

that is, f_1 is an extension of f_0. In this way we build inductively two sequences of elementary extensions and L-embeddings,

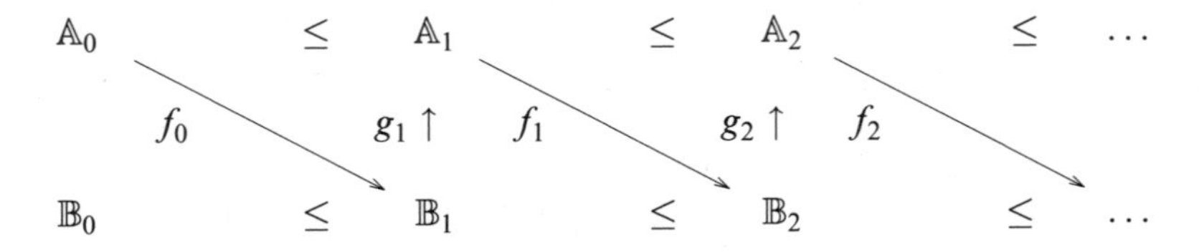

such that, for every $n \in N$, f_{n+1} is an extension of f_n and $f_n g_n(b) = b$, for every $b \in B_n$. Let us define

$$\mathbb{A} = \bigcup\{\mathbb{A}_n\colon\ n \in N\} \quad \text{and} \quad \mathbb{B} = \bigcup\{\mathbb{B}_n\colon\ n \in N\}.$$

Then $\mathbb{A} \in \mathrm{Mod}(T_1)$, $\mathbb{B} \in \mathrm{Mod}(T_2)$, and the function $f = \bigcup_n f_n$ is an L-isomorphism, f: $\mathbb{A} \longrightarrow \mathbb{B}$. Clearly, f: $\mathbb{A} \longrightarrow \mathbb{B}$ is a one-to-one function and if $b \in B$, then $b \in B_n$ for some n and then we have $f g_n(b) = f_n g_n(b) = b$, whence f is onto B. Also, we have

$$\mathbb{A} \models F[a_1, \ldots, a_n] \quad \text{if and only if} \quad \mathbb{B} \models F[f(a_1), \ldots, f(a_n)]$$

for every formula $F \in \mathrm{Fm}(L)$, since all the f_ns are L-embeddings.

We expand model $\mathbb{B}$ of theory T_2 to a model $\mathbb{B}^*$ of theory $T_1 \cup T_2$ as follows: if $r \in R(L_1) \setminus R(L_2)$, then we set $r^{\mathbb{B}^*} = f[r^A]$, that is we transpose the relation r^A by means of the function f onto the universe B. Similarly, we define the additional operations and distinguished elements. Since for $r \in R(L)$ we also have $r^{\mathbb{B}^*} = r^B = f[r^A]$, the reduct of $\mathbb{B}^*$ to L_1 is isomorphic with $\mathbb{A}$. Hence, $\mathbb{B}^* \models T_1$ and finally, $\mathbb{B}^* \models T_1 \cup T_2$, which completes the proof. $\square$

As a corollary we infer the following theorem.

17.6. Craig's Interpolation Theorem (cf. [C3]). *Let F and H be sentences in the languages L_1 and L_2, respectively. If $\vdash (F \to H)$ in $L_1 \cup L_2$, then there is a sentence G in $L_1 \cap L_2$ such that $\vdash (F \to G)$ in L_1 and $\vdash (G \to H)$ in L_2.*

Proof. Let T_0 consist of all the sentences G of the language $L_1 \cap L_2$ for which $\vdash (F \to G)$ holds. Suppose that $\nvdash (G \to H)$, for every $G \in T_0$. Then, we have $T_0 \nvdash H$. Hence, the sentence $\neg H$ is consistent with T_0. Let $\mathbb{A}$ be the reduct to the type of language $L_1 \cap L_2$ of a model of the set $T_0 \cup \{\neg H_0\}$. Clearly, set $T = \mathrm{Th}(\mathbb{A})$ is complete and, as an expansion of T_0 in $L_1 \cap L_2$, consistent with F.

We have

$$T = (T \cup \{F\}) \cap (T \cup \{\neg H\}).$$

By Robinson's theorem, the set $T \cup \{F, \neg H\}$ is consistent, whence $\nvdash (F \to H)$, which contradicts the assumption. □

17.7. Beth's Theorem. *Let T be a set of sentences of a language L. We expand the language L by a new n-argument relation symbol r. We say that the set of sentences $T(r)$ of the language $L\{r\}$ defines r implicitly, if every model $\mathbb{A} \in \mathrm{Mod}(T)$ can be expanded in a unique way to a model $\langle A, r^A \rangle$ of the set $T(r)$.*

17.8. Beth's Definability Theorem (cf. [B1]). *If $T(r)$ defines r explicitly, then there is a formula G of the language L such that $T(r) \vdash \forall x_1, \ldots, x_n[r(x_1, \ldots, x_n) \equiv G]$ (i.e., r is definable in the ordinary sense, described in Chapter 12).*

Proof. Let r_1, r_2 be two different relational symbols, and $T(r_1)$, $T(r_2)$ be the corresponding sets of sentences in the language s $L(r_1)$ and $L(r_2)$ respectively. By the assumption, every model of the set $T(r_1) \cup T(r_2)$ has the form $\langle \mathbb{A};\ r_1^A, r_2^A \rangle$, where $\mathbb{A} \in \mathrm{Mod}(T)$ and $r_1^A = r_2^A$. Hence, we have

17.9 $\quad T(r_1) \cup T(r_2) \vdash \forall x_1, \ldots, x_n[r_1(x_1, \ldots, x_n) \equiv r_2(x_1, \ldots, x_n)].$

Let us introduce new constants $c_1, \ldots, c_n$. We have in particular

$$\big(T(r_1) \cup T(r_2)\big) \vdash \big(r(c_1, \ldots, c_n) \to r_2(c_1, \ldots, c_n)\big),$$

whence

$$\big(F(r_1) \wedge H(r_2)\big) \vdash \big(r_1(c_1, \ldots, c_n) \to r_2(c_1, \ldots, c_n)\big),$$

where $F(r_1)$ is a conjunction of a finite number of sentences from the set $T(r_1)$ and $H(r_2)$ is a conjunction from the set $T(r_2)$. By the rules of the logical calculus we obtain

$$\big(F(r_1) \wedge r_1(c_1, \ldots, c_n)\big) \vdash \big(H(r_2) \to r_2(c_1, \ldots, c_n)\big).$$

By the Craig theorem, there is a sentence $G(c_1, \ldots, c_n)$ of the language $L\{c_1, \ldots, c_n\}$ such that we have

$$F(r_1) \vdash \big(r_1(c_1, \ldots, c_n) \to G(c_1, \ldots, c_n)\big)$$

and

$$H(r_2) \vdash \big(G(c_1, \ldots, c_n) \to r_2(c_1, \ldots, c_n)\big).$$

Therefore

$$T(r_1) \vdash (r(c_1, \ldots, c_n) \equiv G(c_1, \ldots, c_n)),$$

whence, by the theorem on the elimination of constants, we obtain

$$T(r) \vdash \forall x_1, \ldots, x_n[r(x_1, \ldots, x_n) \equiv G(x_1/c_1, \ldots, x_n/c_n)],$$

which completes the proof. □

EXERCISES

17.1. Let $K \subseteq L$ be fields. By a *zero of an ideal* $I \subseteq K[x_1, \ldots, x_n]$, we mean a sequence $a_1, \ldots, a_n \in L$ such that $f(a_1, \ldots, a_n) = 0$, for all $f \in I$. Derive from Theorem 17.2 the Hilbert Nullstellensatz: *if an ideal* $I \subseteq K[x_1, \ldots, x_n]$ *has a zero in some extension* $L \supseteq K$, *then* I *has a zero in* $\bar{K}$*(the algebraic closure of* K*)*. (*Hint*: every ideal in $K[x_1, \ldots, x_n]$ is finitely generated.)

17.2. Show that PA does not admit the elimination of quantifiers.

17.3. Prove that the theory of dense linear orderings without end points admits the elimination of quantifiers.

17.4. Describe the algebra $\mathbb{F}_{\{x,y\}}(T)$, where T is as in Exercise 17.3.

17.5. Show that the Craig interpolation theorem holds for formulas as well.

17.6. In the Robinson theorem, the assumption of the completeness of $T = T_1 \cap T_2$ is essential. Construct a suitable counterexample.

In Exercises 17.7 and 17.8 we define two axiom systems T_1 and T_2 of the real closed fields. Both T_1 and T_2 are considered also in Exercises 17.9 to 17.11.

17.7. Let T_0 be obtained from the usual axioms of the theory of fields (with subtraction) by adding the following axioms F_n:

$$\forall y_0, \ldots, y_n \exists x[y_n \neq 0 \rightarrow y_0 + y_1 x + \cdots + y_n x^n = 0]$$

for odd n. The theory T_1 of real closed fields is obtained from T_0 by adding the following axioms:

$$\forall x_0, \ldots, x_n[x_0^2 + \cdots + x_n^2 = 0 \rightarrow x_0 = 0 \wedge \cdots \wedge x_n = 0]$$

for every n and the axioms $\forall x \exists y[x = y^2 \vee -x = y^2]$.

Show

(a) the formula $x \leq y \equiv \exists z[y - x = z^2]$ defines in T_1 a linear ordering of the universe;

(b) the ordering $\leq$ is congruent with respect to the field operations $x \leq y$ implies $x + z \leq y + z$ and $x \leq y \wedge z \geq 0$ implies $x \cdot z \leq y \cdot z$;

(c) every element greater than or equal to 0 is a square (hence, no negative element is a square).

17.8. The theory T_2 of ordered real closed fields is formulated in the language of field theory with an additional relation symbol $\leq$ and is obtained by adding to T_0 the axioms guaranteeing that $\leq$ is a linear ordering congruent with respect to the field operations and of the axiom $\forall x \exists y [x \geq 0 \rightarrow x = y^2]$. Show that the extension of T_1 by the definition of the ordering (Exercise 17.7) is equivalent with T_2 (i.e., it has the same models).

17.9. Find a simple canonical form of the open formulas of T_1 and of T_2.

17.10. Show that every open formula $F(x)$ of the theory T_1 defines a set belonging to the field $FC(A)$ in any model $\mathbb{A}$ (i.e., it defines a finite set or a set whose complement is finite).

17.11. Find a formula of theory T_1 defining an infinite set with an infinite complement. Hence, theory T_1 does not admit the elimination of quantifiers (see Exercise 11.10).

17.12. If a theory T admits the elimination of quantifiers, then every extension of T obtained by adding new definitions also admits the elimination of quantifiers. The converse theorem is not true: the theory of ordered real closed fields (i.e., the theory T_2 of Exercise 17.8) does admit the elimination of quantifiers; see Tarski's theorem, Chapter 23.

II

SELECTED TOPICS

18

DEFINING FUNCTIONS IN ℕ

In developing syntax (Chapters 5 to 10) we often made use of inductive definitions and proofs. For example, such syntactical notions as terms and formulas, free variables, substitution, and soon were all defined by induction. In this chapter we shall have a closer look at arithmetical functions $f\colon N^k \longrightarrow N$ defined by induction, or—what has the same meaning—by recursion. This leads to the so-called (primitive) recursive functions (introduced by Gödel), which have many applications in mathematical logic (see Chapters 20 and 21). Let us mention here that recursive functions also play (for completely different reasons) an important role in theoretical computer science.

Let $q(a,b)$ and $r(a,b)$ denote the quotient and the remainder of the division of the number a by the number b, for $a,b \in N$. Hence, for an arbitrary integer $b \neq 0$ we have

$$a = q(a,b) \cdot b + r(a,b), \quad \text{where } r(a,b) < b.$$

Additionally, we set $q(a,0) = 0$ and $r(a,0) = a$, for any a. We recall that, by definition, the number theoretic congruence $a \equiv b \bmod m$ holds if and only if $r(a,m) = r(b,m)$ (or, equivalently, if the number $a - b$ is divisible by m). The Chinese remainder theorem states that if the moduli $m_0, \ldots, m_k$ are pairwise relatively prime, then for any numbers $a_0, \ldots, a_k \in N$ the system of congruences

18.1 $$x \equiv a_i \bmod m_i, \quad \text{for } i = 0, \ldots, k,$$

has a solution $x \in N$, $x < m_0 \cdot \cdots \cdot m_k$.

Proof. The proof is by induction on k, that is, on the number of the congruences in the system. For $k = 0$ the solution is the number $x = r(a_0, m_0)$. Assume that we have a solution x of the system 18.1 for a given $k > 0$. It suffices to find a number y satisfying the system

18.2 $$y \equiv x \bmod \bar{m}, \qquad y \equiv a_{k+1} \bmod m,$$

where $\bar{m} = m_0 \cdot \cdots \cdot m_k$ and $m = m_{k+1}$. By the assumption we have $(m, \bar{m}) = 1$, that is, the numbers m and $\bar{m}$ are relatively prime, hence, for some $u, w \in N$ we have $|u\bar{m} - wm| = 1$ (cf. Exercise 13.6). Assume $u\bar{m} - wm = 1$ (the other case is similar). It follows

$$u\bar{m} \equiv 1 \bmod m \quad \text{and} \quad (u\bar{m} - w)m \equiv 1 \bmod \bar{m},$$

and thus, multiplying the first congruence by a_{k+1} and the second by x, we obtain

$$u\bar{m}a_{k+1} \equiv a_{k+1} \bmod m \quad \text{and} \quad (u\bar{m} - w)mx \equiv x \bmod \bar{m}.$$

Hence, the number $y = u\bar{m}a_{k+1} + (u\bar{m} - w)mx$ satisfies system 18.2. Also the number $r(y, \bar{m}m) < m_0 \cdot \ldots \cdot m_{k+1}$ is a solution of this system, which completes the proof. □

Let us define

$$\beta(a, b, i) = r(a, (i+1)b + 1), \qquad \text{for } a, b, i \in N.$$

To every sequence $\langle a_0, \ldots, a_n \rangle$ of natural numbers there corresponds a pair of numbers $\langle a, b \rangle$ in such a way that the initial consecutive values of $\beta(a, b, i)$ are exactly the terms $a_0, \ldots, a_n$ of the sequence. This is put more precisely in the following statement.

18.3. *For every sequence $\langle a_0, \ldots, a_n \rangle$ of natural numbers there exists numbers $a, b \in N$ such that $\beta(a, b, i) = a_i$ for $i = 0, \ldots, n$.*

Proof. Let $m = \max\{n, a_0, \ldots, a_n\}$ and $b = m!$. Then, the numbers $m_i = (i+1)b + 1$ are pairwise relatively prime for $i = 0, \ldots, n$. Suppose that p is the least number greater than 1, such that $p|m_i$ and $p|m_j$ for $i < j$. Hence $p|(m_i - m_j)$, that is, $p|b(j-i)$, whence $p|b$ or $p|(j-i)$ since p is a prime, but $0 < j - i \leq m$, and thus the condition $p|(j-i)$ implies $p|m!$, that is, $p|b$. Hence $p|b$, in either case. But then we also have $p|(i+1)b$ and since $p|m_i$, also $p|\big(m_i - (i+1)b\big)$, that is, $p|1$, a contradiction.

From the Chinese remainder theorem we infer that there is a number $a < \prod_{i \leq n} \big((i+1)b + 1\big)$ such that the following congruences hold:

$$a \equiv a_i \bmod \big((i+1)b + 1\big), \qquad \text{for } i = 0, \ldots, n.$$

Since $a_i \leq m \leq b < (i+1)b + 1$, for every $i \leq n$, we have $r\big(a, (i+1)b + 1\big) = a_i$, for $i \leq n$, that is, $\beta(a, b, i) = a_i$, which completes the proof. □

18.4. Functions of the Class $\Sigma_1(\mathbb{N})$

Every term $t(x_1, \ldots, x_n)$ of PA determines the operation

$$f_t(n_1, \ldots, n_k) = t^{\mathbb{N}}[n_1, \ldots, n_k], \qquad \text{for } n_1, \ldots, n_k \in N.$$

The operation f_t is definable in the model ℕ by the formula $t(x_1, \ldots, x_k) = y$, since we have obviously PA $\vdash \forall x_1, \ldots, x_k \, \exists! y(t = y)$.

As we already know (see Section 5.7), the operations f_t coincide with arithmetical polynomials (i.e., polynomials of several variables with natural number coefficients). In particular, among the operations f_t there are the constant functions

$$C^k_m(n_1, \ldots, n_k) = m, \quad \text{for } m \in N \quad \text{and} \quad n_1, \ldots, n_k \in N$$

and the projections

$$U^k_i(n_1, \ldots, n_k) = n_i, \quad \text{for } 1 \leq i \leq k.$$

Let $\mathcal{F}$ be a family of arithmetical functions (i.e., functions $f\colon N^k \longrightarrow N$, for some k) containing all the arithmetical polynomials. Let $\mathcal{F}_k = \{f \in \mathcal{F}\colon f\colon N^k \longrightarrow N\}$. We assume that the family $\mathcal{F}$ is closed under compositions, if $f \in \mathcal{F}_k$ and $g_1, \ldots, g_k \in \mathcal{F}_l$, then $f(g_1, \ldots, g_k) \in \mathcal{F}_l$, where

$$f(g_1, \ldots, g_k)(n_1, \ldots, n_l) = f\big(g_1(n_1, \ldots, n_l), \ldots, g_k(n_1, \ldots, n_l)\big).$$

The operation of permuting the arguments and of adding redundant arguments does not lead out of class $\mathcal{F}$, Since, if $f \in \mathcal{F}_k$, then

$$f(n_{i_1}, \ldots, n_{i_k}) = f(U^k_{i_1}, \ldots, U^k_{i_k})(n_1, \ldots, n_k)$$

for any permutation $i_1, \ldots, i_k$ of the numbers $1, \ldots, k$. Similarly, if $f \in \mathcal{F}_k$ and $l \geq k$, then the function $g(n_1, \ldots, n_l) = f(n_1, \ldots, n_k)$ belongs to $\mathcal{F}_l$, since $g = f(U^l_1, \ldots, U^l_k)$. By a similar argument, the identification of arguments does not lead out of class $\mathcal{F}$.

It follows also that $\mathcal{F}$ is closed under the function addition and multiplication: $f + g$ and $f \cdot g$ are in $\mathcal{F}$, whenever $f, g \in \mathcal{F}$.

Let us denote

$$z(f) = \{\langle n_1, \ldots, n_k \rangle \in N^k\colon f(n_1, \ldots, n_k) = 0\}$$

and

$$Z(\mathcal{F}) = \{z(f)\colon f \in \mathcal{F}\}.$$

We prove two general properties of the class $Z(\mathcal{F})$.

18.5. *If $\mathcal{F}$ contains the bounded subtraction $x - y$, then $Z(\mathcal{F}_k)$ is, for every $k \geq 1$, a field of subsets of the space N^k containing all the finite sets.*

Proof. We have

$$z(C_0^k) = N^k, \qquad z(C_1^k) = \emptyset, \qquad z(f \cdot g) = z(f) \cup z(g),$$

$$z(C_1^k - f) = N^k \setminus z(f),$$

whence it follows that $Z(\mathcal{F}_k)$ is a field. Since $x - y \in \mathcal{F}$, then also $|x - y| = (x - y) + (y - x)$ belongs to $\mathcal{F}$. If $\langle m_1, \ldots, m_k \rangle \in N^k$ is a fixed point, then

$$z(|U_1^k - C_{m_1}^k| + \cdots + |U_k^k - C_{m_k}^k|) = \{\langle m_1, \ldots, m_k \rangle\},$$

which means that $Z(\mathcal{F}_k)$ contains all the one-element sets, and thus it contains arbitrary finite sets. □

Let $f \in \mathcal{F}_k$. The function

$$(\mathrm{s}\, f)(n_1, \ldots, n_k) = \sum_{i \le n_1} f(i, n_2, \ldots, n_k)$$

is called the *bounded sum of the function* f.

18.6. *If $\mathcal{F}$ contains the bounded subtraction $x - y$ and is closed under the operation of bounded sum ($\mathrm{s}f \in \mathcal{F}$, for any $f \in \mathcal{F}$), then the field $Z(\mathcal{F})$ contains all the sets definable in $\mathbb{N}$ by a bounded formula (i.e., a formula in which all the quantifiers are bounded).*

Proof. Notice that if the set $z(f)$ is definable in $\mathbb{N}$ by a formula F, that is,

$$f(n_1, \ldots, n_k) = 0 \quad \text{if and only if} \quad \mathbb{N} \models F[n_1, \ldots, n_k],$$

then the set $z\big(f(f_{t_1}, \ldots, f_{t_k})\big)$ is also definable in $\mathbb{N}$, for example, by the formula $F(t_1, \ldots, t_k)$; we have

$$\begin{aligned} &\mathbb{N} \models F(t_1, \ldots, t_k)[n_1, \ldots, n_l] \\ &\quad \text{iff} \quad \mathbb{N} \models F[t_1^{\mathbb{N}}(n_1, \ldots, n_l), \ldots, t_k^{\mathbb{N}}(n_1, \ldots, n_l)] \\ &\quad \text{iff} \quad f(f_{t_1}, \ldots, f_{t_k})(n_1, \ldots, n_l) = 0 \end{aligned}$$

The atomic formula $x = y$ defines the set $z(|x - y|) \in Z(\mathcal{F})$, hence the formula $t = s$ defines the set $z(|f_t - f_s|) \in Z(\mathcal{F})$. By 18.5, the sets definable by open formulas belong to $Z(\mathcal{F})$ since they are Boolean combinations of sets of the form $z(|f_t - f_s|)$. Assume that a given formula $F(x, x_1, \ldots, x_k)$ defines a set $z(f)$, where $f \in \mathcal{F}$. Then we have

$$\mathbb{N} \models \forall x \leq yF[n, n_1, \dots, n_k] \quad \text{iff} \quad \forall i \leq n(\mathbb{N} \models F[i, n_1, \dots, n_k])$$

$$\text{iff} \quad \forall i \leq n\big(f(i, n_1, \dots, n_k) = 0\big) \quad \text{iff} \quad \sum_{i \leq n} f(i, n_1, \dots, n_k) = 0.$$

This means that the formula $\forall x \leq y\ F$ defines the set $z(sf)$, which completes the proof. □

By the symbol Σ_1 we denote the class of all the arithmetical formulas of the form $\exists xF$, where F is a bounded formula. Similarly, Π_1 denotes the set of all formulas of the form $\forall xF$, where F is a bounded formula. Hence, the negation of a formula of the class Σ_1 is logically equivalent to a formula of the class Π_1 and the reverse.

The open formula

$$\Lambda_2 \cdot z = (x + y)(x + y + \Lambda_1) + \Lambda_2 \cdot x$$

defines the pairing function $z = J(x, y)$ (see Exercise 13.17). Thus the function J: $N \times N \longrightarrow N$ and maps the set $N \times N$ in a one-to-one way onto N. The bounded formulas

$$\exists y \leq z[\Lambda_2 \cdot z = (x + y)(x + y + \Lambda_1) + \Lambda_2 \cdot x],$$

$$\exists x \leq z[\Lambda_2 \cdot z = (x + y)(x + y + \Lambda_1) + \Lambda_2 \cdot x]$$

define the functions $x = K(z)$ and $y = L(z)$, which are the converse functions of J (Exercise 13.18). Thus, we have

$$J\big(K(a), L(a)\big) = a, \qquad \text{for every } a \in N.$$

Moreover, $K(a) \leq a$ and $L(a) \leq a$. It follows that a formula of the form $\exists x_1, \dots, x_kF$, where F is bounded, is equivalent in $\mathbb{N}$ with a formula of the class Σ_1. In fact, the equivalence

$$\exists x, y\ F \ \equiv\ \exists z[x = K(z) \wedge y = L(z) \wedge F]$$

holds in $\mathbb{N}$ (actually it is provable in PA, but we do not need it here). Of course, $x = K(z)$ and $y = L(z)$ should be replaced by the defining formulas. Then, at the right-hand side we have a formula of the class Σ_1. Similarly, a formula of the form $\exists x, y, zF$ is reduced first to $\exists x, uG$ and next to a formula of the class Σ_1. Let us note that a formula $\forall x \leq yF$, where $F \in \Sigma_1$, is also of the class Σ_1 in $\mathbb{N}$ (see Exercise 13.3).

Let $\Sigma_1(\mathbb{N})$ be the class of all the sets definable in $\mathbb{N}$ by a formula of the class Σ_1. Of course, for a set A to be in $\Sigma_1(\mathbb{N})$ it is sufficient that a formula defining it is equivalent in $\mathbb{N}$ with a Σ_1 one. In particular, if f is a function, then $f \in \Sigma_1(\mathbb{N})$ if the following condition holds:

$$f(n_1, \ldots, n_k) = m \quad \text{if and only if} \quad \mathbb{N} \models F[n_1, \ldots, n_k, m],$$

for some formula $F \in \Sigma_1$ such that $\mathbb{N} \models \forall x_1, \ldots, x_k \exists! y\ F$.

The family of all functions $f \in \Sigma_1(\mathbb{N})$ contains all the polynomials and is closed under compositions. If a formula $F(y_1, \ldots, y_k, y)$ defines a function f and $G_i(x_1, \ldots, x_l, y_i)$ defines g_i for $i = 1, \ldots, k$, then the formula

18.7 $\exists y_1, \ldots, y_k[G_1(x_1, \ldots, x_l, y_l) \wedge \cdots \wedge G_k(x_1, \ldots, x_l, y_k) \wedge F(y_1, \ldots, y_k, y)]$

defines the composition $f(g_1, \ldots, g_k)$. In addition, if $F, G_1, \ldots, G_k \in \Sigma_1$, then the formula 18.7 is equivalent in $\mathbb{N}$ with some formula $H \in \Sigma_1$ and H defines in $\mathbb{N}$ the composition $f(g_1, \ldots, g_k)$.

The function β coding finite sequences is $\Sigma_1(\mathbb{N})$ since it is definable by the bounded formula $\Gamma(a, b, i, c)$:

$$\exists q \leq a\{a = q \cdot [(i + \Lambda_1) \cdot b + \Lambda_1] + c \wedge c < (i + \Lambda_1)b + \Lambda_1\},$$

where the sign $<$ should be replaced by a bounded defining formula, (cf. Exercise 19.11).

18.8. Theorem. *If $g, h \in \Sigma_1(\mathbb{N})$, then the function f defined by induction as follows:*

$$f(0, n_1, \ldots, n_k) = g(n_1, \ldots, n_k),$$

$$f(n + 1, n_1, \ldots, n_k) = h(f(n, n_1, \ldots, n_k), n, n_1, \ldots, n_k)$$

is also in $\Sigma_1(\mathbb{N})$.

Proof. Choose formulas $F_g, F_h \in \Sigma_1$ defining the functions g and h, respectively. The condition $f(n, n_1, \ldots, n_k) = m$ is equivalent with the following one: *There exists a sequence $\langle a_0, \ldots, a_n\rangle$ such that $a_0 = g(n_1, \ldots, n_k)$ and $a_n = m$ and for each $i < n$, $a_{i+1} = h(a_i, i, n_1, \ldots, n_k)$.*

Moreover, we have

18.9
$$\begin{aligned} a_0 = g(n_1, \ldots, n_k) \quad &\text{iff} \quad \mathbb{N} \models F_g[n_1, \ldots, n_k, a_0]z \quad \textit{and} \\ a_{i+1} = h(a_i, i, n_1, \ldots, n_k) \quad &\text{iff} \quad (\mathbb{N} \models F_h[n_1, \ldots, n_k, i, a_i, a_{i+1}]). \end{aligned}$$

The existence of the sequence $a_0, \ldots, a_n$ is, in view of 18.3, equivalent with the existence of two numbers a, b for which $\beta(a, b, i) = a_i$ for $i \leq n$.

Since we have

$$\beta(a, b, i) = c \quad \text{if and only if} \quad \mathbb{N} \models \Gamma(a, b, i, c),$$

condition 18.9 can be expressed by an arithmetical formula $F(x_1, \dots, x_k, x, y)$, that is,

$$f(n, n_1, \dots, n_k) = m \quad \text{if and only if} \quad \mathbb{N} \models F[n_1, \dots, n_k, n, m],$$

where F is the formula

18.10 $$\exists a, b\{\exists z \leq a[F_g(x_1, \dots, x_k, z) \wedge \Gamma(a, b, \Lambda_0, z)] \wedge \Gamma(a, b, x, y)$$
$$\wedge \forall z < x \exists u, w \leq a[\Gamma(a, b, z, u) \wedge \Gamma(a, b, z + \Lambda_1, w) \wedge F_h(x_1, \dots, x_k, z, u, w)]\}$$

By the assumption the formulas F_g and F_h are Σ_1 while Γ is bounded. Hence, the formula F is equivalent in $\mathbb{N}$ with some formula of the class Σ_1, which completes the proof. □

We define by induction a function p as follows: $p(0) = 0$ and $p(n + 1) = n$. Then the bounded subtraction can be defined as

$$x - 0 = x, \qquad x - (y + 1) = p(x - y).$$

Hence, $x - y$ belongs to the class $\Sigma_1(\mathbb{N})$. Also the function sf—the bounded sum—satisfies the conditions

$$\mathrm{s}f(0, n_1, \dots, n_k) = f(0, n_1, \dots, n_k),$$

$$\mathrm{s}f(n + 1, n_1, \dots, n_k) = \mathrm{s}f(n, n_1, \dots, n_k) + f(n + 1, n_1, \dots, n_k).$$

Hence, the class $\Sigma_1(\mathbb{N})$ is closed under the operation of bounded sum. Therefore, the corresponding class of zero sets has the property 18.6.

Now, let us introduce the *minimum* operation: assume that a function g satisfies the condition

18.11 $$\forall n_1, \dots, n_k \exists m \; g(m, n_1, \dots, n_k) = 0,$$

and define the following function f:

$$f(n_1, \dots, n_k) = \min\{m\colon \; g(m, n_1, \dots, n_k) = 0\}.$$

We say that f is the result of the minimum operation on g.

18.12. *The class of functions $\Sigma_1(\mathbb{N})$ is closed under the minimum operation.*

Proof. If $g \in \Sigma_1(\mathbb{N})$ satisfies condition 18.11, then

18.13 $$f(n_1, \dots, n_k) = m \quad \text{iff}$$
$$\mathbb{N} \models (G[m, n_1, \dots, n_k, 0] \wedge \forall x < m \neg G[n_1, \dots, n_k, 0]),$$

where $G \in \Sigma_1(\mathbb{N})$ defines the function g. Now, it suffices to notice that the condition $h \in \Sigma_1(\mathbb{N})$ implies $h \in \prod_1(\mathbb{N})$ for any function h. Clearly,

$$h(n_1, \ldots, n_k) \neq m \quad \text{iff} \quad \mathbb{N} \models (\exists y(F_h(x_1, \ldots, x_k, y) \wedge y \neq z)[n_1, \ldots, n_k, m]),$$

where $F_h \in \Sigma_1$ defines the function h. This means that the complement $N^{k+1} \setminus h$ is a set of the class $\Sigma_1(\mathbb{N})$, and thus $h \in \Sigma_1(\mathbb{N}) \cap \Pi_1(\mathbb{N})$. Coming back to the equivalence 18.13, we note that the condition $\mathbb{N} \models \forall x < m \neg G[n_1, \ldots, n_k, 0]$ is equivalent with $g(n, n_1, \ldots, n_k) \neq 0$ for $n < m$, which can be expressed by a formula of the class Σ_1. Hence, the function f is in $\Sigma_1(\mathbb{N})$, which completes the proof. □

The following definition is due to Kleene [K3].

The class PR, of *primitive recursive* functions, is defined as the least class containing the polynomials and closed under compositions and under defining by induction (as in Theorem 18.8). Closing the class PR under the minimum operation we obtain a larger class—the class of *recursive functions.*

The PR functions were introduced by Gödel [G2]. Gödel's definition of recursiveness was different (but equivalent to the above).

18.14. Corollary. *The recursive functions are precisely the functions of the class* $\Sigma_1(\mathbb{N})$ *(and thus, also* $\Sigma_1(\mathbb{N}) \cap \Pi_1(\mathbb{N})$*).*

Proof. By Theorem 18.8 and the remarks preceding it and by 18.12 it follows that every recursive function is $\Sigma_1(\mathbb{N})$. Conversely, assume that $f \in \Sigma_1(\mathbb{N})$. Thus there is a bounded formula $F(x_1, \ldots, x_k, x, y)$, for which we have

$$f(n_1, \ldots, n_k) = m \quad \text{iff and only if} \quad \mathbb{N} \models \exists x\, F[n_1, \ldots, n_k, m].$$

The set

$$A = \{\langle n_1, \ldots, n_k, l\rangle \in N^{k+1}\colon\ \mathbb{N} \models F[n_1, \ldots, n_k, K(l), L(l)/x]\}$$

as a set definable in ℕ by a bounded formula (recall that K and L are definable by bounded formulas and $K(l), L(l) \leq l$) is of the form $A = z(g)$ for some function $g \in$ PR (see 18.6). Thus we have

$$\forall n_1, \ldots, n_k \, \exists l g(n_1, \ldots, n_k, l) = 0.$$

Hence the function

$$f_1(n_1, \ldots, n_k) = \min\{l\colon\ g(n_1, \ldots, n_k, l) = 0\}$$

is recursive.

The set $B = \{\langle n, l\rangle\colon\ n = L(l)\}$ is the set of zeros of some function h belonging to PR (by 18.6). We have $\forall l \exists n\, h(n,l) = 0$. Hence, the function $L(l) = \min\{n\colon\ h(n,l) = 0\}$ is recursive. But $f(n_1, \ldots, n_k) = L(f_1(n_1, \ldots, n_k))$, which completes the proof. □

The zero sets $z(f)$ for $f \in \mathrm{PR}$ are called *primitive recursive*. Similarly, the zero sets $z(f)$ for recursive functions f are called *recursive*. Instead of $A \in Z(\mathrm{PR})$ we write also $A \in \mathrm{PR}$.

From 18.6 it follows that the sets definable in $\mathbb{N}$ by a bounded formula are in the class PR. We shall prove the following corollary.

18.15. Corollary. *The family of recursive sets coincides with* $\Sigma_1(\mathbb{N}) \cap \Pi_1(\mathbb{N})$.

Proof. If $f \in \Sigma_1(\mathbb{N})$, then also $f \in \Pi_1(\mathbb{N})$ and then

$$f(n_1, \ldots, n_k) = m \quad \text{if and only if} \quad \mathbb{N} \models F[n_1, \ldots, n_k, m],$$

where as F there may occur both a Σ_1-formula and a Π_1-formula. Since

$$z(f) = \{\langle n_1, \ldots, n_k\rangle\colon\ \mathbb{N} \models F[n_1, \ldots, n_k, 0]\},$$

we obtain immediately

$$z(f) \in \Sigma_1(\mathbb{N}) \cap \Pi_1(\mathbb{N}).$$

Conversely, if we have

$$A \in \Sigma_1(\mathbb{N}) \cap \Pi_1(\mathbb{N}),$$

then also

$$N^k \setminus A \in \Sigma_1(\mathbb{N}) \cap \Pi_1(\mathbb{N})$$

and the characteristic function f of the set A is $\Sigma_1(\mathbb{N})$, and thus recursive. □

18.16. Functions of the Class PR

For the class PR of sets and functions there are no simple characterizations such as Corollaries 18.14 and 18.15 for the class of recursive sets and functions. Now we shall prove a few rules allowing us to infer that given functions belong to the class PR.

Obviously, the class PR satisfies the assumptions of propositions 18.5 and 18.6. It is easy to see that the characteristic functions $\mathrm{sg}(x)$ and $\overline{\mathrm{sg}}(x)$ of the sets $\{0\}$ and $N \setminus \{0\}$, respectively, are of the class PR. We can also define them by induction: $\mathrm{sg}(0) = 0$ and $\mathrm{sg}(n+1) = 1$ and, respectively, $\mathrm{sg}(0) = 1$ and $\overline{\mathrm{sg}}(n+1) = 0$. Hence, we obtain [we recall that the symbol $A \in \mathrm{PR}$ means $A \in Z(\mathrm{PR})$],

18.17 $\qquad A \in \mathrm{PR} \quad \text{if and only if} \quad \chi_A \in \mathrm{PR},$

where χ_A denotes the characteristic function of the set $A \subseteq N^k$:

$$\chi_A(n_1, \ldots, n_k) = 0 \text{ on } A \quad \text{and} \quad \chi_A(n_1, \ldots, n_k) = 1 \text{ on } N^k \setminus A.$$

The condition $A \in \text{PR}$ means $A = z(f)$ for some function $f \in \text{PR}$. But then $\chi_A = \text{sg}(f)$ and $\text{sg}(f) \in \text{PR}$. The other implication is obvious.

Let f be defined by "gluing" the functions g_1, g_2:

18.18
$$f(n_1, \ldots, n_k) = \begin{cases} g_1(n_1, \ldots, n_k), & \text{if } \langle n_1, \ldots, n_k \rangle \in A, \\ g_2(n_1, \ldots, n_k), & \text{if } \langle n_1, \ldots, n_k \rangle \notin A. \end{cases}$$

If $A, g_1, g_2 \in \text{PR}$, then also $f \in \text{PR}$, since $A = z(h)$, for some function $h \in \text{PR}$, and then we have

$$f = g_1 \cdot \overline{\text{sg}}(h) + g_2 \cdot \text{sg} \ \ (h),$$

whence $f \in \text{PR}$.

For any set $A \subseteq N^{k+1}$ we define the set $B = \forall x \le n \ A$ as follows:

$$\langle n, n_1, \ldots, n_k \rangle \in B \quad \text{if and only if} \quad \forall i \le n(\langle i, n_1, \ldots, n_k \rangle \in A).$$

If the set A is definable in ℕ,

$$\langle m, n_1, \ldots, n_k \rangle \in A \quad \text{if and only if} \quad (\mathbb{N} \models F[m, n_1, \ldots, n_k]),$$

then the set $B = \forall x \le n \ A$ is also definable:

$$\langle n, n_1, \ldots, n_k \rangle \in B \quad \text{if and only if} \quad \mathbb{N} \models \forall x \le \Lambda_n \ F[n_1, \ldots, n_k].$$

18.19. *If $A \in \text{PR}$, then $\forall x \le n \ A \in \text{PR}$ as well.*

In fact if $A = z(f)$, for a function $f \in \text{PR}$, then

$$\forall x \le n \ A = z(g), \text{ where } g(n, n_1, \ldots, n_k) = \sum_{i \le n} f(i, n_1, \ldots, n_k).$$

We introduce also the *bounded minimum operation*. Let

18.20
$$f(n, n_1, \ldots, n_k) = \min\{m \le n\colon \ g(m, n_1, \ldots, n_k) = 0\},$$

and $f(n, n_1, \ldots, n_k) = n + 1$, if $g(i, n_1, \ldots, n_k) \ne 0$, for every $i \le n$.

It is easy to see that if $g \in \text{PR}$ then also $f \in \text{PR}$; if χ is the characteristic function of the set $\exists x \le m \ z(g)$, then $\chi \in \text{PR}$ and

$$f(n, n_1, \ldots, n_k) = \sum_{i \le n} \chi(i, n_1, \ldots, n_k).$$

In view of 18.17, the condition $g(m, n_1, \ldots, n_k) = 0$ in 18.20 can be replaced by $r(m, n_1, \ldots, n_k)$, where $r \subseteq N^{k+1}$ is an arbitrary PR relation.

Let $h\colon N \longrightarrow N$ be PR. Then, the composition $\varphi(n, n_1, \ldots, n_k) = f(h(n), n_1, \ldots, n_k)$, where f is as in 18.20, is also PR and satisfies

$$\varphi(n, n_1, \ldots, n_k) = \min\{m \leq h(n)\colon\ g(m, n_1, \ldots, n_k) = 0\},$$

and $\varphi(n, n_1, \ldots, n_k) = h(n) + 1$, if $g(i, n_1, \ldots, n_k) \neq 0$, for each $i \leq h(n)$.

Hence, the upper bound $m \leq n$ in 18.20 can be replaced by $m \leq h(n)$, for any $h \in \mathrm{PR}$. A similar remark applies to the relations $\forall x \leq n\ A$ and $\exists x \leq n\ A$, that is the relations $\forall x \leq h(n)\ A$ and $\exists x \leq h(n)\ A$ are PR whenever h and A are such.

Continuing with 18.20, substituting a PR function $n = h(a_1, \ldots, a_k)$, and identifying the arguments a_1 with n_1, ..., a_k with n_k we infer that the function

18.21 $$\varphi(n_1, \ldots, n_k) = \min\{m \leq h(n_1, \ldots, n_k)\colon\ g(m, n_1, \ldots, n_k) = 0\}$$

[and $\varphi(n_1, \ldots, n_k) = h(n) + 1$ if there is no $m \leq h(n_1, \ldots, n_k)$ satisfying $g(m, n_1, \ldots, n_k) = 0$], is PR whenever g and h are PR.

From 18.6 it follows that if a set $A \subseteq N^k$ is definable in $\mathbb{N}$ by a bounded formula, then $A \in \mathrm{PR}$. However, there are functions f definable in $\mathbb{N}$ by a bounded formula which are not primitive recursive. Nevertheless, we can prove the following theorem:

18.22. Theorem. *If a function f is definable in $\mathbb{N}$ by a bounded formula and for some function $g \in \mathrm{PR}$ we have $f(n_1, \ldots, n_k) \leq g(n_1, \ldots, n_k)$ for all $n_1, \ldots, n_k \in N$, then $f \in \mathrm{PR}$ too.*

Proof. By the assumption we have

$$f(n_1, \ldots, n_k) = m \quad \text{if and only if} \quad \mathbb{N} \models F[n_1, \ldots, n_k, m]$$

for some bounded formula F. Hence, the set

$$W_f = \{\langle n_1, \ldots, n_k, m\rangle\colon\ \mathbb{N} \models F[n_1, \ldots, n_k, m]\}$$

belongs to PR. But,

$$f(n_1, \ldots, n_k) = \min\{m \leq g(n_1, \ldots, n_k)\colon\ \langle n_1, \ldots, n_k, m\rangle \in W_f\},$$

so $f \in \mathrm{PR}$ by 18.21. □

18.23. Corollary. *The pairing function J, its converse functions L, K and the function β are of the class* PR.

All these functions are definable by bounded formulas, and are majorized by the following PR functions: $J(x,y) \leq (x+y)(x+y+1)+x$, $K(z) \leq z$, $L(z) \leq z$, and $\beta(a,b,i) \leq a$.

18.24. Remark on Effectivity

Intuitively, a function $f\colon N^k \longrightarrow N$ is effectively computable if there is an algorithm, that is, a mechanical procedure computing each value $f(n_1,\ldots,n_k)$ for all $n_1,\ldots,n_k \in N$. Clearly, the addition $x+y$, multiplication $x \cdot y$ and, more generally, any arithmetical polynomial are computable. More comprehensive considerations lead to the conclusion that computability = recursiveness, that is, recursiveness is a mathematical definition of the intuitive notion of computability. This statement is known as the *Church thesis*, since it was first formulated by Church in 1936.

Given a set $A \subseteq N^k$, the computability of the characteristic function χ_A means that we can effectively (i.e., by means of a mechanical procedure) settle whether $\langle n_1,\ldots,n_k\rangle \in A$ or $\langle n_1,\ldots,n_k\rangle \notin A$, for arbitrary $n_1,\ldots,n_k \in N$. Therefore, the recursive sets $A \subseteq N^k$ serve as a mathematical counterpart of an intuitive notion of effective sets.

EXERCISES

18.1. The function $(\prod f)(n,n_1,\ldots,n_k) = \prod_{i\leq n} f(i,n_1,\ldots,n_k)$ is of the class PR, provided $f \in$ PR.

18.2. Show that the functions $f_n(x_1,\ldots,x_n) = \min\{x_1,\ldots,x_n\}$ and $g_n(x_1,\ldots,x_n) = \max\{x_1,\ldots,x_n\}$, for $n \geq 1$, are of the class PR.

18.3. Show that the functions $q(a,b)$ and $r(a,b)$ (the quotient and the remainder of the division of a by b) are in PR.

18.4. Check that the relations $a|b$ and $d = \mathrm{GCD}(a,b)$ are in PR.

18.5. Check that the function $q(a,b) = [a/b]$—the integer part of the quotient a/b—is in PR.

18.6. Check that the function $f(n) = [\sqrt{n}]$ is primitive recursive.

18.7. The function $f(n) = n!$ is in PR. Show that the set of primes and the function $p(n) =$ the nth prime are in PR as well.

18.8. *Coding Sequences.* The exponential function $f(x,y) = x^y$ is in PR. Let p_n be an increasing enumeration of the set of primes (see Exercise 18.7).

(a) Check that the functions $k_{n+1}(a_0,\ldots,a_n) = p_0^{a_0+1}\cdots p_n^{a_n+1}$, for $n \in N$, are in PR.

(b) Let $K_{n+1} = \{k_{n+1}(a_0,\ldots,a_n)\colon\ a_0,\ldots,a_n \in N\}$ and $K_0 = \{1\}$. Show that the set $K = \bigcup\{K_n\colon\ n \in N\}$ is in PR.

(c) Define functions $l, w \in$ PR so that for $a \in K$ $a > 0$, the equality $a = \prod_{i<l(a)} p_i^{w(a,i)+1}$ will be true.

The number $a \in K$ is called the *code of the sequence* $\langle w(a,0),\ldots,w(a,l(a)-1)\rangle$. Putting $a_i = w(a,i)$, for $i < l(a)$, we write in brief $a = \langle a_0, \ldots, a_n \rangle$, where $n = l(a) - 1$. The number 1 is the code of the empty sequence $\emptyset$.

(d) Define an operation $a * b$ of the class PR so that for $a, b \in K$ the following will hold: if $a = \langle a_0, \ldots, a_n \rangle$ and $b = \langle b_0, \ldots, b_m \rangle$, then $a * b = \langle a_0, \ldots, a_n, b_0, \ldots, b_m \rangle$.

18.9. For an arbitrary function f: $N^{k+1} \longrightarrow N$ we define $\bar{f}(\langle a_0, \ldots, a_k \rangle) = f(a_0, \ldots, a_k)$ and $\underline{f}(a) = 0$, for $a \notin K$ or if $l(a) \neq k + 1$. Check that if $f \in$ PR, then also $\bar{f} \in$ PR and conversely. The contraction operation $\bar{f}$ allows replacing functions of several variables by unary ones. Define an analogous contraction operation $\bar{A}$ for relations $A \subseteq N^k$.

18.10. The function defined by setting

$$f^*(n, n_1, \ldots, n_k) = \langle f(0, n_1, \ldots, n_k), \ldots, f(n, n_1, \ldots, n_k) \rangle$$

is called the *the course-of-values of f*. Let f be defined by the course-of-values induction

$$f(0, n_1, \ldots, n_k) = g(n_1, \ldots, n_k),$$

$$f(n + 1, n_1, \ldots, n_k) = h\big(f^*(n, n_1, \ldots, n_k), n, n_1, \ldots, n_k\big).$$

Show that if $g, h \in$ PR, then also $f \in$ PR.

18.11. The classes of formulas Σ_n and Π_n for $n > 1$ are defined by induction $F \in \Sigma_{n+1}$ if F has the form $\exists x G$ for some formula $G \in \Pi_n$ and the variable x is not bound in G. Similarly, $F \in \Pi_{n+1}$ if F has the form $\forall x G$ for some $G \in \Sigma_n$. With a given formula F we include also in the classes Σ_n and Π_n the formulas logically equivalent with F. Show that the following inclusions hold: $\Sigma_n \subseteq \Sigma_m$ and $\Pi_n \subseteq \Pi_m$, for $n \leq m$. Moreover, $\Sigma_n \subseteq \Pi_{n+1}$ and $\Pi_n \subseteq \Sigma_{n+1}$, for every $n \geq 1$.

18.12. *The Normal Form Theorem.* Show that every arithmetical formula is equivalent in PA with some formula of the class Σ_n or Π_n (Mostowski [M5], Kleene [K3]).

18.13. Let $\mathcal{F}_n = \Sigma_n(\mathbb{N})$, that is, $\mathcal{F}_n$ is the class of all the functions definable in $\mathbb{N}$ by means of a formula of the class Σ_n. Show that $\mathcal{F}_n = \Sigma_n(\mathbb{N}) \cap \Pi_n(\mathbb{N})$ and that the class $\mathcal{F}_n$ is closed under the operations of composition, of defining by induction, and of the minimum operation.

19

TOTAL FUNCTIONS

In this chapter we prove a rather technical result—each primitive recursive function $f\colon N^k \longrightarrow N$ can be syntactically defined in Peano arithmetic. This property will be useful in Chapter 21. As a corollary we obtain a weaker property called representability that, in turn, is valid for all recursive functions. Representability allows us to express enough syntactical properties as theorems of PA to prove the incompleteness theorem in Chapter 20.

We say that a function $f\colon N^k \longrightarrow N$ is *total* (in PA), if there is an arithmetical Σ_1-formula $F_f(x_1, \dots, x_k, y)$ such that $\mathrm{PA} \vdash \forall x_1, \dots, x_k \exists! y F_f$ (formulas with this property will be called total), defining f in $\mathbb{N}$ so that

$$f(n_1, \dots, n_k) = m \quad \text{if and only if} \quad \mathbb{N} \models F_f[n_1, \dots, n_k, m].$$

In particular, a total function must be recursive.

To show that a given function f is total it is sufficient to find a total defining formula of the form $\exists z_1, \dots, z_k\, G$, where G is bounded. The reduction to a single existential quantifier with the help of the evidently total pairing function J (Exercise 13.19) is described in Chapter 18 (see the remarks after the proof of 18.6).

19.1. *Arithmetical polynomials are total.*

We know that every polynomial has the form

$$f_t(n_1, \dots, n_k) = t^{\mathbb{N}}[n_1, \dots, n_k] \quad \text{for } n_1, \dots, n_k \in N,$$

and moreover (cf. Exercise 10.11), $\mathrm{PA} \vdash \forall x_1, \dots, x_k\, \exists! y[t(x_1, \dots, x_k) = y]$.

19.2. *The composition of total functions is a total function.*

Proof. If for the functions $f, g_1, \dots, g_k$ there correspond total defining formulas $F_f, F_{g_1}, \dots, F_{g_k}$, then the composition (see 18.7) $f(g_1, \dots, g_k)$ is definable by the formula $F(x_1, \dots, x_l, y)$:

$$\exists y_1, \dots, y_k[F_{g_1}(x_1, \dots, x_l, y_1) \wedge \cdots \wedge F_{g_k}(x_1, \dots, x_l, y_k) \wedge F_f(y_1, \dots, y_k, y)].$$

To verify that $\mathrm{PA} \vdash \forall x_1, \ldots, x_l \exists! y F$, take an arbitrary model $\mathbb{M}$ and elements $a_1, \ldots, a_l \in M$. The formulas $F_f, F_{g_1}, \ldots, F_{g_k}$ define in $\mathbb{M}$ some functions $f^M, g_1^M, \ldots, g_k^M$. Let $b_i = g_i^M(a_1, \ldots, a_l)$ and $b = f^M(b_1, \ldots, b_k)$. Then we have

$$\mathbb{M} \models F_f[b_1, \ldots, b_k, b] \quad \text{and} \quad \mathbb{M} \models F_{g_i}[a_1, \ldots, a_l, b_i]$$

for $i = 1, \ldots, k$, whence we obtain $\mathbb{M} \models F[a_1, \ldots, a_l, b]$. Thus we have $\mathbb{M} \models \forall x_1, \ldots, x_k \exists y\, F$. Now, if

$$\mathbb{M} \models F[a_1, \ldots, a_l, b'] \quad \text{and} \quad \mathbb{M} \models F[a_1, \ldots, a_l, b''],$$

then

$$b' = f^M\left(g_1^M(a_1, \ldots, a_l), \ldots, g_k^M(a_1, \ldots, a_l)\right) = b'',$$

which completes the proof. □

19.3. Theorem. *Every function $f \in \mathrm{PR}$ is total.*

Proof. Having 19.1 and 19.2, one still has to show that if a function f is defined by induction

$$f(0, n_1, \ldots, n_k) = g(n_1, \ldots, n_k),$$
$$f(n+1, n_1, \ldots, n_k) = h\big(f(n, n_1, \ldots, n_k), n, n_1, \ldots, n_k\big),$$

and the functions g and h are total, then also f is total.

The formula $F(x_1, \ldots, x_k, x, y)$ defining f in $\mathbb{N}$ (see 18.10) contains as its part the formula $\Gamma(a, b, i, c)$ defining the function $\beta(a, b, i) = c$. It is easy to see that β is a total function. We have the theorem on divisibility with a remainder

$$\mathrm{PA} \vdash \forall a, b[b \neq \Lambda_0 \to \exists! q, r(a = q \cdot b + r \wedge r < b)],$$

since if $\mathbb{M}$ is an arbitrary model and $b \in M$, $b \neq 0$, then the parametrically definable set

$$Y = \{a \in M\colon\ \exists! q, r(a = q \cdot b + r \wedge r < b)\}$$

contains 0 and with any a it also contains $a + 1$. Hence, it follows that the function $r = r(a, b)$ is total, and thus also the function β is total as well.

In the proof of the theorem we shall use a property of the function β connected with the Chinese remainder theorem (see 18.3). First, we shall show that a suitable form of that theorem is true in any model of PA. Let $G(x, y)$ be the formula

$$\forall z[z | x \wedge z | y \to z = \Lambda_1],$$

where the sign $z|x$ denotes the formula $\exists u[x = z \cdot u]$. Clearly, we have

$$\mathbb{N} \models G[a, b] \quad \text{if and only if} \quad (a, b) = 1,$$

(a, b are relatively prime).

We shall show that the following property holds:

19.4 $$\text{PA} \vdash \forall x_1, x_2, y[G(x_1, y) \wedge G(x_2, y) \rightarrow G(x_1 \cdot x_2, y)].$$

This means that the so-called fundamental theorem of arithmetic is a theorem of PA. For the proof let us take an arbitrary model $\mathbb{M}$ and let $m_1, m_2 \in M$. The model $\mathbb{M}$ can be extended to an algebraic ring $\mathbb{Z}(\mathbb{M})$ by adding the inverse elements $-a$, for $a \in M$, and extending the operations in the usual way. Then, the parametrically definable set $Y = \{\xi m_1 + \eta m_2 \colon \xi, \eta \in Z(\mathbb{M})\}$ contains (in the ordering $\leq^M$) a least positive element $d = um_1 + wm_2$. The element d is a divisor of every positive element $a \in Y$, since the remainder of the division of a by d belongs to Y, and so it is equal to zero. Hence, we have $d|^M m_1$ and $d|^M m_2$. If $d_1|^M m_1$ and $d_1|^M m_2$, then $d_1|^M (um_1 + wm_2)$, and thus $d_1|d$. Hence, it must be $d = \text{GCD}(m_1, m_2)$. Therefore, it follows that in $\mathbb{Z}(\mathbb{M})$, similarly to the ring of integers, we have

$$(m_1, m_2) = 1 \equiv \big(\text{for some } \xi, \eta \in Z(\mathbb{M}), \xi m_1 + \eta m_2 = 1\big).$$

Now, if $(m_1, m) = 1$ and $(m_2, m) = 1$, that is, $\xi m_1 + \eta m = 1$ and $\bar{\xi} m_2 + \bar{\eta} m = 1$, then, by multiplying side-by-side, we obtain

$$(\xi\bar{\xi})m_1 m_2 + (\eta\bar{\xi} m_2 + \bar{\eta}\xi m_1 + \bar{\eta}\eta m)m = 1$$

that is, $(m_1 m_2, w) = 1$, which yields 19.4.

It is easy to check that the positive combinations $\xi m_1 + \eta m_2$ coincide with the elements of the form $|um_1 - wm_2|$, for $u, w \in M$. In particular, the element $d = \text{GCD}(m_1, m_2)$ has the form

19.5 $$d = |um_1 - wm_2|, \quad \text{for some } u, w \in M.$$

The Chinese remainder theorem can be expressed in any model $\mathbb{M}$ of PA in the following form:

19.6. *If $\varphi, \psi \colon M \longrightarrow M$ are functions parametrically definable in $\mathbb{M}$ such that $(\psi(i), \psi(j))^M = 1$ for $i \neq j$, then for every $l \in M$ there is an element $a \in M$ such that $a \equiv \varphi(i) \bmod \psi(i)$ for all $i \leq^M l$.*

For the proof take the set

$$Y = \{l \in M \colon \exists a \forall i \leq l (a \equiv \varphi(i) \bmod \psi(i))\}.$$

This set is parametrically definable in $\mathbb{M}$, since the functions φ and ψ are such.

The proof of 19.6 is the same as the proof of 18.1. Obviously, we have $0 \in Y$ (with $a = \varphi(0)$). Assume that $l \in Y$ and let a_l be a solution corresponding to l. First, we have to show that there is an element $\bar{m} \in M$ such that

19.7 $\psi(i)|^M \bar{m}$, for all $i \leq^M l$, and $(\bar{m}, m)^M = 1$, where $m = \psi(l+1)$.

To this end we consider the parametrically definable set

$$Z = \{l \in M\colon \ \forall w[\forall i \leq l(\psi(i), w)^M = 1 \rightarrow \exists \bar{m} \forall i \leq l\psi(i)^M|\bar{m} \wedge (\bar{m}, w)^M = 1]\}.$$

Of course, we have $0 \in Z$ (for $\bar{m} = \psi(0)$), and if $l \in Z$, w is given and $\bar{m}_l$ is the $\bar{m}$ corresponding to l, then $\bar{m} = \bar{m}_l \cdot \psi(l+1)$ is divisible in $\mathbb{M}$ by all the $\psi(i)$s, $i \leq l+1$ and $(\bar{m}, w)^M = 1$, in view of 19.4.

So, take $\bar{m} \in M$ for which we have

$$\psi(i)|^M \bar{m}, \quad \text{for} \quad i \leq^M l \quad \text{and} \quad (\bar{m}, m)^M = 1.$$

By the inductive assumption we have $a_l \equiv \varphi(i) \bmod \psi(i)$, for $i \leq^M l$, so that it suffices to find a solution of the system

$$y \equiv a_l \bmod \bar{m},$$

$$y \equiv \varphi(l+1) \bmod m.$$

Since $(\bar{m}, m)^M = 1$, for some $u, w \in M$, we have $|u\bar{m} - wm| = 1$ and the rest of the proof is the same as in 18.1.

Now, we can complete the proof of Theorem 19.3. Recall that $F(x_1, \ldots, x_k, x, y)$ is the formula defining our given function f such as in 18.10. For an arbitrary model $\mathbb{M}$ of PA and for any $d_1, \ldots, d_k \in M$ we have

$$\begin{aligned} \mathbb{M} \models F[d_1, \ldots, d_k, l/x, d/y] & \\ \text{iff} \quad \exists a, b \in M\{ & \beta^M(a, b, 0) = g^M(d_1, \ldots, d_k) \wedge \beta^M(a, b, l) = d \\ & \wedge \forall i < l[\beta^M(a, b, i+1) = h^M(\beta^M(a, b, i), i, d_1, \ldots, d_k)]\}. \end{aligned}$$

We have to prove that

$$\mathbb{M} \models \forall x \exists! y F[d_1, \ldots, d_k].$$

Let us define

$$Y = \{l \in M\colon \ \mathbb{M} \models \exists! y F[d_1, \ldots, d_k, l/x]\}.$$

Of course, $0 \in Y$ [for $a = g^M(d_1, \ldots, d_k)$ and $b = a$]. Assume that $l \in Y$ and let a_l, b_l be the elements corresponding to l. Let $d = h^M(\beta^M(a_l, b_l, l), l, d_1, \ldots, d_k)$.

The function φ defined by the conditions

$$\varphi(i) = \beta^M(a_l, b_l, i), \quad \text{for} \quad i \in M,\ i \neq l+1,$$
$$\varphi(l+1) = d,$$

is parametrically definable in $\mathbb{M}$. From now on the proof is the same as the proof of 18.3. Similarly to the preceding proof (see the remarks following 19.7), we find an element $c \in M$ such that we have $l+1 <^M c$, $d <^M c$ and $\varphi(i) <^M c$, for every $i \leq^M l$, and then we find an element $b \in M$ for which we have $e|^M b$, for every $e \leq^M c$. The function $\psi(i) = (i+1)b + 1$, for $i \in M$, is parametrically definable in $\mathbb{M}$ and we have

$$(\psi(i), \psi(j))^M = 1 \quad \text{for} \quad i <^M j \leq^M l+1.$$

Applying 19.6, we find an element $a \in M$ such that the congruences $a \equiv \varphi(j) \bmod \psi(i)$ hold, for every $i \leq l+1$, that is,

$$\beta^M(a, b, i) = \beta^M(a_l, b_l, i) \quad \text{for} \quad i \leq l$$

and

$$\beta^M(a, b, l+1) = d = h^M(\beta^M(a, b, l), l, d_1, \ldots, d_k).$$

This means that $l+1 \in Y$, which completes the proof of Theorem 19.3. $\square$

Let us note here that the above theorem does not hold for the class $\Sigma_1(\mathbb{N})$—there are recursive functions which are not total (e.g., the function h from the proof of the independence of the Goodstein theorem; see Chapter 22).

19.8. Corollary (Reformulation of the theorem). *For every function $f \in$ PR there is a formula $F_f \in \Sigma_1$ such that we have PA $\vdash \forall x_1, \ldots, x_k\, \exists! y\ F$ and*

$$F(n_1, \ldots, n_k) = m \quad \text{if and only if} \quad \mathbb{N} \models F_f[n_1, \ldots, n_k, m]$$

for any $n_1, \ldots, n_k, m \in N$.

19.9. Representability

It is not difficult to see that a Σ_1-formula true in $\mathbb{N}$ is true in every model of PA. To prove this let us notice first that if $\mathbb{M}$ is an arbitrary model of PA, then the equivalence

19.10 $\qquad \mathbb{N} \models F[n_1, \ldots, n_k] \quad \text{if and only if} \quad \mathbb{M} \models F[n_1, \ldots, n_k]$

holds for every bounded formula. We have $\mathbb{N} \subseteq \mathbb{M}$, hence the value of any term $t[n_1, \ldots, n_k]$ calculated in $\mathbb{N}$ is the same as that calculated in $\mathbb{M}$, whence 19.10

holds for all the atomic formulas and thus also for all the open formulas. Simultaneously, 19.10 is preserved under bounded quantification, since $\mathbb{N}$ is an initial segment of $\mathbb{M}$ (see Exercise 16.16).

Let now F be a Σ_1-formula. Hence, F has the form $\exists xG$, where G is a bounded formula. If $\mathbb{N} \models F[n_1, \ldots, n_k]$ then for some $n \in N$ we have $\mathbb{N} \models G[n_1, \ldots, n_k, n]$ and from 19.10 we obtain $\mathbb{M} \models G[n_1, \ldots, n_k, n]$, and therefore $\mathbb{M} \models F[n_1, \ldots, n_k]$, which proves the claim.

Applying the completeness theorem we obtain the following remark.

19.11. Remark. *If $F(x_1, \ldots, x_k)$ is a Σ_1-formula, then the condition*

$$\mathbb{N} \models F[n_1, \ldots, n_k], \quad \text{for } n_1, \ldots, n_k \in N,$$

implies $\mathrm{PA} \vdash F(\Lambda_{n_1}, \ldots, \Lambda_{n_k})$.

We say that a function $f\colon N^k \longrightarrow N$ is representable (or binumerable) in PA by a formula $H(x_1, \ldots, x_k, y)$, if for any $n_1, \ldots, n_k, m \in N$, the condition $f(n_1, \ldots, n_k) = m$ implies $\mathrm{PA} \vdash H(\Lambda_{n_1}, \ldots, \Lambda_{n_k}, \Lambda_m)$, and also

$$\mathrm{PA} \vdash \exists! y H(\Lambda_{n_1}, \ldots, \Lambda_{n_k}, y).$$

Similarly, we can define representability in a theory T extending PA, $T \supseteq \mathrm{PA}$ (Gödel [G2]).

19.12. Theorem. *Every function $f \in$ PR is representable in PA by a total Σ_1-formula.*

Proof. Using Theorem 19.3, we chose a Σ_1-formula H defining f in PA. The first condition of representability follows from Remark 19.11 above, while the latter follows from the totality of H. □

Representability of sets and relations is defined as follows. We say that a set $A \subseteq N^k$ is representable in PA by a formula H, if the following implications hold for all $n_1, \ldots, n_k \in N$:

$$\langle n_1, \ldots, n_k \rangle \in A \quad \text{implies } \mathrm{PA} \vdash H(\Lambda_{n_1}, \ldots, \Lambda_{n_k});$$

$$\langle n_1, \ldots, n_k \rangle \notin A \quad \text{implies } \mathrm{PA} \vdash \neg H(\Lambda_{n_1}, \ldots, \Lambda_{n_k}).$$

19.13. Theorem. *Every* PR *set $A \subseteq N^k$ is representable in* PA *by a Σ_1-formula.*

Proof. Let $H(x_1, \ldots, x_k, y)$ be a total Σ_1-formula representing the characteristic function $\chi \in$ PR of the set A, as in Theorem 19.12. In particular,

$$n_1, \ldots, n_k \in A \quad \text{implies} \quad \mathrm{PA} \vdash H(\Lambda_{n_1}, \ldots, \Lambda_{n_k}, \Lambda_0).$$

Now, assume $n_1, \ldots, n_k \notin A$. Thus, we have $\mathrm{PA} \vdash H(\Lambda_{n_1}, \ldots, \Lambda_{n_k} \Lambda_1)$. Since H is total, the formula

$$H(x_1, \ldots, x_k.y') \wedge y' \neq y'' \to \neg H(x_1, \ldots, x_k, y'')$$

is provable in PA. But obviously, $\mathrm{PA} \vdash \Lambda_0 \neq \Lambda_1$ and therefore $\mathrm{PA} \vdash \neg H(\Lambda_{n_1}, \ldots, \Lambda_{n_k}, \Lambda_0)$. This proves that the Σ_1-formula$H(x_1, \ldots, x_k, \Lambda_0/y)$ represents the set A in PA. □

The totalness property of formulas defining PR functions is stronger than the uniqueness condition $\mathrm{PA} \vdash \exists! y H(\Lambda_{n_1}, \ldots, \Lambda_{n_k}, y)$ from the definition of representability of functions. In fact, the uniqueness condition can be derived directly from the Σ_1-definability of f in $\mathbb{N}$ (see the exercises for this chapter.). This leads to the following conclusion.

19.14. *Every recursive* [*that is,* $\Sigma_1(\mathbb{N})$] *function or set is representable in* PA *by a* Σ_1*-formula* (in general—not total).

Representability of PR functions will be used in Chapter 20 in the proofs of incompleteness theorems, while Chapter 21 is based on theorem 19.3 on totality of PR functions.

EXERCISES

19.1. The result of the minimum operation applied to a representable (by a Σ_1-formula) function is representable. Hence, each $\Sigma_1(\mathbb{N})$ function is representable (by a Σ_1-formula).

19.2. Show that every set $A \in \Sigma_1(\mathbb{N}) \cap \Pi_1(\mathbb{N})$ is representable in PA by some formula $F_A \in \Sigma_1$.

19.3. Show that the family of the subsets $A \subseteq N^k$ representable in PA by an arbitrary formula is a field.

19.4. Show that if a function f is representable and total in PA, then the set $z(f)$ is representable.

19.5. Show that if a set A is representable, then the characteristic function of A is representable and total.

19.6. *Interpretations*. Let L_1, L_2 be languages, $U(x) \in \mathrm{Fm}(L_1)$ and let I be a function mapping the symbols of language L_1 to the symbols of language L_2 so that $I(x) = x$, for every variable x, $I(c)$ is a constant, for every constant c of the language L_1. Also, for any relation symbols r and function symbols f, the values $I(r)$, $I(f)$ are assumed to be relation and function symbols with the same number of arguments as r and f, respectively. We extend the function I onto $\mathrm{Tm}(L_1)$ by means of the inductive rule

$$I\big(f(t_1, \ldots, t_n)\big) = I(f)\big(I(t_1), \ldots, I(t_n)\big),$$

and next onto $\mathrm{Fm}(L_1)$ by means of the rules:

$$I(t = s) = [I(t) = I(s)],$$

$$I\big(r(t_1, \ldots, t_n)\big) = I(r)\big(I(t_1), \ldots, I(t_n)\big),$$

$$I(\neg F) = \neg I(F), \qquad I(F \to G) = (I(F) \to I(G))$$

and

$$I(\forall x F) = \forall x [U(x) \to I(F)].$$

For any $F(x_1, \ldots, x_n) \in \mathrm{Fm}(L_1)$ we denote by F^I the formula $U(x_1) \wedge \cdots \wedge U(x_n) \to I(F)$, called the translation of F under I.

Let T_1, T_2 be sets of sentences in the languages L_1 and L_2, respectively. We say that the system $\langle U, I\rangle$ is an *interpretation of the theory* T_1 *in* T_2, if the following conditions are fulfilled:

1. $T_2 \vdash \exists x\, U(x)$,
2. $T_2 \vdash \big(U(x_1) \wedge \ldots \wedge U(x_n) \to U\big(I(f)(x_1, \ldots, x_n)\big)\big)$ for every function symbol f of language L_1,
3. $T_2 \vdash F^I$ for every sentence $F \in T_1$.

Assume that $\langle U, I\rangle$ is an interpretation of T_1 in T_2. Show that:

(a) if $T_1 \vdash F$, then $T_2 \vdash F^I$,

(b) if T_2 is consistent, then T_1 is consistent;

(c) every model $\mathbb{B} \in \mathrm{Mod}(T_2)$ determines a model $\mathbb{A} = \mathbb{B}^I \in \mathrm{Mod}(T_1)$ with the universe $A = \{b \in B\colon\ \mathbb{B} \models U[b]\}$;

(d) if $\mathbb{B}_1 \leq \mathbb{B}_2$, then $\mathbb{B}_1^I \leq \mathbb{B}_2^I$.

19.7. Define a more general type of an interpretation $\langle U, I\rangle$ in which $I(r), I(f)$ may be some formulas of the language L_2, such that the formulas $I(f)$, $I(c)$ define functions and constants, respectively, in T_2. Show that if $\langle U, I\rangle$ is an interpretation T_1 in a definable extension of the theory T_2, then there is an interpretation T_1 in T_2 in the above generalized sense. Construct interpretations of $\mathrm{Th}(\mathbb{Z})$ and $\mathrm{Th}(\mathbb{Q})$ in $\mathrm{Th}(\mathbb{N})$.

19.8. Prove that PA is interpretable in the Zermelo–Fraenkel set theory.

19.9. Let $n = c_k^n \cdot 2^k + \cdots + c_1^n \cdot 2 + c_0^n$, where $c_i^n = 0, 1$, be the digital representation in base 2 of the number n. We set

$$n \epsilon m \quad \text{if and only if} \quad c_n^m = 1.$$

Show that the structures $\langle R_\omega, \in\rangle$ and $\langle N, \epsilon\rangle$ are isomorphic.

19.10. Prove that set theory without the infinity axiom is interpretable in PA. (Use Exercise 19.9).

19.11. Let t be an arbitrary arithmetical term.

(a) If t has one free variable z, then there are integers $k_0, \ldots, k_m \in N$

such that

$$\mathrm{PA} \vdash t(z) = \Lambda_{k_0} + \Lambda_{k_1} \cdot z + \cdots + \Lambda_{k_m} \cdot z^m.$$

(b) Let $k = \max\{k_0, \ldots, k_m\}$. The formula $z > \Lambda_k \to t(z) < z^{m+1}$ is provable in PA.

(c) For an arbitrary formula F, the formula $\exists x(x < t(z) \wedge F)$ is equivalent in PA with

$$\bigvee_{0 \le i \le k} (z = \Lambda_i \wedge \bigvee_{0 \le j < t(i)} F(\Lambda_j)) \vee \exists x_0, \ldots, x_m$$

$$< z(z > \Lambda_k \wedge \bigvee_{0 \le i \le m} [(x_i < \Lambda_{k_i}) \wedge \bigwedge_{i < j \le m} (x_j = \Lambda_{k_j})]$$

$$\wedge F(x_0 + x_1 z + \cdots + x_m z^m / x)).$$

(Use Exercise 13.20 and 13.21.).

Here, the formula $x_i < \Lambda_{k_i}$ can be replaced by the open (bounded) one $\bigvee_{0 \le l < k_i} x_i = \Lambda_l$ and the formula $z > \Lambda_k$ by $\bigwedge_{0 \le l \le k} z \ne \Lambda_l$. Hence it follows that the formula $\exists x(x < t(z) \wedge F)$, where F is open, is equivalent in PA with a bounded formula. Thus, any formula F in which every quantifier is bounded by a term is equivalent in PA with a bounded one. Prove it by induction on the subformulas of F.

(d) Generalize property (c) for terms $t(z_1, \ldots, z_m)$ of several variables.

Hence, the quantifiers of the form $\exists x < t$ or $\forall x < t$, where t is an arbitrary term, are equivalent to ordinary bounded quantifiers (i.e., bounded by variables).

19.12. Let $f\colon N^k \longrightarrow N$ be a recursive function and $H(x_1, \ldots, x_k, y)$ be its Σ_1 definition. Thus, H is of the form $\exists z G$, where G is a bounded formula. Prove that the formula

$$\exists z\{H \wedge \forall y', z'[J(y', z') < J(y, z) \to \neg G(y'/y, z'/z)]\},$$

where J is a pairing function defined by a bounded formula (see Exercise 13.19) represents f in PA. (Use Exercise 19.11 above).

Hence, the representability of recursive functions follows directly from their Σ_1-definability.

20

INCOMPLETENESS OF ARITHMETIC

In this chapter we present the celebrated incompleteness theorem, the most startling discovery in mathematical logic. The theorem says, in particular, that such a mathematical theory as Peano arithmetic or Zermelo–Fraenkel set theory is incomplete; that is, there are undecidable statements. At present, we know numerous problems of set theory, topology, or functional analysis that have been shown to be undecidable (the independence proofs are highly nontrivial). The continuum hypothesis is perhaps the most familiar example. An arithmetical sentence independent of PA will be constructed here and also in Chapter 22.

The incompleteness theorem, proved by Gödel in 1931 and strengthened by Rosser in 1936, is quite general—it holds for any consistent theory T having an effectively given set of axioms and containing Peano arithmetic (actually, a weak fragment of PA suffices). Therefore, the incompleteness theorem also has a valuable philosophical significance, since it reveals essential limitations of the deductive power of axiomatic theories—the impossibility of a complete axiomatic description of the integers and other mathematical objects.

First, we shall describe a method of coding finite sequences of natural numbers which is more adequate than that used before. Simultaneously, we shall be checking that sets and functions connected with this coding are in PR.

By induction we define the generalized product

$$\prod_{i<0} f(i, n_1, \ldots, n_k) = 1,$$

$$\prod_{i<n+1} f(i, n_1, \ldots, n_k) = \prod_{i<n} f(i, n_1, \ldots, n_k) \cdot f(n, n_1, \ldots, n_k).$$

Hence, the function $\prod_{i<n} f(i, n_1, \ldots, n_k)$ is in PR, provided $f \in$ PR. The set of primes $E \subseteq N$ is defined by the bounded formula

$$x > 1 \wedge \forall y \leq x[y|x \rightarrow (y = 1 \vee y = x)].$$

Hence, the function p defined by induction,

$$p_0 = 2$$

$$p_{n+1} = \min\{a \leq p_n! + 1\colon\ a > p_n \wedge a \in E\},$$

enumerating the consecutive prime numbers, is in PR, since the function $f(n) = n!$ is in PR $[0! = 1$ and $(n+1)! = n! \cdot (n+1)]$. Also, the exponential function $f(x, y) = x^y$ is in PR $(x^0 = 1, x^{y+1} = x^y x)$. It follows that the functions

20.1 $$\langle a_0, \ldots, a_n \rangle = p_0^{a_0+1} \cdot \cdots \cdot p_n^{a_n+1}$$

are in PR, for every $n \in N$. The number $a = \langle a_0, \ldots, a_n \rangle$ is called the *code of the sequence* (or simply the *sequence*) $\langle a_0, \ldots, a_n \rangle$ and is denoted by the same symbol as the sequence itself.

Let K be the set of all the codes of the sequences. Also, the number 1, the code of the empty sequence $\emptyset$, is included in K. Thus, we have

$$a \in K \quad \text{if and only if} \quad \big(a = 1 \vee (a > 1 \wedge \forall i \leq a[p_i | a \rightarrow \forall j < i\ p_j | a])\big),$$

whence $K \in \text{PR}$ as defined by a bounded formula.

Define the following functions l and w:

$$l(a) = \min\{n \leq a\colon\ p_n \nmid a\},$$

$$w(a, i) = \min\{k \leq a\colon\ p_i^{k+2} \nmid a\}.$$

Obviously, we have $l, w \in \text{PR}$. It is clear that $l(\langle a_0, \ldots, a_n \rangle) = n + 1$ and $w(\langle a_0, \ldots, a_n \rangle, i) = a_i$, for $i < l(a)$. Therefore,

$$a = \prod_{i<l(a)} p_i^{w(a,i)+1}, \qquad \text{for each } a \in K,\ a > 1.$$

Now, we shall define the *concatenation*

$$a * b = a \cdot \prod_{j<l(b)} p_{l(a)+j}^{w(b,j)+1}.$$

Clearly, the function $f(a, b) = a * b$ is in PR. If $a, b \in K$, then also $a * b \in K$ and $l(a * b) = l(a) + l(b)$. Moreover, if $a = \langle a_0, \ldots, a_n \rangle$ and $b = \langle b_0, \ldots, b_n \rangle$, then $a * b = \langle c_0, \ldots, c_{n+m} \rangle$ where $c_i = a_i$, for $i < l(a)$, and $c_{l(a)+j} = b_j$, for $j < l(b)$.

The function f^* defined inductively

$$f^*(0, n_1, \ldots, n_k) = 1,$$

$$f^*(n+1, n_1, \ldots, n_k) = f^*(n, n_1, \ldots, n_k) * \langle f(n, n_1, \ldots, n_k) \rangle$$

is called the *course-of-values* of the function f. Thus we have

$$f^*(n, n_1, \ldots, n_k) = \langle f(0, n_1, \ldots, n_k), \ldots, f(n-1, n_1, \ldots, n_k)\rangle,$$

for $n \geq 1$.

20.2 (Induction on the course-of-values). *If* $h \in \mathrm{PR}$ *and the function* f *is defined inductively*

20.3 $$f(n, n_1, \ldots, n_k) = h\big(f^*(n, n_1, \ldots, n_k), n, n_1, \ldots, n_k\big),$$

then also $f \in \mathrm{PR}$.

Proof. If f satisfies equation 20.3, then f^* satisfies the equations

$$f^*(0, n_1, \ldots, n_k) = 1,$$

$$f^*(n+1, n_1, \ldots, n_k) = f^*(n, n_1, \ldots, n_k) * \langle h\big(f^*(n, n_1, \ldots, n_k), n, n_1, \ldots, n_k\big)\rangle,$$

whence $f^* \in \mathrm{PR}$, and since

$$f(n, n_1, \ldots, n_k) = w\big(f^*(n+1, n_1, \ldots, n_k), n\big),$$

then also $f \in \mathrm{PR}$. □

20.4. *If* $B \in \mathrm{PR}$, $B \subseteq N^{k+2}$ *and for an* $A \subseteq N^{k+1}$ *we have*

$$\langle n, n_1, \ldots, n_k\rangle \in A \quad \textit{if and only if} \quad \langle \chi_A^*(n, n_1, \ldots, n_k), n, n_1, \ldots, n_k\rangle \in B,$$

then also $A \in \mathrm{PR}$.

It is sufficient to observe that

$$\chi_A(n, n_1, \ldots, n_k) = \chi_B\big(\chi_A^*(n, n_1, \ldots, n_k), n, n_1, \ldots, n_k\big),$$

and to apply 20.2.

20.5 Defining a Language in $\mathbb{N}$

We have defined a logical language as the system of mutually disjoint sets $\langle R, F, C, X, S\rangle$, where the set of variables X is countable and the set of logical signs $S = \{=, \neg, \rightarrow, \forall\}$ is a four-element set. If also the set of relation symbols R, function symbols F, and the set of constants C are at most countable, then we may assume that they all are some subsets of the set N.

We use the following decomposition of the set N into mutually disjoint infinite sets,

$$N = \bigcup\{N_m\colon\ m \in N\}, \quad \text{where} \quad N_m = \{\langle m, n\rangle\colon\ n \in N\},$$

where the symbol $\langle\ ,\ \rangle$ denotes the code of a sequence, as defined in 20.1.

Assume that the relation symbol r_i is the number $\langle 0, \langle i, n_i\rangle\rangle$, where n_i denotes the number of arguments of the symbol r_i; similarly f_j is the number $\langle 1, \langle j, m_j\rangle\rangle$, where f_j is an m_j-argument function symbol, and the constant c_k is the number $\langle 2, k\rangle$. The set of variables is defined as $X = \{\langle 3, n\rangle\colon\ n \in N\}$, and the logical symbols are defined as follows: the equality sign is the number $\langle 4, 0\rangle$, the negation sign is the number $\langle 4, 1\rangle$, the implication sign is the number $\langle 4, 2\rangle$, and the quantifier sign is the number $\langle 4, 3\rangle$. Since we have obviously,

$$x \in X \quad \text{iff} \quad \exists n \leq x[x = \langle 3, n\rangle] \quad \text{iff} \quad \exists n \leq x[|x - \langle 3, n\rangle| = 0],$$

the set of variables, under this definition, is of class PR. Assume that the sets R, F, C are finite. Then, evidently, all the sets of symbols of the language are in PR.

The terms and the formulas are constructed by means of the operation of finite sequences. If the sequences are replaced by their codes (by a sequence we mean the number $\langle a_0, \ldots, a_n\rangle$ defined in 20.1), then also terms and formulas are numbers. Of course, the whole construction is so arranged that the set of formulas is disjoint from the set of terms and from the set of function symbols and soon, so that no collision will occur. Similarly, the syntactical operations (such as adding the negation or the substitution of terms for variables) are functions from some N^k into N. In view of further applications we have to show that certain sets of formulas (e.g., the set of logical axioms) and certain syntactical operations are sets and functions of class PR.

20.6. *The set of terms* Tm *is* PR.

Proof. Since t is a term if t is a variable or a constant or if t has the form $t = \langle f, t_1, \ldots, t_m\rangle$, we have

$$\begin{aligned} t \in \mathrm{Tm} \quad \text{iff} \quad & \big(t \in X \vee t \in C \\ & \vee \{t \in K \wedge \exists f \leq t[f \in F \wedge \big(l(t) = w\big(w(f, 1), 1\big) + 1\big) \wedge w(t, 0) = f] \\ & \wedge \forall i < l(t) - 1\ w\big(\chi^*_{\mathrm{Tm}}(t), w(t, i+1)\big) = 0\}\big), \end{aligned}$$

whence $\mathrm{Tm} \in \mathrm{PR}$ in view of 20.4. (Here and below χ is the characteristic function; cf. 18.7). □

20.7. *The set of atomic formulas A and the set of all the formulas* Fm *are in* PR.

Proof. The atomic formulas have the form $\langle t, =, s\rangle$ or $\langle r, t_1, \ldots, t_n\rangle$, hence

$$\begin{aligned} F \in A \quad \text{iff} \quad & \big(F \in K \wedge \{\exists t, s \leq F(F = \langle t, =, s\rangle \wedge t, s \in \mathrm{Tm}) \\ & \vee \exists r \leq F[r \in R \wedge \big(l(F) = w\big(w(r,1),1\big) + 1 \wedge w(F,0) = r \\ & \wedge \forall i < l(F) - 1\big(w(F, i+1) \in \mathrm{Tm}\big)\big]\}\big), \end{aligned}$$

and assuming that the other formulas have the form $\langle \neg, F\rangle, \langle F, \to, G\rangle, \langle\langle\forall, x\rangle, F\rangle$, we have

$$\begin{aligned} F \in \mathrm{Fm} \quad \text{iff} \quad & F \in K \wedge \{F \in A \vee \big[l(F) = 2 \wedge \big(w(F,0) = \neg\big) \\ & \vee \exists x \leq F(x \in X \wedge w(F,0) = \langle\forall, x\rangle)\big) \wedge w\big(\chi^*_{\mathrm{Fm}}(F), w(F,1)\big) = 0\big] \\ & \vee \big[l(F) = 3 \wedge w(F,1) = \to \\ & \wedge w\big(\chi^*_{\mathrm{Fm}}(F), w(F,0)\big) = 0\big) \wedge w\big(\chi^*_{\mathrm{Fm}}(F), w(F,2)\big) = 0\big]\}, \end{aligned}$$

whence $A, \mathrm{Fm} \in \mathrm{PR}$. (Here and below $\neg$ and $\to$ are the just defined integers). □

Let us write explicitly the operations creating new formulas:

$$\mathrm{neg}(F) = \langle \neg, F\rangle,$$

$$\mathrm{imp}(F, G) = \langle F, \to, G\rangle,$$

$$\mathrm{gen}(x, F) = \langle\langle\forall, x\rangle, F\rangle.$$

It is immediate that these functions are in PR.

It is sometimes more convenient to use the symbols occurring on the left hand sides of the above equalities.

20.8. *The relations $V(t, x)$ and $V_f(F, x)$ (x is a free variable in the term t and in the formula F, respectively) are in* PR.

Proof. We have

$$\begin{aligned} V(t,x) \quad \text{if and only if} \quad & \big(t \in \mathrm{Tm} \wedge x \in X \wedge (t \in X \wedge t = x) \\ & \vee \exists i < l(t) - 1\big[w\big(\chi^*_V(t,x), w(t, i+1)\big) = 0\big]\big). \end{aligned}$$

We check without difficulty that $V_f(F, x)$ is in PR. □

Since the condition $V(F, x)$ implies $x \leq F$, the set of all the sentences $\{F \in \mathrm{Fm}\colon \forall x \leq F \text{ not } V(F, x)\}$ is in PR. Also, the function V defined by the conditions $V(t) = 0$, if $t \notin \mathrm{Tm}$; $V(t) = \langle t\rangle$, if $t \in X$; $V(t) = 1$, if $t \in C$; and finally,

$$V(t) = \min\{a \leq t\colon\ a \in K \wedge \forall x \leq t\big[V(t,x) \equiv \exists i < l(a)\big(x = w(a,i)\big)\big]\}$$

otherwise, is in PR and the value $V(t)$ is the sequence of all the free variables of the term t.

In a similar way we check that the extension of the function V onto the set Fm (defined in an obvious way) remains in PR. Clearly, we set $V(a) = 0$ if $a \notin \mathrm{Tm}$ or $a \notin \mathrm{Fm}$.

Similarly, the functions $Z(F)$ =(the sequence of the bound variables of F) and $C(a)$ =(the sequence of the constants occurring in a) are in PR.

20.9 *The function* $q(a, x, t) = a(t/x)$ *(the substitution of* t *for* x*) is in* PR.

Proof. Define first the function $h \in \mathrm{PR}$:

$$h(d, a) = \min\{b \leq a \cdot l(a) \cdot p_{l(a)}^{d+1} : b \in K \wedge l(b) = l(a)$$

$$\wedge w(b, 0) = w(a, 0) \wedge \forall i < l(a) - 1[w(b, i+1) = w(d, w(a, i+1))]\},$$

which replaces the terms a_{i+1} of the sequence a by the terms $d_{w(a,i+1)}$. Then, we have

$$q(s, x, t) = \begin{cases} s, & \text{if } s \in C \text{ or } s \in X \text{ and } s \neq x, \\ t, & \text{if } s = x, \end{cases}$$

and

$$q(a, t, x) = h(q^*(a, x, t), a), \quad \text{for } a \notin C \cup X,$$

hence $q \in \mathrm{PR}$. □

Similarly, we check that the substitution

$$q(a, v, t) = a(t_1/y_1, \ldots, t_n/y_n),$$

where $v = \langle y_1, \ldots, y_n\rangle$ is a sequence of variables and $t = \langle t_1, \ldots, t_n\rangle$ is a sequence of terms, is in PR.

In a similar manner we can prove that the functions $s(t/c)$ and $F(t/c)$ (substitution of a term for a constant) are of class PR.

Following the definition of proper substitution (see Chapter 7) in the case of the one-element domain Y, we can easily prove that the relation $\mathrm{prop}(F, t, x)$, saying that t is properly substituted for x in F, is PR.

20.10. *The set of logical axioms is in* PR.

Proof. The set LOG consists of universal closures of formulas of the form $\forall y_1, \ldots, y_n H$, where H runs over the formulas of the five sets $Z_1, \ldots, Z_5$. Each of these sets is in PR. For instance, Z_1 contains three schemes, one of which is $F \to (G \to F)$. Hence, H is a formula in this scheme if and only if $\exists F, G \leq H[H = \mathrm{imp}\ (F, \mathrm{imp}(G, F))]$. Similarly, we check that the two remaining schemes are in PR, hence the set Z_1 is in PR.

For an illustration, we shall check that $Z_3, Z_5 \in \mathrm{PR}$. We have

$$H \in Z_3 \equiv \exists F, x \leq H \exists t \leq H\big[x \in X \wedge t \in \mathrm{Tm} \wedge H = \mathrm{imp}(\mathrm{gen}(x,F), F(t/x)) \wedge \mathrm{prop}(F,t,x)\big],$$

$$H \in Z_5 \equiv \exists F, x \leq H\big[x \in X \wedge \text{ not } V(F,x) \wedge H = \mathrm{imp}\ \ (F, \mathrm{gen}\ \ (x,F))\big].$$

Hence, $Z = Z_1 \cup \cdots \cup Z_5$ is a set in PR.

The function of restricting a sequence $a \restriction j = \langle a_0, \ldots, a_{j-1} \rangle$, for $j < l(a)$ is in PR

$$a \restriction j = \min\{b \leq a\colon\ b \in K \wedge l(b) = j \wedge \forall i < j[w(b,i) = w(a,i)]\}.$$

Now, we define inductively the function $g(a,F)$ as $g(0,F) = g(1,F) = F$ and

$$g(a,F) = \mathrm{gen}\big(w(a, l(a)-1), w(g^*(a,F), a \restriction l(a)-1)\big), \quad \text{for } a > 1.$$

Hence, $g \in \mathrm{PR}$. If a is a sequence of variables, $a = \langle y_1, \ldots, y_n \rangle$, then $g(a,F) = \forall y_1, \ldots, y_n\ F$. Now, from the obvious equivalence

$$H \in \mathrm{LOG} \quad \text{if and only if} \quad \exists a, F \leq H[F \in Z \wedge H = g(a,F)],$$

we obtain $\mathrm{LOG} \in \mathrm{PR}$. This completes the proof of 20.10. □

20.11. *If $T \subseteq \mathrm{Fm}$ and $T \in \mathrm{PR}$, then the relation $D_T(d,F) \equiv T \vdash_d F$ (d is a proof of the formula F in the theory T) is of class* PR.

We have the following equivalence

$$D_T(d,F) \quad \text{iff} \quad d \in K \wedge F = w(d, l(d)-1) \wedge \forall i < l(d)\big(w(d,i) \in \mathrm{Fm}\big) \wedge \forall i < l(d)\big[w(d,i) \in \mathrm{LOG} \vee w(d,i) \in T \vee \exists k,j < i\big[w(d,j) = \mathrm{imp}\ \ (w(d,k), w(d,i))\big]\big],$$

from which we get $D_T \in \mathrm{PR}$.

The results obtained above can be summed up as follows: every finite language (i.e., a language having finitely many relation and function symbols) is arithmetically definable in the model $\mathbb{N}$ in such a way that the syntactical relations and operations (possibly except D_T) are of class PR. If the set T is arithmetically definable, then also the relation $D_T(d,F)$ is arithmetically definable. If T is $\Sigma_1(\mathbb{N})$ or PR, then also D_T is $\Sigma_1(\mathbb{N})$ or PR, respectively.

In the sequel we shall use sometimes the assumption $T \in \mathrm{PR}$ or $T \in \Sigma_1(\mathbb{N})$ which means, intuitively, that the axiom system T is effective (cf. Remark 18.24).

Example. Every finite axiomatic system is PR. PA is PR (in any finite expansion of the language of arithmetic). The Zermelo or Zermelo–Fraenkel set theory is also PR.

20.12. Incompleteness Theorems

Let T be a theory (a set of sentences) in a language $L(T)$. Assume that $L(T)$ is a finite expansion of the language of PA. Let $N(x)$ be a formula in $L(T)$. The relativization F^N of a sentence F to $N(x)$ can be obtained from F by replacing every subformula of F of the form $\forall x\, F_1$ by $\forall x\big(N(x) \to F_1\big)$.

We say that a theory T contains PA if for some formula $N(x)$ the following sentences are theorems of T:

$$N(\Lambda_0),\ \ N(\Lambda_1),\ \forall x, y(N(x) \wedge N(y) \to N(x+y),$$
$$\forall x, y(N(x) \wedge N(y) \to N(x \cdot y).$$

We assume also that $\mathrm{PA}^N = \{F^N\colon\ F \in \mathrm{PA}\} \subseteq T$, that is, every axiom of arithmetic relativized to $N(x)$ is an axiom of the theory T.

Example. Let $L(T) = L(\mathrm{PA})$ and let $T \supseteq \mathrm{PA}$ be an arbitrary extension of PA. Take as $N(x)$ the formula $x = x$. Then $F^N \equiv_l F$, that is, the relativization F^N is logically equivalent with F, so we can identify PA^N with PA.

The next example is connected with interpretations; see Exercises 19.6 and 19.7.

Example. Let us assume that PA is interpretable in T, for example that T is a system of set theory. Then the arithmetical operations and constants are definable in T and a certain formula $N(x)$ defines in T the natural numbers. Hence, after a suitable modification, we may assume that T contains PA. Namely, we replace $L(T)$ by the union $L(T) \cup L(\mathrm{PA})$, and enlarge T by PA^N and by the axioms defining the arithmetical operations and constants. If, in addition, T was PR in $L(T)$, then T remains PR after the above modification.

For theories T containing PA we assume that the language $L(T)$ is defined in $\mathbb{N}$ as described in Section 20.5.

Consequently, formulas of T are some integers and since T contains PA, some formulas of T express properties of formulas of T. It may happen that a given sentence F expresses, in this way, a property of itself. For obvious reasons, we say that F is a self-referential sentence in this case. The forthcoming lemma states the existence of self-referential sentences, relative to an arbitrary, given in advance, property.

20.13. The Diagonal Lemma (Gödel). *If a theory T contains* PA, *then for every formula $G(y)$ there is a sentence F such that we have $T \vdash \big(F \equiv G(\Lambda_F/y)\big)$.*

Proof. Let x be a variable distinct from y. The arithmetical terms Λ_n are defined inductively, and so the function $g(a) = \Lambda_a$ is PR. In view of 20.9, the function $f(a) = a(\Lambda_a/x)$ is PR as well. By representability (see Theorem 19.12) there is a formula of arithmetic (and so of the theory T) $\Gamma_f(x, y)$ for which we have

$$T \vdash \exists! y \ \Gamma_f(\Lambda_a/x) \quad \text{for each } a \in N,$$

and if $f(a) = b$, then $T \vdash \Gamma_f(\Lambda_a, \Lambda_b)$ for $a, b \in N$, since T contains PA.

Now, let $H(x)$ be the formula $\exists y[\Gamma_f(x, y) \wedge G(y)]$. Intuitively, $H(x)$ expresses the following: the formula $x(\Lambda_x)$ has the property G. If F is the sentence $H(\Lambda_H)$, then we have $F = f(H)$, and hence also $T \vdash \Gamma_f(\Lambda_H, \Lambda_F)$. Thus, we obtain $T, F \vdash G(\Lambda_F)$, because we have $T \vdash \exists! y \ \Gamma_f(\Lambda_H)$. On the other hand, we have $T, G(\Lambda_F) \vdash F$, which completes the proof. □

The sentence F in Lemma 20.13 is sometimes called the diagonal sentence for the formula $G(y)$. Now, we are about to prove the incompleteness of Peano arithmetic. The diagonal lemma is the crucial point of the proof.

20.14. The Incompleteness Theorem (Gödel [G3]). *Peano arithmetic is incomplete, that is, there is an arithmetical sentence F such that* PA $\nvdash F$ *and* PA $\nvdash \neg F$.

(More generally, every arithmetically definable extension T of PA is incomplete, if only $\mathbb{N}$ is a model of T.)

Proof. If T is definable in $\mathbb{N}$, then the relation $D_T(d, F)$ is definable as well (see Section 20.12). Therefore, there is an arithmetical formula $\Gamma_D(x, y)$, for which the following equivalence holds:

$$D_T(d, F) \quad \text{if and only if} \quad \mathbb{N} \models \Gamma_D(d, F), \quad \text{for all } d, F \in N.$$

It follows immediately that

20.15 $\qquad T \vdash F \quad \text{if and only if} \quad \mathbb{N} \models \exists x \ \Gamma_D[F], \quad \text{for } F \in N.$

Now, let F be the diagonal sentence for the formula $\neg \exists x \Gamma_D$. Thus, by Lemma 20.13 we have

$$(*) \qquad \mathbb{N} \models F \quad \text{if and only if} \quad \mathbb{N} \models \neg \exists x \ \Gamma_D(\Lambda_F).$$

Hence, using 20.15, we obtain at once

$$T \vdash F \quad \text{if and only if} \quad (\mathbb{N} \models \neg F).$$

By the assumption $\mathbb{N}$ is a model for T, whence $T \vdash F$ implies $\mathbb{N} \models F$, while $T \vdash \neg F$ implies $T \vdash F$, via $(*)$. In either case we get a contradiction which proves that the sentence F is undecidable in T. □

Notice that the sentence F in the above proof expresses the following: "the sentence F has no proof in T." A direct application of the diagonal lemma leads also to the following interesting theorem of Tarski [T2].

20.16. Tarski's Theorem on the Undefinability of Truth. *If T contains the arithmetic and $\mathbb{A}$ is an arbitrary model of the theory T, then the set $T(\mathbb{A}) = \{F\colon\ \mathbb{A} \models F\}$ of the sentences true in $\mathbb{A}$ is not definable in $\mathbb{A}$.*

Proof. Suppose, on the contrary, that $T(\mathbb{A}) = \{F\colon\ \mathbb{A} \models H[F]\}$, for some formula $H(x)$. If F is the diagonal sentence for the formula $\neg H(x)$, then we must have

$$\mathbb{A} \models F \quad \text{if and only if} \quad \mathbb{A} \models \neg H[F],$$

which contradicts the definition of the formula H. $\square$

Now, we prove the incompleteness of theories T containing PA. We shall assume that T is a recursive set, which, intuitively, means that T is an effective set of axioms. Obviously, we assume also that T is consistent. In his original proof Gödel used a stronger assumption (so called ω-consistency) that was eliminated later by Rosser [R3] by an artful modification. For example, in contrast to Theorem 20.14, this includes some extensions of PA having nonstandard models only. Thus, we shall prove the following theorem.

20.17. Incompleteness Theorem (Gödel, Rosser). *If a consistent theory T contains* PA *and T is recursive, then T is incomplete.*

Proof. From the assumption it follows that the relation

$$R(d, F) \quad \text{if and only if} \quad D_T(d, F) \wedge \forall z \leq d \text{ not } D_T\big(z, \operatorname{neg}(F)\big)$$

is recursive as well. By 19.14 on representability there is an arithmetical Σ_1-formula $\Gamma_D(x, y)$ representing the relation D_T:

$$\text{if } D_T(d, F), \text{ then } T \vdash \Gamma_D(\Lambda_d, \Lambda_F);$$

and

$$\text{if not } D_T(d, F), \text{ then } T \vdash \neg\Gamma_D(\Lambda_d, \Lambda_F).$$

We also have a formula Γ_n representing the function neg F. It is easy to check that the formula $\Gamma_R(x, y)$ [we omit the relativization of the quantifiers to $N(x)$], $\Gamma_D(x, y) \wedge \forall z \leq x \neg\Gamma_D\big(z, \operatorname{neg}(y)\big)$, where $\Gamma_D\big(z, \operatorname{neg}(y)\big)$ stands for the formula $\exists w[\Gamma_D(z, w) \wedge \Gamma_n(y, w)]$, represents in T the relation R.

Let F be the diagonal sentence for the formula $\neg\exists x \Gamma_R$. Thus, we have

20.18 $$T \vdash \big(F \equiv \forall x\big[\Gamma_D(x, \Lambda_F) \rightarrow \exists z \leq x\ \Gamma_D\big(z, \operatorname{neg}(\Lambda_F)\big)\big]\big).$$

We easily check that F is an undecidable sentence on the basis of the theory T. Assume that $T \vdash F$. Hence for some number $d \in N$ we have $D_T(d, F)$ and also not $D_T((e, \mathrm{neg}(F))$, for every $e \leq d$, since T is consistent. Hence it follows that $R(d, F)$ and, by the representability of R, we have $T \vdash \Gamma_R(\Lambda_d, \Lambda_F)$, whence $T \vdash \exists x\, \Gamma_R(\Lambda_F)$, and consequently $T \vdash \neg F$, a contradiction.

Similarly, if $T \vdash \neg F$, then for some number $e \in N$ we have$D_T(e, \mathrm{neg}(F))$ and not $D_T(d, F)$, for every $d \in N$. Hence, by the representability of D_T:

20.19 $$T \vdash \neg\Gamma_D(\Lambda_d, \Lambda_F), \quad \text{for every } d \in N$$

and

$$T \vdash \neg\Gamma_D(\Lambda_e, \mathrm{neg}\ \ (\Lambda_F)), \quad \text{for some } e \in N.$$

Let $\mathbb{A}$ be an arbitrary model of the theory T. The formula $N(x)$ defines in $\mathbb{A}$ a model of PA:

$$M = \{a \in A\colon\ \mathbb{A} \models N[a]\}.$$

Let $a \in M$. Then, by 20.19 the condition $\mathbb{A} \models \Gamma_D[a, F]$ implies $a^M > N$, that is, a is a nonstandard number. In particular, we have $a^M > e$. Thus, we have

$$\mathbb{A} \models (\Gamma_D(x, \Lambda_F) \to \exists z \leq x\ \Gamma_D(z, \mathrm{neg}\ \ (\Lambda_F))]\,[a],$$

for every $a \in M$. Hence, it follows (omitting the relativization)

$$T \vdash \forall x[\Gamma_D(x, \Lambda_F) \to \exists z \leq x\ \Gamma_D(z, \mathrm{neg}\ \ (\Lambda_F))],$$

and thus, by 20.18, we obtain $T \vdash F$, which contradicts the consistency of the theory T. □

EXERCISES

We say that a given language L is (of class) PR (or $\Sigma_k(\mathbb{N})$ or $\Pi_k(\mathbb{N})$ (cf. Exercise 17.11), if the sets of its symbols and the syntactical relations and functions considered in Section 20.5, except possibly of the relation D_T, are definable in $\mathbb{N}$ and are PR (or $\Sigma_k(\mathbb{N})$ or $\Pi_k(\mathbb{N})$, respectively).

20.1. If a given language L has countably many relation symbols r_n or function symbols f_n, then L is definable in $\mathbb{N}$ provided that the functions $\arg(r_n)$ and $\arg(f_n)$ are definable in $\mathbb{N}$. Check that if $\arg(r_n)$, $\arg(f_n)$ are PR, then also L is PR.

20.2. Show that Lemma 20.13, the diagonal lemma, and Theorems 20.14 and 20.16 hold also for a theory T in a countable language $L(T)$ of class $\Sigma_1(\mathbb{N}) \cap \Pi_1(\mathbb{N})$ (see Exercise 19.1).

20.3. Show that Theorem 20.17 holds for theories $T \in \Sigma_1(\mathbb{N}) \cap \Pi_1(\mathbb{N})$, even if language $L(T)$ is countable of class $\Sigma_1(\mathbb{N}) \cap \Pi_1(\mathbb{N})$.

20.4. Let T be consistent, $T \supseteq \text{PA}$. If $T \in \Sigma_1(\mathbb{N}) \cap \Pi_1(\mathbb{N})$, then the relation

$$E_H(d, n_1, \ldots, n_k) \quad \text{if and only if} \quad D_T\big(d, H(\Lambda_{n_1}, \ldots, \Lambda_{n_k})\big)$$

for a fixed formula $H(x_1, \ldots, x_k)$, is of class $\Sigma_1(\mathbb{N}) \cap \Pi_1(\mathbb{N})$.

Show that every function f representable in T is $\Sigma_1(\mathbb{N})$.

20.5. If $T, L(T) \in \Sigma_1(\mathbb{N}) \cap \Pi_1(\mathbb{N})$, then the set of theorems $T^* = \{F\colon\ T \vdash F\}$ is of class $\Sigma_1(\mathbb{N})$.

20.6. Show that if T is complete and $T, L(T) \in \Sigma_1(\mathbb{N})$, then $T^* = \{F\colon\ T \vdash F\} \in \Sigma_1(\mathbb{N}) \cap \Pi_1(\mathbb{N})$.

20.7. Prove that $\text{PA}^* \in \Sigma_1(\mathbb{N}) \setminus \Pi_1(\mathbb{N})$. Hence, by Exercise 20.6, PA is not complete. (*Hint*: Assume that $\text{PA}^* \in \Sigma_1(\mathbb{N}) \cap \Pi_1(\mathbb{N})$ and apply the diagonal lemma to the negation of a formula representing PA^*).

20.8. *Coding Sets*. By a *code* of the set $\{k_0, \ldots, k_n\} \subseteq N$ we mean the number $2^{k_0} + \cdots + 2^{k_n}$. Hence, every number $m \in N$ codes a set $m^* = \{k < m\colon\ c_k^m = 1\}$, where c_k^m is a coefficient of 2^k in the digital representation of m at base 2 (cf. Exercise 19.9). Consider the relation $e(k, m)$ defined as $k < m \wedge c_k^m = 1$. Check that

(a) $m^* = \{k\colon\ e(k, m)\}$, and e is of class PR;

(b) the equivalences

$$f(a, b) = c \quad \text{iff} \quad \forall k\big[e(k, c) \equiv \big(e(k, a) \vee e(k, b)\big)\big],$$

$$g(a, b) = c \quad \text{iff} \quad \forall k\big[e(k, c) \equiv \big(e(k, a) \wedge e(k, b)\big)\big]$$

define functions of class PR.

(c) Denote $a \cup b = f(a, b)$ and $a \cap b = g(a, b)$. If $\mathbb{M}$ is a model of PA, then $(a \cup b)^* = a^* \cup b^*$ and $(a \cap b)^* = a^* \cap b^*$, for any $a, b \in M$.

20.9. Show that, for an arbitrary model $\mathbb{M}$, a formula F and an element $b \in M$, there exists an $a \in M$ such that

$$a^* = \{k \in M\colon\ k \leq^M b \text{ and } \mathbb{M} \models F[k]\}$$

and

$$\mathbb{M} \models \forall x\big[\Gamma_e(x, a) \equiv \big(x \leq b \wedge F(x)\big)\big],$$

where Γ_e defines the relation e following the definition in Exercise 20.8.

20.10. We say that a number $a \in N$ *codes* a function, if for all $k, l \in a^*$ the condition $K(k) = K(l)$ implies $L(k) = L(l)$, that is, if the set $\{\langle K(k), L(k)\rangle\colon\ e(k, a)\}$ is a function. (Here, K and L are the converses of the pairing function J). Check that

(a) The set of the codes of functions is PR.

Let Fn be the formula defining this set in the natural way with the help of the formula Γ_e.

(b) For any model $\mathbb{M}$ and any $a \in M$, if $\mathbb{M} \models Fn[a]$, then the set $\{\langle k, l\rangle: J^M(k, l) \in a^*\}$ is a function.

20.11. Show that the function $f(a) =$ (the number of elements of the set a^*), for $a \in N$ is PR. Hence, f is a total function in PA.

Show that for any model $\mathbb{M}$ and element $a \in M$ (under a suitable choice of the formula representing f):

(a) $f^M(a) = n$ iff a^* has n elements;

(b) if $f^M(a) >^M n$, for every $n \in N$, then the set a^* is infinite.

20.12. If the formulas $F_1, \ldots, F_n$ represent in PA the sets $A_1, \ldots, A_n \subseteq N^k$, then any Boolean combination of the formulas $F_1, \ldots, F_n$ represents the corresponding Boolean combination of the sets $A_1, \ldots, A_n$.

20.13. If the formula F represents in PA the set $A \subseteq N^{k+1}$, then the formula $\exists x \leq y\ F$ represents the set $\exists x \leq y\ A$.

20.14. If a is a code of a sequence of length less than or equal to n with terms less than or equal to n, or a code of a subset of the interval $[0, n]$, or a code of a partial function from $[0, n]$ to $[0, n]$, then $a \leq 2^{4n^2+1}$.

21

ARITHMETICAL CONSISTENCY

In this chapter we prove three major theorems all concerning Peano arithmetic. First, we shall convince ourselves that the logical calculus of a finite language can be treated as a part of PA so that even a suitable form of the completeness theorem can be proved in PA. This is the Hilbert–Bernays theorem. Next, we present a beautiful proof, due to Kreisel, of another famous incompleteness theorem of Gödel resulting in a new independent sentence of a given T. This time, it is a sentence expressing the consistency of T, $\mathrm{Cons}(T)$. It should be noted that T is supposed to be consistent, effectively axiomatizable, and containing PA, as before. Finally, we construct a so called Σ_1-universal formula, that is, a formula defining the truth relation for Σ_1-formulas. Besides being of interest on their own, all the theorems mentioned above serve as a powerful tool in various branches of logic and mathematics.

Let T be a consistent theory in a finite language $L(T) = \langle R, F, C, X, S\rangle$. As we already know (see Section 20.5), the language $L(T)$ can be arithmetically defined in the model $\mathbb{N}$ in such a way that the syntactical relations and operations are of class PR. We shall check now that theorems of the logical calculus for language $L(T)$ can also be expressed and proved in PA. In this way we shall obtain the arithmetical version of the completeness theorem.

First, we fix arithmetical formulas defining the logical notions. If $R = \{r_0, \ldots, r_{n_R}\}$ and the symbols r_i are defined as the numbers $r_i = \langle 0, \langle i, \arg(r_i)\rangle\rangle$, then the set R is defined by the formula

$$R(x):\ \ x = \Lambda_{r_0} \vee \ldots \vee x = \Lambda_{r_{n_R}}.$$

In the same way we build the formulas $F(x)$, $C(x)$, and $S(x)$ defining the sets of function symbols F, constants C, and logical symbols S, respectively. The set of variables $X = \{\langle 3, n\rangle\colon\ n \in N\}$, is defined by the formula $X(x)$: $\exists y \le x[x = \langle \Lambda_3, y\rangle]$, where the symbol $\langle\ ,\ \rangle$ is replaced by a Σ_1 formula defining in PA the total function of forming a two-element sequence. Hence, for an arbitrary model $\mathbb{M}$ of PA, the set $X^M = \{a \in M\colon\ \mathbb{M} \models X[a]\}$ is identical with the set $\{\langle 3, m\rangle^M\colon\ m \in M\}$, while the standard part of X^M, that is, the set $X^M \cap (N \times N)$ is equal to X. The remaining sets of symbols defined in $\mathbb{M}$ are the same as defined in $\mathbb{N}$.

By Theorem 19.3 there exist formulas of class $\Sigma_1(\mathbb{N})$ defining in PA the total functions: $\langle a_0, \ldots, a_m\rangle$, of forming a sequence; $l(a)$, of length of a sequence; $w(a, i)$, of terms of a sequence; $a * b$, of concatenating sequences; and the formula $K(x)$ defining the set K, of codes of sequences. From those formulas we construct definitions in PA of the syntactical notions. For the set of terms Tm we have the equivalence (cf. 20.6)

$$t \in \mathrm{Tm} \quad \text{if and only if} \quad B\big(t, \chi^*_{\mathrm{Tm}}(t)\big),$$

where $B(x, y)$ is the relation

$$x \in X \vee x \in C \vee \{x \in K \wedge \exists f \leq x\big[f \in F \wedge \big(l(x) = w\big(w(f, 1), 1\big) + 1\big) \\ \wedge\, w(x, 0) = f\big] \wedge \forall i < l(x) - 1\big[w\big(y, w(x, i+1)\big) = 0\big]\}.$$

We replace the sets and functions occurring here by their already fixed definitions, preserving the structure of the connectives and of the bounded quantifiers, for example, the condition $x \in X$ is replaced by the formula $X(x)$, $x \in C$, by the formula $C(x)$, and so on. Thus, we obtain an arithmetical formula $\Gamma_B(x, y)$ representing the relation $B(x, y)$ in PA (see Exercises 20.12 and 20.13). The formula $\mathrm{Tm}(x)$ will be chosen in such a way that the condition

21.1 $$\mathbb{M} \models \mathrm{Tm}[t] \quad \text{if and only if} \quad \mathbb{M} \models \Gamma_B[t, \chi^*_{\mathrm{Tm}}(t)]$$

is fulfilled, for every model $\mathbb{M}$ and for each $t \in M$. Condition 20.1 ensures that the terms in the sense of the model $\mathbb{M}$ (i.e., the elements of the set $\mathrm{Tm}^M = \{t \in M\colon\ \mathbb{M} \models \mathrm{Tm}[t]\}$) are exactly those elements $t \in M$ that are variables in the sense of $\mathbb{M}$, or constants, or elements of the form $t = \langle f, t_1, \ldots, t_m\rangle$, where $f \in F$, $m = \arg(f) \in M$ and $t_1, \ldots, t_m$ are terms $m \in M$.

To this end, notice that the course of values function χ^* of the characteristic function χ of the set $\mathrm{Tm} \subseteq N$ is a function of class PR satisfying the inductive conditions

21.2 $$\chi^*(0) = 1, \qquad \chi^*(t+1) = h\big(\chi^*(t), t\big),$$

where $h(a, b) = a * \langle \chi^*_B(b, a)\rangle$ is a function of class PR.

Hence, by Theorems 19.3 and 19.12, χ^* is a total function definable in PA by the formula $\Gamma_{\chi^*}(x, y) \in \Sigma_1$, and moreover the equations

21.3 $$(\chi^*)^M(0) = 1, \\ (\chi^*)^M(t+1) = h^M\big((\chi^*)^M(t), t\big)$$

are satisfied in every model $\mathbb{M}$. From Γ_{χ^*}, taking into account the condition $\chi(t) = w(\chi^*(t+1), t)$, we obtain the formula $\Gamma_\chi(x, y) \in \Sigma_1$ defining χ in PA, and moreover from 21.2 and 21.3 we have $\chi^M(t) = \chi_{B^M}(t, \chi^*(t))$ for any model $\mathbb{M}$. Let $\text{Tm}(x)$ be the formula $\Gamma_\chi(x, \Lambda_0)$. Of course, we have

$$\mathbb{M} \models \text{Tm}[t] \quad \text{iff} \quad B^M(t, \chi^*(t)) \quad \text{iff} \quad \mathbb{M} \models \Gamma_B[t, \chi^*(t)],$$

and thus the formula $\text{Tm}(x)$ satisfies the required condition 20.1. In particular, $\text{Tm} \subseteq \text{Tm}^M$, that is, the ordinary terms remain terms in any model $\mathbb{M}$. In fact, $\text{Tm} = \text{Tm}^M \cap N$, that is, Tm is the standard part of Tm^M.

In a similar way, applying 20.7 to 20.10, we obtain arithmetical formulas representing the remaining syntactical notions connected with the language $L(T)$. In particular, we obtain the formulas $A(x)$ and $\text{Fm}(x)$, so that, for any model $\mathbb{M}$, the set $A^M = \{F \in M\colon \ \mathbb{M} \models A[F]\}$ (of atomic formulas in the sense of $\mathbb{M}$) consists of the sequences $\langle t, =, s\rangle$, where $t, s \in \text{Tm}^M$, and the sequences $\langle r, t_1, \ldots, t_n\rangle$, where $r \in R$, $n = \arg(r)$ and $t_1, \ldots, t_n \in \text{Tm}^M$. Similarly, the set Fm^M is the closure in $\mathbb{M}$ of the set A^M under the total functions neg, imp, gen. The substitution operations $s(t/x)$ and $F(t/x)$ are total functions in PA, and in any model $\mathbb{M}$ they have their usual properties. The formula $\text{LOG}(x)$ defines in $\mathbb{M}$ the set of logical axioms in the sense of $\mathbb{M}$. We assume that the set $T \subseteq \text{Fm}$ is definable in $\mathbb{N}$ by an arithmetical formula $T(x)$. By 20.11, we construct the formula $D_T(x, y)$, for which, in any model $\mathbb{M}$, we have $\mathbb{M} \models D_T[d, F]$ if and only if d is a proof in the sense of $\mathbb{M}$ in T of the formula F, which means that every element $w^M(d, i)$, for $i <^M l(d)$, is a formula in the sense of $\mathbb{M}$, and that $w^M(d, i) \in T^M$ or $w^M(d, i) \in \text{LOG}^M$, or there are $k, j <^M i$ such that

$$w^M(d, k) = \text{imp}^M(w^M(d, j), w^M(d, i)).$$

Note that, for any nonstandard $\mathbb{M}$, there are always nonstandard (or infinite) proofs d, that is, nonstandard $d >^M N$ satisfying $\mathbb{M} \models D_T[d, F]$ even for a finite $F \in M$. This follows easily from the overspill principle (Exercise 13.13). In contrast to finite $d \in N$, the nonstandard proof in $\mathbb{M}$ cannot be interpreted as ordinary proofs out of $\mathbb{M}$. It may happen that a nonstandard $d \in M$ is a proof in $\mathbb{M}$ of the false statement $\Lambda_0 = \Lambda_1$.

Hence, the formula $W_T(y)$: $\exists x\ D_T(y)$ defines in $\mathbb{M}$ the set of the theorems of the theory $T^M = \{F \in \text{Fm}^M\colon \ \mathbb{M} \models T[F]\}$. If $T \in \text{PR}$ and as $T(x)$ we take the corresponding representing formula, then $W_T(y)$ is of class Σ_1.

Now, we shall formulate arithmetical counterparts of theorems of logic. We have

21.4 $$\text{PA} \vdash [(\text{LOG}(F) \vee T(F)) \rightarrow W_T(F)].$$

For any model $\mathbb{M}$, if $F \in \text{LOG}^M \cup T^M$, then the element $d = \langle F\rangle$ is in M and d is a proof in T^M of the formula F.

We have

21.5 $$\mathrm{PA} \vdash [W_T(F) \wedge W_T(F \to G) \to W_T(G)].$$

If $d, e \in M$ are proofs in $\mathbb{M}$ of the formulas F and $F \to G$, respectively, then the sequence $d * e * \langle G \rangle \in M$ is a proof of the formula G.

The theorem on the induction with respect to a proof can be expressed in PA as follows: for any arithmetical formula $S(x)$ let Γ_S be the sentence

$$\forall F, G\big[\mathrm{Fm}(F) \wedge \mathrm{Fm}(G) \to \big(S(F) \wedge S(F \to G) \to S(G)\big)\big].$$

Hence, the following equivalence holds:

$$(\mathbb{M} \models \Gamma_S) \quad \text{if and only if} \quad \big(\textit{the set } S^M \textit{ is closed under modus ponens}\big).$$

It follows that

21.6 $$\mathrm{PA} \vdash \big(\Gamma_S \wedge \forall x[T(x) \wedge \mathrm{LOG}(x) \to S(x)] \to \forall x[W_T(x) \to S(x)]\big).$$

If $d \in M$ is a proof in T^M of a formula F, then the definable set

$$\{i \in M\colon\ i <^M l^M(d) \to w(d, i) \in S^M\}$$

satisfies the premise of the suitable induction axiom.

The syntactical compactness: "if $T \vdash F$, then $T_0 \vdash F$ for some finite subset $T_0 \subseteq T$" can be expressed as follows: let $T_z(x)$ be the formula $T(x) \wedge x \leq z$. Clearly, we have

21.7 $$\mathrm{PA} \vdash [W_T(F) \to \exists z W_{T_z}(F)].$$

If $d \in M$ is a proof in T^M of a formula F, then since $w(d, i) <^M d$, for $i < l(d)$, d is a proof in T_d^M. (All the notions and theorems considered so far can be generalized to the case where the set T^M is parametrically definable, that is, where the formula T has, besides the variable x, other free variables.)

Let $(T; y)(x)$ denote the formula $T(x) \vee x = y$. The deduction theorem takes the form

21.8 $$\mathrm{PA} \vdash [W_{(T;F)}(G) \equiv W_T(F \to G)].$$

We have $\mathrm{PA}, W_T(F \to G) \vdash W_{(T;F)}(F \to G)$ and $\mathrm{PA}, W_T(F \to G) \vdash W_{(T;F)}(F)$ by 21.4, hence we get $\mathrm{PA}, W_T(F \to G) \vdash W_{(T;F)}(G)$ by 21.5, whence the implication, from the right to the left follows. To prove the converse implication we apply 21.6 to the formula $W_T(F \to x)$ in place of $S(F, x)$ and the theory $(T; F)$ in place of T and we proceed as in the proof of Theorem 9.1.

The proofs of the counterparts of the remaining logical laws differ from the proofs given in Chapter 9 only typographically. As an illustration we shall formulate the generalization rule (see Theorem 10.1).

21.9 $$\mathrm{PA} \vdash \big(X(y) \wedge \forall F[T(F) \to \neg V(F,y)] \to \forall F\big[W_T(F) \to W_T\big(\mathrm{gen}(y,F)\big)\big]\big).$$

The proof can be obtained by applying 21.6 to the formula $S(y,x)$: $W_T\big(\mathrm{gen}(y,x)\big)$.

On the other hand, to formulate the counterpart of Theorem 9.3, let us denote by $\mathrm{Cons}(T)$ the sentence

$$\forall F\big[\neg W_T(F) \vee \neg W_T\big(\mathrm{neg}(F)\big)\big].$$

Then, from Theorem 9.3, we obtain the following relationships:

21.10 $$\mathrm{PA} \vdash [\neg\mathrm{Cons}(T) \equiv \forall F W_T(F)]$$

and

21.11 $$\mathrm{PA} \vdash [\mathrm{Cons}(T) \equiv \neg W_T(\Lambda_{F_0} \wedge \Lambda_{\mathrm{neg}\ F_0})],$$

for any formula $F_0 \in \mathrm{Fm}$.

Next, Theorem 9.11 (*reductio ad absurdum*) yields

21.12 $$\mathrm{PA} \vdash [\neg\mathrm{Cons}(T;F) \equiv W_T(\neg F)],$$

whence

21.13 $$\mathrm{PA} \vdash [\mathrm{Cons}(T) \wedge W_T(F) \to \mathrm{Cons}(T;F)]$$

and

21.14 $$\mathrm{PA} \vdash \big[\mathrm{Cons}(T) \to \forall F\big(\mathrm{Cons}(T;F) \vee \mathrm{Cons}(T;\neg F)\big)\big]$$

The use of the logical axiom of substitution gives the theorem on the change of variables,

21.15 $$\mathrm{PA} \vdash \big[X(z) \wedge F < z \wedge z \neq x \to \big(\forall x\, F \equiv \forall z F(z/x)\big)\big].$$

Heading for the proof of the completeness theorem, we define the Henkin axioms. Let $z(x,F) = \langle 3, \langle x, F\rangle\rangle$. Evidently, this is a total function. The arithmetical formula $h(F)$;

$$\exists x,z\ G \leq F\big[z = z(x,G) \wedge \mathrm{Fm}(G) \wedge F = \mathrm{imp}\big(\mathrm{neg}\big(\mathrm{gen}\big(x,\mathrm{neg}(G)\big)\big), G(z/x)\big)\big]$$

determines in an arbitrary model $\mathbb{M}$ the set of formulas of the form

$\neg\forall x \neg G \rightarrow G(z/x)$, where the pair $\langle x, G\rangle$ runs over $X^M \times \mathrm{Fm}^M$, and moreover the variable z is chosen in such a way that it does not occur in the pairs $\langle y, H\rangle$, for $y \leq x$ and $H \leq G$. Now, let $T_h(x)$ be the formula $T(x) \vee h(x)$. We have then

21.16 $$\mathrm{PA} \vdash [\mathrm{Cons}(T) \rightarrow \mathrm{Cons}(T_h)]$$

for any set of sentences T, which can be easily proved by applying 21.7, 21.9, and 21.15. Of course, if $T \in \mathrm{PR}$, then also $T \cup h^N \in \mathrm{PR}$.

It still remains to determine arithmetically a maximal consistent extension of the theory T_h. Let $K_0 \subseteq K$ consist of the codes of the zero–one sequences:

$$a \in K_0 \quad \text{if and only if} \quad \big(a \in K \wedge \forall i < l(a)\big[w(a, i) = 0 \vee w(a, i) = 1\big]\big).$$

This condition determines an arithmetical formula $K_0(x)$ such that the set $K_0^M = \{a \in M\colon\ \mathbb{M} \models K_0[a]\}$ consists of all the sequences $a \in M$ all of whose terms are 0 or 1. The set K_0^M is partially ordered by the definable relation

$$a \leq_0 b \quad \text{if and only if} \quad (l(a) \leq l(b) \wedge \forall i < l(a)[w(a, i) = w(b, i)].$$

The pair $\langle K_0^M, \leq_0^M\rangle$ is called the *binary tree of* M and the elements $a \in K_0^M$ are called the *vertices* of that tree.

The function f defined inductively by the equations

21.17
$$f(0) = \min \mathrm{Fm}$$
$$f(n+1) = \min\{F \leq \mathrm{neg}(f(n))\colon\ F \in \mathrm{Fm} \wedge F > f(n)\}$$

is total in PA and in any model $\mathbb{M}$ it enumerates all the formulas: $\mathrm{Fm}^M = \{f^M(m)\colon\ m \in M,\ m > 0\}$ [m can be referred to as the number of the formula $f^M(m)$]. Hence, the total function g defined by setting,

$$g(a) = \begin{cases} f(l(a)), & \text{if } w\big(a, l(a) - 1\big) = 0, \\ \mathrm{neg}(f(l(a)), & \text{if } w\big(a, l(a) - 1\big) = 1 \end{cases}$$

and

$$g(a) = 0, \quad \text{for } a \notin K_0 \ \text{ and } \ g(1) = 1,$$

assigns to a vertex $a \in K_0$ the formula F with the number $l(a)$ or its negation $\neg F$, according to whether the last term of the sequence a is 0 or 1.

The arithmetical formula $\Gamma(x)$ is called a *branch* for a given model $\mathbb{M}$, if the set $\Gamma^M = \{a\colon\ \mathbb{M} \models \Gamma[a]\}$ is a linearly ordered by $\leq_0^M$ subset of K_0^M containing sequences of all possible lengths. A branch $\Gamma(x)$ is called *consistent* (in the sense of $\mathbb{M}$), if the set $S_\Gamma^M = \{g^M(a)\colon\ a \in \Gamma^M \wedge a > 1\}$ is consistent. In this case S_Γ^M is, of course, a maximal consistent set of formulas and is definable in $\mathbb{M}$ by the formula $S_\Gamma(x)$, $\exists a(\Gamma(a) \wedge \Gamma_g(a,x))$, where Γ_g is the formula defining the total function g in PA.

Therefore, it is sufficient to determine a consistent branch $\Gamma(x)$, such that $T_h^M \subseteq S_\Gamma^M$.

The function

$$\bigwedge(a) = \bigwedge_{i<l(a)} w(a,i) = w(a,0) \wedge \cdots \wedge w(a, l(a)-1),$$

assigning to a sequence of formulas their conjunction, is PR and so, it is total in PA. As $\Gamma(a)$ we take the formula

$$K_0(a) \wedge \{a = 1 \vee [a > 1 \wedge \neg W_{T_h}(\neg \bigwedge_{i<l(a)} g(a\restriction(i+1)))$$

$$\wedge \forall j < l(a)[\neg W_{T_h}(\bigwedge_{i<j} g(a\restriction(i+1)) \to \neg f(j)) \to w(a,j) = 0]]\}.$$

Hence, the set Γ^M consists of those sequences $a \in K_0^M$ for which the conjunction of the formulas corresponding to the restrictions $a \restriction i$ is consistent with T_h^M, and moreover, for $j < l(a)$, we take $w(a,j) = 0$ unless the formula F with the number j is inconsistent with T_h^M and the preceding formulas. Hence, directly from the definition of the formula Γ, the following condition is fulfilled:

if $a,b \in M$ and $l(a) \leq^M l(b)$, then $a \leq_0^M b$.

Applying 21.14, we see that if $a \in \Gamma^M$, then either $a * \langle 0 \rangle \in \Gamma^M$ or $a * \langle 1 \rangle \in \Gamma^M$. Hence, from 21.16 we obtain by induction

$$\text{PA}, \text{Cons}(T) \vdash \forall x \exists a[l(a) = x \wedge \Gamma(a)],$$

that is, under the assumption $\text{Cons}(T)$, the formula Γ is a consistent branch. Of course, $T_h^M \subseteq S_\Gamma^M$, for any model $\mathbb{M}$ such that $\mathbb{M} \models \text{Cons}(T)$. Thus, we have shown the following lemma

21.18. Lemma. *For every arithmetical formula $T(x)$ there is an arithmetical formula Γ such that if $\mathbb{M} \models \text{Cons}(T)$, then Γ is a consistent branch in $\mathbb{M}$ and the formula S_Γ defines in $\mathbb{M}$ a maximal consistent extension of the theory T_h^M.*

If $T \in \text{PR}$, then we can find a formula $T(x) \in \Sigma_1$, representing T in PA. In this case $T = T^M \cap N$, and hence $T \subseteq S_\Gamma^M$ and thus S_Γ^M is an extension of the theory T.

Let a structure

$$\mathbb{U} = \langle U;\ \{r^U\}_{r\in R}, \{f^U\}_{f\in F}, \{c^U\}_{c\in C}\rangle$$

be a model of the theory T. We say that $\mathbb{U}$ is definable in the given model $\mathbb{M}$ of PA, if the universe U as well as the relations r^U, operations f^U, and constants c^U are arithmetically definable in $\mathbb{M}$ by means of some formulas H_U, H_r, for $r \in R$; H_f, for $f \in F$; and H_c, for $c \in C$.

We shall prove the following version of the completeness theorem:

21.19. Hilbert–Bernays Theorem (cf. [HB]). *Let T be a consistent set in a finite language $L(T)$. If $T \in$ PR, then there is a system of arithmetical formulas*

$$H = \langle H_U, \{H_r\}_{r\in R}, \{H_f\}_{f\in F}, \{H_c\}_{c\in C}\rangle$$

such that in any model $\mathbb{M}$ of PA *satisfying* Cons(T), *H defines a model $\mathbb{U}$ for T.*

Proof. If $\mathbb{M} \models \text{Cons}(T)$, then from Lemma 21.18 we find a formula S defining in $\mathbb{M}$ a maximal consistent extension of the theory T_h^M. Following the proof of the completeness theorem we define a model on variables $x \in X^M$ (cf. Exercise 11.6). Let $H_U(x)$ be the formula

$$X(x) \wedge \forall y[X(y) \wedge S(x = y) \rightarrow x \leq y]$$

and let u be the function

$$u(t) = \min\{y \in X^M\colon\ \mathbb{M} \models S[(t = y)]\} \quad \text{for } t \in \text{Tm}^M.$$

The function u is well defined since $\exists y(t = y)$ is in LOG^M, for a variable $y \geq t$, hence $\exists y(t = y)$ belongs to S^M and it remains only to apply a suitable Henkin axiom. It follows that

$$U = \{m \in M\colon\ \mathbb{M} \models H_U[m]\} = \{u(t)\colon\ t \in \text{Tm}^M\}.$$

Moreover, we have

21.20 $\quad u(t) = u(s)$ if and only if $(t = s) \in S^M$, for all $t, s \in \text{Tm}^M$.

If $u(t) = u(s)$, then for $z = u(t)$ the formulas $t = z$ and $z = s$ belong to S^M, and thus also the formula $t = s$ is in S^M.

We define the relations r^U as follows:

$$r^U(z_1, \ldots, z_n) \quad \text{if and only if} \quad \langle r, z_1, \ldots, z_n\rangle \in S^M, \quad \text{for } z_1, \ldots, z_n \in U.$$

Then we have

21.21 $\quad r^U(u(t_1), \ldots, u(t_n))$ if and only if $\langle r, t_1, \ldots, t_n\rangle \in S^M$, for $t_1, \ldots, t_n \in \text{Tm}^M$.

The formulas $u(t_1) = t_1, \ldots, u(t_n) = t_n$ are in S^M. From the logical axioms of equality follows the equivalence

$$\langle r, u(t_1), \ldots, u(t_n)\rangle \in S^M \quad \text{if and only if} \quad \langle r, t_1, \ldots, t_n\rangle \in S^M.$$

Hence, we obtain 21.21.

The operations f^U are defined as follows:

$$f^U(z_1, \ldots, z_m) = u(\langle f, z_1, \ldots, z_m\rangle), \quad \text{for } z_1, \ldots, z_m \in U.$$

Then we have

21.22 $\quad f^U(u(t_1), \ldots, u(t_m)) = u(\langle f, z_1, \ldots, z_m\rangle) \;$ for $\; t_1, \ldots, t_m \in \mathrm{Tm}^M$.

We prove this in the same way as 21.21.

Finally, the constants c^U are defined by the equalities $c^U = u(c)$, for arbitrary c. The formula S and the function u determine in an obvious way formulas H_r, H_f, and H_c defining relations, operations, and constants, respectively. So we have defined in $\mathbb{M}$ the system

$$\mathbb{U} = \langle U, \{r^U\}, \{f^U\}, \{c^U\}\rangle.$$

A term $t \in \mathrm{Tm}^M$ will be called *standard* if t has the form $s(y_1/x_1, \ldots, y_n/x_n)$, where $s(x_1, \ldots, x_n) \in \mathrm{Tm} \subseteq N$ and $y_1, \ldots, y_n \in X^M$, that is, t is obtained from an ordinary term $s \in \mathrm{Tm}$ by a change of variables. It is easy to see that standard terms are built as ordinary terms except that the ground set of variables is X^M, not X. Similarly, a formula $F \in \mathrm{Fm}^M$ will be called *standard* if F is obtained from some formula $G \in \mathrm{Fm}$ by a change of variables (free or bound). Hence, standard formulas are built in the same way as ordinary formulas (i.e., as elements of the set Fm) except that the atomic formulas have the form $t = s$ or $r(t_1, \ldots, t_n)$, where the terms $t, s, t_1, \ldots, t_n$ are standard.

Now, we shall prove the following equivalence:

21.23 $\quad \mathbb{U} \models F[u] \;$ if and only if $\; F \in S^M, \;$ for standard formulas F.

The function u is treated here as an assignment in $\mathbb{U}$ (more precisely, u should be replaced by the restriction $u|X^M$).

First, we check the equality $t[u] = u(t)$, for standard terms t.If $x \in X^M$, then $x[u] = u(x)$. If $c \in C$, then $c[u] = c^U = u(c)$, by the definition.

The inductive step

$$\begin{aligned}\langle f, t_1, \ldots, t_m\rangle[u] &= f^U(t_1[u], \ldots, t_m[u]) \\ &= f^U(u(t_1), \ldots, u(t_m)) = u(\langle f, t_1, \ldots, t_m\rangle),\end{aligned}$$

follows by 21.21.

We check 21.23 for atomic formulas

$$\mathbb{U} \models (t = s)[u] \quad \text{iff} \quad t[u] = s[u] \quad \text{iff} \quad u(t) = u(s) \quad \text{iff} \quad (t = s) \in S^M,$$

from 21.20;

$$\mathbb{U} \models \langle r, t_1, \ldots, t_n \rangle [u] \quad \text{iff} \quad r^U(t_1[u], \ldots, t_n[u])$$

$$\text{iff} \quad r^U(u(t_1), \ldots, u(t_n)) \quad \text{iff} \quad \langle r, t_1, \ldots, t_n \rangle \in S^M,$$

because of 21.21.

Since S^M is a maximal consistent set, for any formulas $F, G \in \mathrm{Fm}^M$ we have

$$\neg F \in S^M \quad \text{if and only if} \quad F \notin S^M,$$

$$F \to G \in S^M \quad \text{if and only if} \quad \textit{either } F \notin S^M \textit{or } G \in S^M.$$

Hence, we infer immediately that 21.23 is preserved under adding a negation and under forming an implication. It remains to check the case of the quantifier

$$\mathbb{U} \not\models \forall x\, F[u] \quad \text{iff} \quad \exists z \in U(\mathbb{U} \not\models F[u(z/x)])$$

$$\text{iff} \quad \exists z \in U(\mathbb{U} \not\models F[u(z[u]/x)]) \quad \text{iff} \quad \exists z \in U(\mathbb{U} \not\models F(z/x)[u])$$

$$\text{iff} \quad \exists z \in U(\neg F(z/x) \in S^M) \quad \text{iff} \quad \forall x\, F \notin S^M.$$

In the last equivalence the implication to the right follows immediately from the logical axiom of substitution; the formula $\forall x F \to F(z/x)$ is in S^M. The converse implication follows from the Henkin condition: if $\forall x F \notin S^M$, then $\neg \forall x F \in S^M$, and since $(\neg \forall x F \to \neg F(z/x)) \in S^M$, for some variable z [we may assume that $z = u(z)$], also $\neg F(z/x) \in S^M$.

In this way we have proved equivalence 21.23.

Since every sentence $F \in T$ is standard and $T \subseteq S^M$, $\mathbb{U}$ is a model of the theory T, which finishes the proof. □

Taking into account the equivalence 21.23 we may express the Hilbert–Bernays theorem in a more precise form, preserving the same notation.

21.24. Corollary. *If $T \in \mathrm{PR}$ is a consistent set of sentences (in a finite language $L(T)$), then there is an arithmetical formula $S(x)$ such that every model $\mathbb{M}$ of* PA *in which* $\mathrm{Cons}(T)$ *holds, determines a definable in $\mathbb{M}$ model $\mathbb{U}$ of the theory T for which we have*

$$\mathbb{U} \models F[u(t_1), \ldots, u(t_n)] \quad \text{if and only if} \quad F(t_1, \ldots, t_n) \in S^M,$$

for every standard formula F, any standard terms $t_1, \ldots, t_n$ and the assignment u.

21.25. Gödel's Theorem on Consistency

First, we shall prove the following lemma

Lemma. *Let* $\mathbb{M}_1$, $\mathbb{M}_2$ *be models of* PA. *If* $\mathbb{M}_2$ *is definable in* $\mathbb{M}_1$ (*see the passage preceding* Theorem 21.19), *then* $\mathbb{M}_1$ *can be isomorphically embedded into* $\mathbb{M}_2$ *as an initial segment of* $\mathbb{M}_2$.

Proof. From the assumption it follows in particular that the operation $x +^{M_2} y$ (the addition in $\mathbb{M}_2$) and the elements 0^{M_2} and 1^{M_2} are definable in $\mathbb{M}_1$. Thus there is a definable function $f\colon\ M_1 \longrightarrow M_1$ satisfying the equations

21.26
$$f(0^{M_1}) = 0^{M_2},$$
$$f(a+1) = f(a) +^{M_2} 1^{M_2}, \quad \text{for } a \in M_1.$$

Hence, $f\colon\ M_1 \longrightarrow M_2$ and f is monotone: if $a <^{M_1} b$, then $f(a) <^{M_2} f(b)$, since, for any $a \in M_1$, the set

$$Z(a) = \{c \in M_1\colon\ f(a) <^{M_2} f(a+c+1)\}$$

is definable in $\mathbb{M}_1$ with the parameter a, contains 0^{M_1} and, with any c, it contains also $c+1$, by 21.26. In particular f is a one-to-one function. Of course, $f(1^{M_1}) = 1^{M_2}$, which immediately follows from 21.26. Applying induction to the definable set

$$Z(a) = \{b \in M_1\colon\ f(a +^{M_1} b) = f(a) +^{M_2} f(b)\},$$

we infer that f preserves the addition and similarly we find that

$$f(a \cdot^{M_1} b) = f(a) \cdot^{M_2} f(b).$$

To see that the image $f[M_1] \subseteq M_2$ is an initial segment of M_2 assume that $b \in M_2$ and $b \leq^{M_2} f(a)$, for an $a \in M_1$. Then, the set

$$Z(b) = \{a \in M_1\colon\ b \leq^{M_2} f(a)\}$$

is parametrically definable in $\mathbb{M}_1$, since both the relation $\leq^{M_2}$ and the function f are definable in $\mathbb{M}_1$. Let $a_0 = \min Z(b)$. If $a_0 = 0^{M_1}$, then $b \leq^{M_2} f(0^{M_1}) = 0^{M_2}$, and thus $b = 0^{M_2}$, whence $b = f(0^{M_1})$. If $a_0 >^{M_1} 0$, then

$$f(a_0 - 1) \leq^{M_2} b <^{M_2} f(a_0 - 1) +^{M_2} 1^{M_2},$$

that is, either $b = f(a_0 - 1)$ or $b = f(a_0)$, which finishes the proof. □

Let T be a set of sentences in a finite language $L(T)$. We assume that the language $L(T)$ is defined in PA as we have described before. If T is definable, then there exists the sentence $\mathrm{Cons}(T)$ expressing the consistency of T for which we have (see 21.10 and 21.11)

$$(T \text{ is consistent}) \quad \text{if and only if} \quad (\mathbb{N} \models \mathrm{Cons}(T)).$$

If $T \in \mathrm{PR}$, then $D_T \in \Sigma_1$, hence $\mathrm{Cons}(T)$ is a Π_1-formula.

Using the Hilbert–Bernays theorem we shall prove now the following famous theorem of Gödel on consistency.

21.27. Gödel's Theorem [G3]. *If T is a consistent set of sentences, $T \in \mathrm{PR}$ and T contains* PA, *then* $T \nvdash \mathrm{Cons}(T)$.

Proof (Kreisel). Since T contains the arithmetic, the arithmetical part (the reduct to the language of PA) of any model for T is a model of PA. Applying Lemma 21.18 we fix a formula $S_\Gamma(x)$ defining a maximal consistent extension of T such that the branch Γ is defined as in the proof of Lemma 21.18, that is, Γ is the earliest possible branch in the lexicographical ordering. Let $\mathbb{A}_0 \in \mathrm{Mod}(T)$ be an arbitrary model of the theory T and $\mathbb{M}_0$ be its arithmetical part. Assume, tending to a contradiction, that $T \vdash \mathrm{Cons}(T)$. Hence, we have $\mathbb{A}_0 \models \mathrm{Cons}(T)$, and thus also $\mathbb{M}_0 \models \mathrm{Cons}(T)$, since $\mathrm{Cons}(T)$ is an arithmetical sentence. By the Hilbert–Bernays theorem there is a model $\mathbb{A}_1$ of the theory T definable in $\mathbb{M}_0$ and such that the formula S_Γ determines all the sentences true in $\mathbb{A}_1$. In particular, the arithmetical part $\mathbb{M}_1$ of the model $\mathbb{A}_1$ is definable in $\mathbb{M}_0$. Since $\mathbb{A}_1 \models \mathrm{Cons}(T)$, also $\mathbb{M}_1 \models \mathrm{Cons}(T)$ and again we have a model $\mathbb{A}_2$ of the theory T definable in $\mathbb{M}_1$, so its arithmetical part is definable in $\mathbb{M}_1$. Proceeding in this way, we obtain inductively a sequence $\{\mathbb{A}_n\colon\ n \in N\}$ of models of the theory T and a sequence $\{\mathbb{M}_n\colon\ n \in N\}$ of their arithmetical parts such that $\mathbb{M}_{n+1}$ is definable in $\mathbb{M}_n$, for every $n \in N$. Moreover, the truth in $\mathbb{M}_n$ is determined by the formula S_Γ in $\mathbb{M}_{n-1}$.

Applying Lemma 20.13, the diagonal lemma, we find a sentence F satisfying the condition

$$\mathrm{PA} \vdash \big(F \equiv \neg S_\Gamma(\Lambda_F)\big).$$

The sentence F is alternately true and false in the models $\mathbb{M}_n$; if, for example, $\mathbb{M}_0 \models F$, then $\mathbb{M}_1 \models \neg F$, $\mathbb{M}_2 \models F$, and so on. Fix an m for which $F = f(m)$ under the enumeration 21.17. Let

$$\Gamma_m^{M_n} = \{a \in K_0^M\colon\ l(a) \le m \text{ and } \mathbb{M}_n \models \Gamma[a]\}$$

be the initial part of the branch Γ in the model $\mathbb{M}_n$ consisting of the vertices of length less than or equal to m. Recall that to the longest vertex in $\Gamma_m^{M_n}$ there corresponds (via the function g) the sentence F or its negation. Hence, it follows immediately that $\Gamma_m^{M_{n+1}} \neq \Gamma_m^{M_n}$, for every $n \in N$, since, as we have mentioned, the condition $F \in S_\Gamma^{M_n}$ implies $\neg F \in S_\Gamma^{M_{n+1}}$ and conversely. The sequences $a \in \Gamma_m^{M_n}$ determine a single zero–one sequence of length m, denoted by γ_n, which is their common extension. We shall show that γ_{n+1} is lexicographically later than γ_n. So, let $i < m$ be the least index for which $\gamma_{n+1}(i) \neq \gamma_n(i)$. If it were $\gamma_n(i) = 1$, then this would mean that the formula $G = f(i)$ is inconsistent with

$$T_i = T_h^{M_n} \cup \{g(\gamma \restriction j)\colon\ j \leq i\}.$$

Thus, we may write $\mathbb{M}_n \models \neg\,\mathrm{Cons}(T_i \cup \{G\})$. Since $\mathbb{M}_{n+1}$ is definable in $\mathbb{M}_n$, $\mathbb{M}_n$ can be isomorphically embedded into $\mathbb{M}_{n+1}$ as an initial segment, see the lemma in section 21.25. But then we also have $\mathbb{M}_{n+1} \models \neg\,\mathrm{Cons}(T_i \cup \{G\})$ since the Σ_1 sentences are absolute upwards (see Exercise 16.17). This means that also $\gamma_{n+1}(i) = 1$ holds, contrary to the definition of the number i. Thus, $\gamma_n(i) = 0$ and $\gamma_{n+1}(i) = 1$, that is, $\gamma_n < \gamma_{n+1}$ lexicographically, which proves the claim. In particular, it follows that all the sequences $\gamma_0, \gamma_1, \gamma_2, \ldots$ are different, which is impossible, since they all have a constant length m. In this way the proof has been completed. □

For another proof see the Exercises in this chapter.

21.28. A Universal Formula

Assume now that $T = \mathrm{PA}$, in particular, the arithmetical formulas defining in PA the syntactical notions refer now to the language of PA. In view of the Tarski theorem, 20.16, the relation

$$r(F, a) \quad \text{if and only if} \quad \mathbb{M} \models F[a]$$

is not arithmetically definable in $\mathbb{M}$ (for any model $\mathbb{M}$). However, if the range of the variable F is restricted, for example, to the class Σ_1, then—as we shall prove—the satisfaction relation is arithmetically definable.

We start with defining the values of terms $t[a] = u$.

21.29. Lemma. *There exists an arithmetical formula $W(x, y, z) \in \Sigma_1$ such that for any model $\mathbb{M}$ of arithmetic, any term $t \in Tm$, any sequence $a \in K^M$, which is defined at least for the free variables of the term t, and any element $u \in M$, we have*

$$t^M[a] = u \quad \text{if and only if} \quad \mathbb{M} \models W[t, a, u].$$

Proof. The condition $t \in \mathrm{Tm}$ is equivalent with the following condition.

21.30 *There is a sequence $\tau = \langle t_0, \ldots, t_n\rangle$ of terms such that every term t_i is a variable or a constant or has the form $t_i = \langle +, t_{j_1}, t_{j_2}\rangle$ or $t_i = \langle \cdot, t_{j_1}, t_{j_2}\rangle$, for some $j_1, j_2 < i$, and $t_n = t$.*

The shortest such sequence τ is called the *sequence of the subterms* of the term t. The property 21.30 can be expressed arithmetically by the formula $H_0(\tau, t)$ of class Σ_1 of the form

$$\begin{aligned} &K(\tau) \wedge w(\tau, l(\tau)-1) = t \wedge \forall i < l(\tau)\{\mathrm{Tm}(w(\tau, i)) \\ &\wedge \big[X(w(\tau, i)) \vee w(\tau, i) = \Lambda_{c_0} \vee w(\tau, i) = \Lambda_{c_1} \\ &\vee \exists j_1, j_2 < i\big(w(\tau, i) = \langle +, w(\tau, j_1), w(\tau, j_2)\rangle \\ &\vee w(\tau, i) = \langle \cdot, w(\tau, j_1), w(\tau, j_2)\rangle\big)\big]\}, \end{aligned}$$

where c_0, c_1 are the numbers corresponding to the constants Λ_0, Λ_1. Hence, we have $\mathrm{PA} \vdash (\mathrm{Tm}(t) \equiv \exists \tau H_0(\tau, t))$, by 21.1.

To calculate the value $t^M[a]$ we compute consecutively the values of the subterms $t_i^M[a] = b_i$. Next, we arrange the so-obtained values $b_i \in M$ in a single sequence $b \in M$. This process is described by the formula $H_1(\tau, a, b) \in \Sigma_1$ of the form

$$\begin{aligned} &K(b) \wedge l(b) = l(\tau) \\ &\wedge \forall i < l(\tau)\{[X(w(\tau, i)) \to w(\tau, i) < l(a) \wedge w(b, i) = w(a, w(\tau, i))] \\ &\wedge [(w(\tau, i) = \Lambda_{c_0} \to w(b, i) = \Lambda_0) \wedge (w(\tau, i) = \Lambda_{c_1} \to w(b, i) = \Lambda_1)] \\ &\wedge \forall j_1, j_2 < i(w(\tau, i) = \langle +, w(\tau, j_1), w(\tau, j_2)\rangle \to w(b, i) = w(b, j_1) + w(b, j_2)) \\ &\wedge \forall j_1, j_2 < i(w(\tau, i) = \langle \cdot, w(\tau, j_1), w(\tau, j_2)\rangle \to w(b, i) = w(b, j_1) \cdot w(b, j_2))\}. \end{aligned}$$

Now, let $W(t, a, u)$ be the formula

$$\exists \tau, b[H_0(\tau, t) \wedge H_1(\tau, a, b) \wedge w(b, l(b) - 1) = u].$$

Clearly, W is of class Σ_1. We have to check that the following holds:

$$(t^M[a] = u) \quad \text{if and only if} \quad (\mathbb{M} \models W[t, a, u]), \quad \text{for } t \in \mathrm{Tm}.$$

So let $t(a) = u$ in $\mathbb{M}$ and let $\tau = \langle t_0, \ldots, t_n\rangle$ be the sequence of the subterms of t. Putting $b_i = t_i^M[a]$ and $b = \langle b_0, \ldots, b_n\rangle$ we obtain $u = b_n$, and moreover

21.31 $$\mathbb{M} \models H_0[\tau, t] \quad \text{and} \quad \mathbb{M} \models H_1[\tau, a, b],$$

whence it follows that $\mathbb{M} \models W[t, a, u]$. Conversely, if we have $\mathbb{M} \models W[t, a, u]$, then let us take $\tau, b \in M$, for which we have 21.31. We may assume that τ is the sequence of the subterms of t, since rejecting from τ the redundant values and the corresponding values of the sequence b, we shall obtain the sequence of the subterms τ_0 and a sequence b_0 satisfying 21.31, too. From the condition $\mathbb{M} \models H_1[\tau, a, b]$ we get, by induction on $i < l(\tau)$, the equations $b_i = w(\tau, i)[a]$. Since the last term of the sequence b is u, we must have $u = t^M[a]$, which completes the proof. □

A similar method will be applied to define the satisfaction relation. We shall use the method of coding sequences described in Exercises 20.8 to 20.11.

21.32. Theorem. *There exists an arithmetical formula $\Phi(x, y) \in \Sigma_1$ such that for every model $\mathbb{M}$, every $F \in$ Fm of class Σ_1 of language $L(\mathrm{PA})$, and every sequence $a \in M$, defined on the free variables of the formula F, we have*

$$\mathbb{M} \models F[a] \quad \text{if and only if} \quad \mathbb{M} \models \Phi[F, a].$$

Proof. Let us denote by Fm_0 the family of bounded formulas. The condition $F \in \mathrm{Fm}_0$ is equivalent with the following:

21.33 *There is a sequence of formulas $\varphi = \langle F_0, \ldots, F_n \rangle$ such that $F_n = F$ and every term F_i is an atomic formula or $F_i = \mathrm{neg}\ (F_j)$, for some $j < i$; or $F_i = \mathrm{imp}\ (F_{j_1}, F_{j_2})$, for some $j_1, j_2 < i$; or $F_i = \mathrm{gen}\ (x \leq y, F_j)$, for some $j < i$; and for some variables $x, y \leq F$.*

Here $\mathrm{gen}(x \leq y, F)$ denotes $\mathrm{gen}(x, x \leq y \to F)$.

The shortest sequence φ with the property 21.33 is called the *sequence of the subformulas* of the formula F. Similarly as for terms, we build an arithmetical formula $H_0(x, y) \in \Sigma_1$ for which we have

$$\mathrm{PA} \vdash [\mathrm{Fm}_0(F) \;\equiv\; \exists \varphi H_0(\varphi, F)].$$

In addition, modifying H_0 in an obvious way, we may ensure that no free variable in F is bound in F.

If F is a bounded formula, then the relation $\mathbb{M} \models F[a]$ is uniquely determined by the relations $\mathbb{M} \models F_i[b]$, where F_i runs over the subformulas of F, and the bs are some sequences defined for both the free and bound variables of F and with terms not exceeding a. By Exercise 20.14, the codes of such sequences are $\leq 2^{4(aF)^2+1}$. Let us put

$$K_F(a) = \{b \in K\colon\ l(b) = F \text{ and } b \leq 2^{4(aF)^2+1}\},$$

and

$$r(t, s) = \{b \in K_F(a)\colon\ t[b] = s[b]\}.$$

Identifying the sets $Z \subseteq K_F(a)$ with their arithmetical codes (Exercises 20.8 and 20.9) and applying Lemma 21.29 we infer that the function r is of class Σ_1. We have

$$u = r(t,s) \quad \text{iff} \quad \forall b \in K_F(a)\{[e(b,u) \to \exists v(W(t,b,v) \wedge W(s,b,v)))]$$
$$\wedge [\neg e(b,u) \to \exists v, w(W(t,b,v) \wedge W(t,b,w) \wedge v \neq w)]\}$$

(see Exercise 20.8) and for any model $\mathbb{M}$ we have

$$r^M(t,s) = \{b \in K_F^M(a)\colon \ \mathbb{M} \models (t = s)[b]\},$$

for $t, s \in \mathrm{Tm}$. Also the functions

$$n(Z) = K_F(a) \setminus Z, \quad \text{for } Z \subseteq K_F(a),$$

$$i(Z_1, Z_2) = (K_F(a) \setminus Z_1) \cup Z_2, \quad \text{for } Z_1, Z_2 \subseteq K_F(a),$$

$$g(x \leq y, Z) = \{b \in K_F(a)\colon \forall u \leq w(b,y)[b(u/x) \in Z]\},$$

$$\text{for } Z \subseteq K_F(a) \quad \text{and } x, y \leq F$$

are of class PR. If, for example,

$$Z = \{b \in K_F^M(a)\colon \ \mathbb{M} \models G[b]\},$$

then

$$n^M(Z) = \{b \in K_F^M(a)\colon \ \mathbb{M} \models \neg G[b]\}$$

and the functions i and g have analogous properties. Hence, the condition $\mathbb{M} \models F[a]$ is equivalent with the existence of sequences $\varphi = \langle F_0, \ldots, F_n \rangle$, $z = \langle Z_0, \ldots, Z_n \rangle \in M$ having the following properties (here, we use the boundedness of F): φ is the sequence of the subformulas of F, z is the sequence of the codes of the subsets of the set $K_F(a)$ whose every term Z_i has the form $r^M(t,s)$, $n^M(Z_j)$, $i^M(Z_{j_1}, Z_{j_2})$, or $g^M(x \leq y, Z_j)$, according to the form of the subformula F_i; Z_n contains some extension of the sequence a.

The above properties determine a formula $H_1(\varphi, z, a)$ of class Σ_1. Thus, we have

$$\mathbb{M} \models F[a] \quad \text{if and only if} \quad \mathbb{M} \models U[F,a], \quad \text{for } F \in \mathrm{Fm}_0,$$

where $U(F,a)$ is the formula

$$\exists \varphi, Z(H_0(\varphi, F) \wedge H_1(\varphi, z, a)).$$

If $\mathbb{M} \models F[a]$ holds, then for the sequence of the subformulas $\varphi = \langle F_0, \ldots, F_n \rangle$ we have $\mathbb{M} \models H_0[\varphi, F]$ and putting

$$Z_i = \{b \in K_F^M(a)\colon\ \mathbb{M} \models F_i[b]\} \quad \text{and} \quad z = \langle Z_0, \ldots, Z_n \rangle,$$

we obtain $\mathbb{M} \models H_1[\varphi, z, a]$, whence $\mathbb{M} \models U[F, a]$.

Conversely, if $\mathbb{M} \models U[F, a]$, that is, $\mathbb{M} \models H_0[\varphi, F]$ and $\mathbb{M} \models H_1[\varphi, z, a]$, then (similarly as in the proof of Lemma 21.29) we may assume that $\varphi = \langle F_0, \ldots, F_n \rangle$ is the sequence of the subformulas for F and then, by induction on i, we show that

$$Z_i = \{b \in K_F^M(a)\colon\ \mathbb{M} \models F_i[b]\} \quad \text{for} \quad i \leq n,$$

whence we obtain $\mathbb{M} \models F[a]$.

Hence U is a so called *universal formula* for the class of bounded formulas. Taking as $\Phi(G, a)$ the formula

$$\exists z, y[U(h(G), y) \wedge y = a(z/x)],$$

where h is a PR function satisfying the condition $h(\exists xF) = F$, we obtain

$$\mathbb{M} \models G[a] \quad \text{if and only if} \quad \mathbb{M} \models \Phi[G, a],$$

which means that Φ is a universal formula for the class Σ_1.

Clearly, $\Phi \in S_1$, which finishes the proof. □

The universal formula Φ defined above determines the formulas $\Phi_n(F, x_1, \ldots, x_n)$ universal for $F \in \Sigma_1$, where F has n free variables. Namely, we take as Φ_n the formula

$$\exists x, y_1, \ldots, y_n \big[\Phi(F, x) \wedge \big(w(x, y_1) = x_1\big) \\ \cdots \wedge \big(w(x, y_n) = x_n\big) \wedge (V_f(F) = \langle y_1, \ldots, y_n \rangle)\big].$$

Therefore, we have the following corollary.

21.34. Corollary. *For every $n \in N$ there is an arithmetical formula $\Phi_n(y, x_1, \ldots, x_n)$ of class Σ_1 such that for every formula $F \in \Sigma_1$ with n free variables we have*

$$\mathbb{M} \models F[a_1, \ldots, a_n] \quad \text{if and only if} \quad \mathbb{M} \models \Phi_n[F, a_1, \ldots, a_n]$$

for any model $\mathbb{M}$ and elements $a_1, \ldots, a_n \in M$.

EXERCISES

21.1. Show that if $T \in \mathrm{PR}$ [or, more generally, $T \in \Sigma_1(\mathbb{N})$] then we have

$$T \vdash F \quad \text{if and only if} \quad \mathrm{PA} \vdash W_T(\Lambda_F), \quad \text{for every sentence } F.$$

21.2. Show that if T contains PA and $T \in \mathrm{PR}$, then

$$T \vdash F \quad \text{if and only if} \quad T \vdash (W_T(\Lambda_F) \to F) \quad \text{(Löb).}$$

Hence, for every independent from T sentence F there is a model in which the negation $\neg F$ is true and, simultaneously, there is a (nonstandard) proof of the sentence F.

21.3. The diagonal sentence for the formula $W_T(y)$ is a theorem of theory T (under the assumptions of Exercise 21.2).

21.4. If $T \in \mathrm{PR}$, then the formula $S_\Gamma(x)$ defining a maximal consistent extension of the theory T is of class Σ_2 and, simultaneously, Π_2 (see Exercise 18.11).

21.5. Check that the Hilbert–Bernays theorem holds under a more general assumption $T \in \Sigma_1(\mathbb{N}) \cap \Pi_1(\mathbb{N})$.

21.6. From the Hilbert–Bernays theorem derive the theorem on the incompleteness of any consistent $T \in \mathrm{PR}$ containing the arithmetic.

21.7. Check that the universal formula $U(F, a)$ for the class of bounded formulas is equivalent in PA with a formula of class Π_1.

21.8. Show that for every $m, n \geq 1$ there is an arithmetical Σ_m-formula $\Phi_n^m(F, x_1, \ldots, x_n)$, universal for the class of formulas $F \in \Sigma_m$ with n free variables. Define the Π_m-formulas $\Phi_n^m(F, x_1, \ldots, x_n)$, universal for the class Π_m (see Exercise 18.11).

21.9. Show that for any model $\mathbb{M}$ of PA the set $u_m = \{a \in M\colon\ \mathbb{M} \models \Phi_1^m[a]\}$ (see Exercise 21.8) does not belong to the class $\Pi_m(\mathbb{M})$. Hence, in any model $\mathbb{M}$, the hierarchy of definable sets is strictly increasing; if $n < m$, then $\Sigma_n(\mathbb{M}) \subsetneq \Sigma_m(\mathbb{M})$ and $\Pi_n(\mathbb{M}) \subsetneq \Pi_m(\mathbb{M})$ (see Exercise 18.11).

21.10. Let $I\Sigma_n$ denote the fragment of PA in which we assume the induction axioms only for the formulas of class Σ_n. Show that $I\Sigma_n$ is a finitely axiomatizable theory.

21.11. Let T, S be sets of sentences definable in $\mathbb{N}$ by formulas $T(x)$, $S(x)$, respectively. Show that the sentence

$$\forall x[T(x) \to S(x)] \to [\mathrm{Cons}(S) \to \mathrm{Cons}(T)]$$

is a theorem of PA.

21.12. Show that if T is a consistent theory that is arithmetically definable in $\mathbb{N}$ and contains PA, then $T \nvdash \mathrm{Cons}(T)$.

21.13. The Mostowski theorem states that in PA one can prove the consistency of any finite fragment of PA: PA $\vdash$ Cons(PA_n), for every n ($\mathrm{PA}_n(\varphi)$ is $\mathrm{PA}(\varphi) \wedge \varphi \leq \Lambda_n$). The formula $T(x)$: $\mathrm{PA}(x) \wedge \forall y \leq x\, \mathrm{Cons}(\mathrm{PA}_y)$, defines PA in $\mathbb{N}$. Show that we have PA $\vdash$ Cons(T) (Feferman [F1]). Hence, the choice of the formula defining the theory has an essential significance for the Gödel consistency theorem (cf. Exercise 21.12).

21.14. Let $T \in \mathrm{PR}$ and $T \supseteq \mathrm{PA}$. Prove by induction on F that for every sentence $F \in \Sigma_1$ we have $\mathrm{PA} \vdash \big(F \to W_T(\Lambda_F)\big)$. (The proof is not difficult but lengthy).

Hence follows the property

$$\mathrm{PA} \vdash [W_T(\Lambda_F) \to W_T(\Lambda_{W_T(\Lambda_F)})],$$

for arbitrary F.

21.15. Under the assumptions of Exercise 21.14 show, applying that exercise and Exercise 21.1, that we have $T \vdash \big(\mathrm{Cons}(T) \equiv G\big)$, where G is the diagonal sentence for the formula $\neg W_T(y)$. Hence, Theorem 21.27 follows from Theorem 20.14.

21.16. Let $\mathbb{M} \in \mathrm{Mod}(T)$. By a *satisfaction class* for $\mathbb{M}$ we mean a set $S \subseteq M$ consisting of the pairs $J^M(F, a)$ such that $F \in \mathrm{Fm}^M$ and the sequence $a \in K^M$ is defined in $\mathbb{M}$ at least on $V^M(F)$, satisfying the following conditions:

(a) $J^M(\langle r, t_1, \ldots, t_n\rangle, a) \in S$ if and only if $r^M\big(t_1^M(a), \ldots, t_n^M(a)\big)$, for $t_1, \ldots, t_n \in \mathrm{Tm} \subseteq N$, $a \in M$;

(b) $J^M(F \to G, a) \in S$ if and only if $J^M(F, a) \notin S$ or $J^M(G, a) \in S$, for $F, G \in \mathrm{Fm}^M$, $a \in M$;

(c) $J^M(\forall x\, F, a) \in S$ if and only if for every $b \in M$, $J^M\big(F, a(b/x)\big) \in S$, for $F \in \mathrm{Fm}^M$, $a \in M$.

Assume that $\mathbb{M} \models \mathrm{PA}$ and that the assumptions of the Hilbert–Bernays theorem are fulfilled for T. Let $\mathbb{U}$ be defined as in the proof of the Hilbert–Bernays theorem. Then there is a set $S \subseteq U$ definable in $\mathbb{M}$ which is a satisfaction class for $\mathbb{U}$.

21.17. Show that if S is a satisfaction class for $\mathbb{M}$ and $F \in \mathrm{Fm} \subseteq N$ and a sequence $a \in M$ is defined on $V(F)$, then $J^M(F, a) \in S$ if and only if $\mathbb{M} \models F[a]$.

21.18. Show that if $\mathbb{M} \models \mathrm{PA}$ and there is a system $\mathbb{U} \in \mathrm{Mod}(T)$ definable in $\mathbb{M}$, and a satisfaction class S for $\mathbb{U}$ definable in $\mathbb{M}$ such that $\mathbb{M} \models \forall F\big(F \in T \to J(F, \emptyset) \in S\big)$, then $\mathbb{M} \models \mathrm{Cons}(T)$ (for a definable T).

22

INDEPENDENCE OF GOODSTEIN'S THEOREM

A theorem of Goodstein, although proved with the help of a set-theoretical apparatus, expresses a natural property of integers and can be formulated by an arithmetical formula. In this chapter we prove that it is, in fact, independent of Peano arithmetic. Thus, we obtain another example of an undecidable mathematical statement. The proof is somewhat complicated and uses a sophisticated technique of rapidly growing functions.

We have proved the incompleteness of the arithmetic PA by constructing arithmetical sentences F, such that both non-PA $\vdash F$ and non-PA $\vdash \neg F$ hold. In Theorem 20.14, F is the diagonal sentence for the formula $\neg W_{\mathrm{PA}}(y)$. Thus F has the meaning: "F has no proof in PA". From Theorem 21.27 we get non PA $\vdash$ Cons(PA), and since the relation $\mathbb{N} \models$ Cons(PA) holds, it follows that Cons(PA) is also undecidable in PA. Thus, in both cases, the sentence F, although formulated in the language of arithmetic, is of a logical origin.

In this chapter we shall prove that a theorem of Goodstein (which can be expressed in PA by an arithmetical sentence G) is undecidable in PA. In the proof of Goodstein's theorem given here (in which we make use of some properties of ordinals) we show $\mathbb{N} \models G$. It is a bit more difficult to prove non-PA $\vdash G$, that is, the independence of G. To this end we shall show that the provability of G in PA is equivalent with the totality in PA of a certain recursive function h. Then we shall prove an interesting theorem of Wainer from which it follows that the function h cannot be total in PA. The independence of Goodstein's theorem from PA was proved by and Kirby and Paris, by a slightly different method (cf. [KP]). We begin with showing a proof of Goodstein's theorem.

22.1. *Definition.* Let $g, p \in N, p > 1$. Consider the following process:

1. We represent g in base p. We get

$$g = a_{k_1}p^{k_1} + a_{k_2}p^{k_2} + \cdots + a_{k_n}p^{k_n}, \quad \text{where each } a_i < p.$$

Let $f_1(x) = a_{k_1}x^{k_1} + a_{k_2}x^{k_2} + \ldots + a_{k_n}x^{k_n}$.

2. We represent $k_1, \ldots, k_n$ in base p. We get

$$g = a_{k_1} p^{b_{l_1^1} p^{l_1^1} + \cdots + b_{l_{m_1}^1} p^{l_{m_1}^1}} + \cdots + a_{k_n} p^{b_{l_1^n} p^{l_1^n} + \cdots + b_{l_{m_n}^n} p^{l_{m_n}^n}} \quad \text{where all } b_{l_j^i} < p.$$

Let $f_2(x) = a_{k_1} x^{b_{l_1^1} x^{l_1^1} + \cdots + b_{l_{m_1}^1} x^{l_{m_1}^1}} + \cdots + a_{k_n} x^{b_{l_1^n} x^{l_1^n} + \cdots + b_{l_{m_n}^n} x^{l_{m_n}^n}}$.

We continue the process by representing the exponents of the powers of x occurring in f_n. After finitely many steps, say after K steps, the exponents occurring in f_K will all be smaller than p. The function f_K will be called *the representation of g in pure basis p* and will be denoted by $f_{g,p}(x)$.

To define the function $f_{g,p}(x)$ in a more formal way, we may proceed as follows. We let $f_{0,p}(x) = 0$ and if $f_{k,p}$ is defined for numbers $k < g$, then to define $f_{g,p}$ we represent g in base p as

$$g = p^{k_1} \cdot a_{k_1} + p^{k_2} \cdot a_{k_2} + \cdots + p^{k_n} \cdot a_{k_n},$$

where each $0 \leq a_i < p,\ g > k_1 > k_2 > \ldots k_n$, and we let

$$f_{g,p}(x) = x^{f_{k_1,p}(x)} \cdot a_{k_1} + x^{f_{k_2,p}(x)} \cdot a_{k_2} + \cdots + x^{f_{k_n,p}(x)} \cdot a_{k_n}.$$

The value $f_{g,p}(x)$ can be computed, in a straightforward way, also for $x = \omega$.

Also inductively, we define the notion of a *coefficient* of the representation. If this notion is defined for $k < g$ then the coefficients of the representation of g in pure basis p are the numbers $a_{k_1}, \ldots, a_{k_n}$ and the coefficients of the representations $f_{k_1,p}(x), \ldots, f_{k_n,p}(x)$.

Note the following remark—its proof is straightforward by induction on g.

22.2. Remark.

1. *If k is the largest integer such that $p^k | g$ and $j = g - p^k$, then*

$$f_{g,p}(x) = f_{j,p}(x) + x^{f_{k,p}(x)}.$$

2. *If $x \geq p$ then $f_{g,p}(x) < f_{g+1,p}(x)$.*

Now we are ready to define the Goodstein sequence.

22.3. *Definition*. Let $g, p \in N, p > 1$. By the *Goodstein sequence* in base p starting from g we shall mean the sequence $g_0, g_1, \ldots$, where $g_0 = g$ and, if $g_n \neq 0$, then $g_{n+1} = f_{g_n, p+n}(p + n + 1) - 1$. If $g_n = 0$ the sequence ends.

22.4. Goodstein's Theorem [G1]. *Let $g, p \in N$, $p > 1$. Let g_n be the Goodstein sequence in base p starting from g. Then there is an n such that $g_n = 0$.*

Example (A Goodstein sequence). Let $p = 2$, $g = 8$. We have

$$\begin{aligned} g_0 &= 2^3 = 2^{2+1}, \\ g_1 &= 3^{3+1} - 1 = 3^4 - 1 = 2 \cdot 3^3 + 2 \cdot 3^2 + 2 \cdot 3 + 2, \\ g_2 &= 2 \cdot 4^4 + 2 \cdot 4^2 + 2 \cdot 4 + 1, \\ g_3 &= 2 \cdot 5^5 + 2 \cdot 5^2 + 2 \cdot 5, \\ g_4 &= 2 \cdot 6^6 + 2 \cdot 6^2 + 6 + 5, \\ g_5 &= 2 \cdot 7^7 + 2 \cdot 7^2 + 7 + 4, \end{aligned}$$

and so on.

Proof. To prove the Goodstein theorem assume that $g_0, g_1, \ldots$ is the Goodstein sequence in base p starting from g. We shall associate with every g_n an ordinal α_n. Let $\alpha_n = f_{g_n, p+n}(\omega)$. Then, if $g_n \neq 0$, we have

$$\alpha_n = f_{g_{n+1}+1, p+n+1}(\omega),$$

for every n. This follows immediately from the inductive definition of g_{n+1} (Definition 22.3). On the other hand,

$$\alpha_{n+1} = f_{g_{n+1}, p+n+1}(\omega).$$

Hence, $\alpha_{n+1} < \alpha_n$, by Remark 22.2, 2). By the well-foundedness of the ordinals, the sequence is finite. □

Example (A sequence g_n, α_n). Let $p = 2$, $g = 4$. We have

$$\begin{aligned} g_0 &= 2^2, & \alpha_0 &= \omega^\omega, \\ g_1 &= 3^3 - 1 = 2 \cdot 3^2 + 2 \cdot 3 + 2, & \alpha_1 &= \omega^2 \cdot 2 + \omega \cdot 2 + 2, \\ g_2 &= 2 \cdot 4^2 + 2 \cdot 4 + 1, & \alpha_2 &= \omega^2 \cdot 2 + \omega \cdot 2 + 1, \\ g_3 &= 2 \cdot 5^2 + 2 \cdot 5, & \alpha_3 &= \omega^2 \cdot 2 + \omega \cdot 2, \\ g_4 &= 2 \cdot 6^2 + 6 + 5, & \alpha_4 &= \omega^2 \cdot 2 + \omega + 5, \\ g_5 &= 2 \cdot 7^2 + 7 + 4, & \alpha_5 &= \omega^2 \cdot 2 + \omega + 4, \\ g_6 &= 2 \cdot 8^2 + 8 + 3, & \alpha_6 &= \omega^2 \cdot 2 + \omega + 3, \\ g_7 &= 2 \cdot 9^2 + 9 + 2, & \alpha_7 &= \omega^2 \cdot 2 + \omega + 2, \\ g_8 &= 2 \cdot 10^2 + 10 + 1, & \alpha_8 &= \omega^2 \cdot 2 + \omega + 1, \\ g_9 &= 2 \cdot 11^2 + 11, & \alpha_9 &= \omega^2 \cdot 2 + \omega, \\ g_{10} &= 2 \cdot 12^2 + 11, & \alpha_{10} &= \omega^2 \cdot 2 + 11, \\ g_{11} &= 2 \cdot 13^2 + 10, & \alpha_{11} &= \omega^2 \cdot 2 + 10, \\ &\vdots & &\vdots \\ g_{21} &= 2 \cdot 23^2, & \alpha_{21} &= \omega^2 \cdot 2, \\ g_{22} &= 24^2 + 23 \cdot 24 + 23, & \alpha_{22} &= \omega^2 + \omega \cdot 24 + 23, \\ &\vdots & &\vdots \\ g_{45} &= 47^2 + 23 \cdot 47, & \alpha_{45} &= \omega^2 + \omega \cdot 23, \end{aligned}$$

and so on.

It can be shown that $\alpha_n = 0$ for

$$n = 2^{2^{2^3 \cdot 3} \cdot 2^3 \cdot 3} \cdot 2^{2^3 \cdot 3} \cdot 2^3 \cdot 3 - 3.$$

Now, we shall show how to express Goodstein's theorem in the language of arithmetic.

We define the function $f(p, u, p')$, which expresses the change of the basis p to the basis p' in the representation of a number u in pure basis p. Let

$$f(p, u, p') = u, \quad \text{if } u < p,$$

and

$$f(p, u, p') = \sum_{k \leq u} c(u, p, k) \cdot (p')^{f(p, k, p')},$$

where $c(u, p, k)$ denotes the kth digit (the coefficient of p^k) in the representation of u in the basis p. All the notions occurring in the above definition can be easily replaced by their arithmetical definitions. Since the values $f(p, a, p')$ depend on $f(p, a', p',)$ for $a' < a$, f is well defined and recursive.

We have $f(p, g, p') = f_{g,p}(p')$, where $f_{g,p}$ is such as in 22.1.

Now, we may define the Goodstein sequence in base p starting from g:

$$g_0(g, p) = g,$$

and

$$g_{i+1}(g, p) = f(p + i, g_i(g, p), p + i + 1) - 1.$$

The function $g_i(g, p)$ as a function of the triple i, g, p is recursive. Let $F(i, g, p, y)$ be a Σ_1-formula representing this function in PA. Then, the Goodstein theorem can be expressed by the formula

$$\forall g, p(p > 1 \rightarrow \exists m F(m, g, p, 0)).$$

Now, we shall define a few notions concerning countable ordinals. We shall need most of them not earlier than in Theorem 22.26.

Let ε_0 be the limit of the following sequence $\{\omega_n \colon n \in N\}$ of ordinals: $\omega_0 = 1$, $\omega_{n+1} = \omega^{\omega_n}$. Each ordinal number $\lambda < \varepsilon_0$ can be presented in the Cantor normal form

$$\lambda = \omega^{\lambda_1} + \omega^{\lambda_2} + \cdots + \omega^{\lambda_m},$$

where

$$\lambda > \lambda_1 \geq \lambda_2 \geq \cdots \geq \lambda_m \geq 0.$$

For each ordinal number $\lambda < \varepsilon_0$ we may define the representation of λ in pure basis ω and its coefficients repeating Definition 22.1, where instead of g we place λ and instead of p we place ω, and the numbers k_i are ordinals, not necessarily natural numbers. We shall denote the suitable function by $f_{\lambda,\omega}(x)$.

22.5. *Definition*. Let λ be a limit ordinal number less than ε_0. Let us define the *canonical sequence* $\{\lambda_n\colon\ n \in N\}$ of numbers less than λ converging to λ. Let

$$\lambda = \omega^{\lambda_1} + \omega^{\lambda_2} + \ldots + \omega^{\lambda_m}, \quad \text{where} \quad \lambda > \lambda_1 \geq \lambda_2 \geq \cdots \geq \lambda_m \geq 0$$

be the representation of λ in Cantor's normal form. Then $\lambda = \beta + \omega^{\lambda_m}$.

Let $n \in \omega$ and

$$\{\lambda\}(n) = \begin{cases} \beta + \omega^{\lambda_m - 1} \cdot n, & \text{if } \lambda_m \text{ is a successor,} \\ \beta + \omega^{\{\lambda_m\}(n)}, & \text{if } \lambda_m \text{ is limit.} \end{cases}$$

Let $\{0\}(n) = 0$. Then $\{\lambda_n = \{\lambda\}(n)\colon\ n \in N\}$ is a sequence converging to λ.

22.6. *Definition* (cf. [C1]). Consider the following operation on ordinal numbers $\alpha < \varepsilon_0$. Let $x \in N$ and let

$$P_x(\alpha) = \begin{cases} \alpha - 1, & \text{if } \alpha \text{ is a successor,} \\ 0, & \text{if } \alpha = 0, \\ P_x(\{\alpha\}(x)), & \text{if } \alpha \text{ is a limit.} \end{cases}$$

22.7. Lemma. *Assume that* $g, p \in N,\ p > 1$. *Then*

$$P_p(f_{g+1,p}(\omega)) = f_{g,p}(\omega).$$

Proof. We first prove by induction on λ that

$$P_x(\omega^\lambda) = \omega^{P_x(\lambda)}(x-1) + P_x(\omega^{P_x(\lambda)}).$$

Then, by induction on α we show that if $\alpha = \beta + \omega^\lambda$ then

$$P_x(\alpha) = \beta + P_x(\omega^\lambda) = \beta + \omega^{P_x(\lambda)}(x-1) + P_x(\omega^{P_x(\lambda)}).$$

Now the proof of the lemma is by induction on g. Assume the result for $g' < g$. Let k be the largest integer such that $p^k|(g+1)$ and let $j = g + 1 - p^k$. If $k = 0$, then $f_{g+1,p}(\omega) = f_{g,p}(\omega) + 1$ and $P_p(f_{g+1,p}(\omega)) = P_p(f_{g,p}(\omega) + 1) = f_{g,p}(\omega)$. If $k > 0$, then $g = j + p^{k-1}(p-1) + (p^{k-1} - 1)$. Then, in view of Remark 22.2, item 1,

$$\begin{aligned} P_p(f_{g+1,p}(\omega)) &= P_p(f_{j,p}(\omega) + \omega^{f_{k,p}(\omega)}) \\ &= f_{j,p}(\omega) + \omega^{P_p(f_{k,p}(\omega))}(p-1) + P_p(\omega^{P_p(f_{k,p}(\omega))}), \end{aligned}$$

and

$$f_{g,p}(\omega) = f_{j,p}(\omega) + \omega^{f_{k-1,p}(\omega)}(p-1) + f_{p^{k-1}-1,p}(\omega).$$

Hence, in view of the inductive assumption, the lemma follows. □

Now, we have

22.8 $$\alpha_n = f_{g_{n+1}+1,p+n+1}(\omega),$$

for every n—see the proof of the Goodstein theorem.

By Lemma 22.7 we infer

$$P_{p+n+1}(\alpha_n) = P_{p+n+1}(f_{g_{n+1}+1,p+n+1}(\omega)) = f_{g_{n+1},p+n+1}(\omega),$$

and hence we have the following corollary.

22.9. Corollary

$$\alpha_{n+1} = P_{p+n+1}(\alpha_n).$$

22.10. Remark. *By Definition of the α_ns we have $\alpha_n < \varepsilon_0$.*

22.11. *Definition*. We shall define a sequence of functions G_α: $N \longrightarrow N$ for $\alpha < \varepsilon_0$. Let

$$G_0(x) = 0,$$
$$G_{\alpha+1}(x) = G_\alpha(x) + 1,$$
$$G_\lambda(x) = G_{\{\lambda\}(x)}(x) \quad \text{for a limit } \lambda.$$

The following remark gives the intuition lying behind this definition. Its second part will be needed just at the very end of the chapter.

22.12. Remark. *For every α we may obtain $G_\alpha(x)$ as follows. Consider the representation of α in pure basis ω and substitute x for ω in this representation, that is, $G_\alpha(x) = f_{\alpha,\omega}(x)$. On the other hand, if $g = G_\alpha(x)$ is given, then $\alpha = f_{g,x}(\omega)$, provided that x is large enough—it has to be larger than the coefficients in the representation of α in pure basis ω.*

Proof. The proof of the first part is by induction on α. As for the second part this is intuitively clear; to obtain $g = f_{\alpha,\omega}(x)$ we substitute x for ω in the representation of α in pure basis ω. To obtain $f_{g,x}(\omega)$ we substitute again ω for x in the representation of g in pure basis x. Both representations have the same coefficients since the coefficients are less than x. Thus, at the end we get again α. A formal proof can be again by induction on α. □

Note also the following

22.13. Remark. *If $g_0, g_1, \ldots$ is the Goodstein sequence in base p starting from g and $\alpha_n = f_{g_n, p+n}(\omega)$, then for every n we have $g_n = G_{\alpha_n}(p+n)$.*

The proof follows immediately from Remark 22.12.

To show that Goodstein's theorem is independent of the arithmetic PA we need a characterization of the class of recursive functions which are total in PA (whose totality is provable in PA—see Chapter 19). In this class there are, for instance, the primitive recursive functions, by Theorem 19.3. We shall show that such functions do not grow too fast; they do not grow faster than the functions from the so-called Hardy's sequence. Then we shall show that the recursive function determined by the Goodstein sequence

$$h(g, p) = \min\{m\colon \; g_m(g, p) = 0\}$$

grows faster, whence we shall derive the independence of Goodstein's theorem of PA.

Now, we shall define the Hardy sequence. Let for $\alpha < \varepsilon_0$:

$$H_0(x) = x,$$

$$H_{\alpha+1}(x) = H_\alpha(x+1),$$

$$H_\lambda(x) = H_{\{\lambda\}(x)}(x) \quad \text{for a limit } \lambda.$$

This sequence is defined in a similar way as the sequence G_α; however it grows much faster in the ordering 'less almost everywhere' of functions (we say that a function f is "less almost everywhere" than a function g if $f(n) < g(n)$ for all, except possibly finitely many, n's).

The sequence $\{H_\alpha\}_{\alpha<\varepsilon_0}$ is called *Hardy's sequence*. It can be shown that for $\alpha < \beta$ we have $H_\alpha(x) < H_\beta(x)$ for almost all x. For example

$$\begin{aligned} H_1(x) &= x+1, & G_1(x) &= 1,\\ H_n(x) &= x+n, & G_n(x) &= n,\\ H_\omega(x) &= 2x, & G_\omega(x) &= x,\\ H_{\omega+n}(x) &= 2(x+n), & G_{\omega+n}(x) &= x+n,\\ H_{\omega\cdot 2}(x) &= 4x, & G_{\omega\cdot 2}(x) &= 2x,\\ H_{\omega\cdot n}(x) &= 2^n x, & G_{\omega\cdot n}(x) &= xn,\\ H_{\omega^2}(x) &= 2^x \cdot x, & G_{\omega^2}(x) &= x^2. \end{aligned}$$

We shall need the following property of Hardy's sequence:

22.14. *For every $n \in N$ and any α, x we have*

$$H_{\omega^\alpha \cdot n}(x) = H^{(n)}_{\omega^\alpha}(x),$$

where $H^{(n)}_{\omega^\alpha}$ denotes the nth iteration of the function H_{ω^α}.

Proof. We apply induction on n. We have $H_0(x) = x = H^{(0)}_{\omega^\alpha}(x)$. Assume 22.14 for n. It is enough to prove

$$H_{\omega^\alpha \cdot (n+1)}(x) = H_{\omega^\alpha \cdot n}\big(H_{\omega^\alpha(x)}\big).$$

To this end we shall show by induction on β that

$$H_{\omega^\alpha \cdot n+\beta}(x) = H_{\omega^\alpha \cdot n}\big(H_\beta(x)\big) \quad \text{for} \quad \beta \leq \omega^\alpha.$$

For $\beta = 0$ the above equality holds. Assume that it holds for $\beta = \gamma$ and we shall prove it for $\beta = \gamma + 1$. We have

$$H_{\omega^\alpha \cdot n+\gamma+1}(x) = H_{\omega^\alpha \cdot n+\gamma}(x+1) = H_{\omega^\alpha \cdot n}\big(H_\gamma(x+1)\big) = H_{\omega^\alpha \cdot n}\big(H_{\gamma+1}(x)\big).$$

Now consider a limit β and assume the above equality for smaller numbers. We have

$$H_{\omega^\alpha \cdot n+\beta}(x) = H_{\omega^\alpha \cdot n+\{\beta\}(x)}(x) = H_{\omega^\alpha \cdot n}\big(H_{\{\beta\}(x)}(x)\big) = H_{\omega^\alpha \cdot n}\big(H_\beta(x)\big).$$

Hence we obtain

$$H_{\omega^\alpha \cdot (n+1)}(x) = H_{\omega^\alpha \cdot n+\omega^\alpha}(x) = H_{\omega^\alpha \cdot n}\big(H_{\omega^\alpha}(x)\big),$$

which completes the proof. □

The most important property of the Hardy sequence is expressed by the following theorem:

22.15. Wainer's Theorem (cf. [W_1, W_2]). *Let f: $N \longrightarrow N$ be a recursive function. Let $F(x, y)$ be a Σ_1 formula representing f in* PA. *Assume that* PA $\vdash \forall x \exists ! y\, F(x, y)$. *Then, there is an $\alpha < \varepsilon_0$ and an $n_0 \in N$ such that for $x \in N$, $x \geq n_0$, $f(x) \leq H_\alpha(x)$.*

To prove this theorem we need some new notions.

22.16. *Definition.* Let f: $N \longrightarrow N$ be a partial function (i.e., dom $(f) \subseteq N$) with a finite domain and let $S \subseteq N$. We say that S is an *approximation* of f if

$$\forall x \in S \setminus \{\max S\} \forall y < x - 2\big(y \in \operatorname{dom} f \rightarrow f(y) < x^+ \vee f(y) \geq \max S\big),$$

where x^+ denotes the least element of S greater than x, $\max S$ denotes the greatest element of S, and $\operatorname{dom} f$ denotes the domain of f.

Remark. *If $S \subseteq N$ is infinite, $f\colon N \longrightarrow N$ is such as in Definition 22.16, then there is an infinite subset $S_1 \subseteq S$ such that S_1 is an approximation of f.*

Proof. Let $a_0, a_1, \ldots$ be defined as follows: $a_0 =$ the least element of S, and

$$a_{n+1} = \min\{a \in S\colon \forall x < a_n - 2(f(x) < a \wedge a_n < a)\}.$$

Let $S_1 = \{a_0, a_1, \ldots\}$.

22.17. *Definition*. Assume that S is a finite set. Let $A_n^k(S)$ for $k, n \in N$ denote the property

$$\forall f_1 \exists S_1 \subseteq S \forall f_2 \exists S_2 \subseteq S_1 \cdots \forall f_n \exists S_n \subseteq S_{n-1}$$

$$\Big(\bigwedge_{i=1}^{n} S_i \text{ is an approximation of } f_i \wedge \operatorname{card} S_n > k\Big),$$

where f_i run over functions with finite domains. This property expresses the iteration of a property similar to the one stated in the remark following Definition 22.16.

22.18. Remark. *We have*

$$A_0^k(S) \quad \text{if and only if} \quad \operatorname{card} S > k$$

and

$$A_{n+1}^k(S) \equiv \forall f_1 \exists S_1 (S_1 \subseteq S \wedge A_n^k(S_1) \wedge S_1 \text{ is an approximation of } f_1).$$

Using the coding of sets and functions we may express the formula $A_n^k(S)$ in the language of arithmetic—see Exercises 20.8, 20.10, 20.11.

The proof of Wainer's theorem splits into proofs of the following lemmas.

Lemma 1. *Let ω_n^k denote $\{\omega_{n+1}\}(k)$. Let $a, b \in N$, $a > 0$. Let $I(a, b)$ denote the code of the set $\{x\colon a \leq x \leq b\}$. Assume that $H_{\omega_n^k}(a) < b$. Then $\mathbb{N} \models A_n^k[I^N(a, b)]$.*

Lemma 2. *Let f, F be as in the assumptions of Wainer's theorem. Assume that for every pair $n, k \in N$ we have*

$$\mathbb{N} \models \exists x \exists y [A_n^k(I(x, y)) \wedge F(x, y)].$$

Then, there is a model $\mathbb{M}$ *of the arithmetic* PA *and* $a, b \in M$ *such that*

$$\mathbb{M} \equiv \mathbb{N}, \qquad \mathbb{M} \models F[a, b]$$

and, for any $n, k \in N$, $\mathbb{M} \models A_n^k[I^M(a, b)]$.

Lemma 3. *If* $\mathbb{M} \models \mathrm{PA}$, $\mathbb{M} \equiv \mathbb{N}$, $a, b \in M$, $\mathbb{M}$ *is countable, and* $\mathbb{M} \models A_n^k[I^M(a, b)]$ *for all* $n, k \in N$, *then there is an initial segment* $\mathbb{I} \subseteq \mathbb{M}$ *such that* $\mathbb{I} \models \mathrm{PA}$, $a \in I$, $b \in M \setminus I$.

Now we shall show how the theorem follows from the lemmas.

Let f, F be given. Suppose that the conclusion of the Wainer theorem is not true. Then for any pair $n, k \in N$ there is an $x \in N$ such that $x >^N 0$ and $f(x) >^N H_{\omega_n^k}(x)$. By Lemma 1, we obtain

$$\mathbb{N} \models \exists x \exists y \big[A_n^k\big(I(x, y)\big) \wedge F(x, y)\big].$$

By Lemma 2, it follows that there is a model $\mathbb{M}$ of the arithmetic PA and $a, b \in M$ such that $\mathbb{M} \equiv \mathbb{N}$, $\mathbb{M} \models F[a, b]$ and for any $n, k \in N$ $\mathbb{M} \models A_n^k\big[I^M(a, b)\big]$. We have $\mathbb{M} \models \forall x \exists! y\; F(x, y)$, since $\mathbb{M} \equiv \mathbb{N}$.

By Lemma 3 it follows that there is an initial segment $\mathbb{I} \subseteq \mathbb{M}$ such that $\mathbb{I} \models \mathrm{PA}$, $a \in I$, and $b \in M \setminus I$. Since, as we have assumed, $\mathrm{PA} \vdash \forall x \exists y\; F(x, y)$ holds, we have $\mathbb{I} \models \exists y F[a]$.

Let $b' \in I$ be such that $\mathbb{I} \models F[a, b']$. By the fact that $\mathbb{I}$ is an initial segment of $\mathbb{M}$ and that F is Σ_1 it follows (see Exercise 16.16) that $\mathbb{M} \models F[a, b']$. But $\mathbb{M} \models F[a, b]$ and $b' \neq b$ since $b' \in I$ and $b \notin I$. This is a contradiction with $\mathbb{M} \models \exists! y\; F[a]$, and so the Wainer theorem has been proved.

Proof of Lemma 1. We need a certain generalization of the definition of Hardy's functions.

22.19. *Definition.* Let $S \subseteq N$. We define for $x \in S$:

$$H_0^S(x) = x, \qquad H_1^S(x) = x^+,$$

where, as in Definitions 22.16, x^+ is the least element of S greater than x. If S is finite, then H_1^S is a partial function from S into S defined on $S \setminus \{\max S\}$. Let us define

$$H_{\alpha+1}^S(x) = H_\alpha^S\big(H_1^S(x)\big) \quad \text{for} \quad \alpha < \varepsilon_0,$$

$$H_\lambda^S(x) = H_{\{\lambda\}(x)}^S(x) \quad \text{for a limit } \lambda < \varepsilon_0.$$

Each of the above functions from S into S can be partial (i.e., defined at a subset of S) and we should read the definition as follows: the left-hand side is defined provided that the right-hand side is.

We shall say that a set S is α-large if $S \neq \emptyset$, and $H_\alpha^S(\min S) = \max S$, where $\min S$, $\max S$ are the least and the largest element of S, respectively.

Example. If $k \in N$, then S is k-large if and only if card $S = k+1$. S is ω-large if and only if card $S = \min S + 1$.

22.20. Remark. *Let $S_1 \subseteq S_2$ be such that S_1 is an interval in S_2, that is, there are a, b such that $S_1 = S_2 \cap [a, b]$. Then*

$$H_\alpha^{S_1} = H_\alpha^{S_2} \restriction S_1 \quad \text{for} \quad \alpha < \varepsilon_0.$$

Here by the restriction we mean $H_\alpha^{S_2} \cap S_1^2$. Hence ($[a, b]$ is α-large) $\equiv$ $\big(H_\alpha(a) = b\big)$. This follows from the fact that

$$H_\alpha^{[a,b]} = H_\alpha^N \restriction [a, b] = H_\alpha \restriction [a, b].$$

22.21. Remark. *If S is ω^α-large,* $\min S > 0$ *and f is a function with a finite domain, then there exists $S' \subseteq S$ such that S' is α-large and S' is an approximation of f and* $\min S' = \min S$.

This remark is crucial for the proof of Lemma 1. Before we prove Remark 22.21, let us consider the following auxiliary remark.

22.22. Remark. *Let S be $\omega^{\alpha+1}$-large and f be a function with a finite domain. Assume that* $\min S = a_0 > 0$. *Then there is an $a \in S$ such that $a > a_0$ and an $S' \subseteq S$ such that* $\min S' = a$, *S' is ω^α-large, and*

$$\forall x < a_0 - 2\big(x \in \operatorname{dom} f \rightarrow f(x) < a \ \text{ or } f(x) \geq \max S'\big).$$

That is, $\{a_0, a\}$ is an approximation of f with a certain strong property, namely for $x < a_0 - 2$, $x \in \operatorname{dom} F$ if $f(x) \geq a$, then $f(x)$ is much larger than a.

Proof of Remark 22.22. Let S be $\omega^{\alpha+1}$-large, $S \subseteq N$, and $\min S > 0$. Let $a_0 = \min S$, $b_0 = \max S$. We have $H_{\omega^{\alpha+1}}^S(a_0) = b_0$. Hence, $H_{\omega^\alpha \cdot a_0}^S(a_0) = b_0$. The left-hand side is, by the relativized to S version of 22.14, equal to $(H_{\omega^\alpha}^S)^{(a_0)}(a_0)$, where $(H_{\omega^\alpha}^S)^{(a_0)}$ is the a_0-fold iteration of the function $H_{\omega^\alpha}^S$.

Let $a_k = (H_{\omega^\alpha}^S)^{(k)}(a_0)$ for $k = 0, \ldots, a_0$. We have $a_0 < a_1 < \cdots < a_{a_0} = b_0$ since $a_0 > 0$, $\omega^\alpha > 0$.

Notice that the image of the interval $[0, a_0 - 3]$ under the function f has at most $a_0 - 2$ elements. There are $a_0 - 1$ intervals $[a_j, a_{j+1})$ for $j \geq 1$ and thus, there is one among them that does not contain any value $f(x)$ for $x < a_0 - 2$. Let it be $[a_{j_0}, a_{j_0+1})$. Let

$$a = a_{j_0}, \qquad S' = S \cap [a_{j_0}, a_{j_0+1}).$$

We have

$$H^{S'}_{\omega^\alpha}(a_{j_0}) = H^{S}_{\omega^\alpha}(a_{j_0}) = a_{j_0+1}$$

by Remark 22.20. Hence S' is ω^α-large. If $x < a_0 - 2$, $x \in \operatorname{dom} f$, then $f(x) \notin [a_{j_0}, a_{j_0+1})$ and thus $f(x) < a_{j_0}$ or $f(x) \geq a_{j_0+1} = \max S'$. Hence a, S' are as required in the Remark 22.22.

Proof of Remark 22.21. We apply induction on α. Let $\alpha = 0$ and let S be $\omega^\alpha = 1$-large. Let f be given. Then $S = \{a_0, a_1\}$ and $S' = \{a_0\}$ is 0-large and is an approximation of f.

The nonlimit step. Assume that the remark holds for α. Assume that S is $\omega^{\alpha+1}$-large and let f be given. Let $a_1 \in S$, $S' \subseteq S$ be such as in the conclusion of Remark 22.22, that is S' is ω^α-large, $\min S' = a_1$, and

$$\forall x < \min S' - 2\big(x \in \operatorname{dom} f \rightarrow f(x) < a_1 \text{ or } f(x) \geq \max S'\big).$$

By the inductive assumption there is an $S'' \subseteq S'$ which is α-large and such that $\min S'' = \min S' = a_1$ and S'' is an approximation of f. We shall show that $\{a_0\} \cup S''$ is an approximation of f which is $\alpha + 1$-large, where $a_0 = \min S$.

We have $\min(\{a_0\} \cup S'') = a_0 = \min S$. Assume that $S'' = \{a_1, a_2, \ldots, a_n\}$. Then we have

$$H^{\{a_0\}\cup S''}_{\alpha+1}(a_0) = H^{\{a_0\}\cup S''}_{\alpha}\big(H^{\{a_0\}\cup S''}_{1}(a_0)\big) = H^{\{a_0\}\cup S''}_{\alpha}(a_1) = H^{S''}_{\alpha}(a_1) = a_n.$$

Thus, $\{a_0\} \cup S''$ is $\alpha + 1$-large. Let now

$$a \in \{a_0\} \cup S'', \quad a < \max(\{a_0\} \cup S'') \;\; x < a - 2.$$

If $a = a_j$ for a $j \geq 1$, then $f(x) < a_{j+1}$ or $f(x) \geq \max S''$. If $a = a_0$, then $f(x) < a_1$ or $f(x) \geq \max S''$. Thus $\{a_0\} \cup S''$ is an approximation of f with the required properties.

The limit case. Let α be limit and assume that for $\beta < \alpha$ the remark holds. Let S be ω^α-large, $a_0 = \min S$ and let f be given. Then S is $\omega^{\{\alpha\}(a_0)}$-large, since

$$\max S = H^{S}_{\omega^\alpha}(a_0) = H_{\omega^{\{\alpha\}S(a_0)}}(a_0).$$

By the inductive assumption, there is an $S' \subseteq S$ which is $\{\alpha\}(a_0)$-large, satisfies $\min S' = a_0$ and is an approximation of f. Then S' is α-large since

$$H^{S}_{\alpha}(a_0) = H^{S'}_{\{\alpha\}(a_0)}(a_0) = \max S'.$$

Hence S' is as required in the remark, which completes the proof of Remark 22.21.

We continue the proof of Lemma 1. Assume $a > 0$, $H_{\omega_n^k}(a) \leq b$. Let $S = [a, H_{\omega_n^k}(a)] \subseteq [a, b]$. Then S is ω_n^k-large.

Note the following property of the numbers ω_n^k which explains the point of their definition:

$$\omega_0^k = k, \ \omega_n^k = \omega^{\omega_{n-1}^k}.$$

Let f be given. By Remark 22.21 we find an $S_1 \subseteq S$ which is ω_{n-1}^k-large and is an approximation of f. Iterating this procedure n times we obtain

$$\forall f_1 \exists S_1 \subseteq [a, b] \forall f_2 \exists S_2 \subseteq S_1 \ldots \forall f_n \exists S_n \subseteq S_{n-1}$$

$$\Big(\big(\bigwedge_{i=1}^{n} S_i \text{ is an approximation of } f_i\big) \wedge \big(\bigwedge_{i=1}^{n} S_i \text{ is } \omega_{n-1}^k\text{-large}\big) \wedge (S_n \text{ is } k\text{-large})\Big).$$

Hence we infer $A_n^k([a, b])$. So we have finished the proof of Lemma 1.

Proof of Lemma 2. Let us fix $n, k \in N$. We have

$$\mathbb{N} \models \exists x, y\big(F(x, y) \wedge \bigwedge_{n', k' \leq n, k} A_{n'}^{k'}(I(x, y))\big).$$

Let T be the following theory in the language $L\{\underline{a}, \underline{b}\}$:

$$\text{Th }(\mathbb{N}) \cup \{F(\underline{a}, \underline{b})\} \cup \{A_n^k\big(I(\underline{a}, \underline{b})\big): \ n, k \in N\}.$$

Since every finite fragment of T has a model (namely $\langle \mathbb{N}; a, b\rangle$ for suitably large a, b), T has a model $\langle \mathbb{M}; a, b\rangle$, and the proof of the lemma has been completed.

Proof of Lemma 3. We shall use in the proof the following notion.

22.23. *Definition*. Let $\mathbb{M} \models$ PA and let $I \subseteq M$ be an initial segment in M. We say that I is *strong* in $\mathbb{M}$ if for every function $f\colon \ M \longrightarrow M$ that is definable in $\mathbb{M}$ there is an $e \in M \setminus I$ such that for $x \in I$ we have $f(x) \in I$ or $f(x) \geq^M e$ (by definability we mean, here and later, the parametric definability).

The proof of Lemma 3 consists first of showing that if $a, b \in M$ satisfy the assumptions, then there is an initial segment $I \subseteq M$ which is strong in $\mathbb{M}$ and satisfies $a \in I$ and $b \in M \setminus I$; next, we show that a strong initial segment is a universe of a model $\mathbb{I} \subseteq \mathbb{M}$ of PA.

Let $\mathbb{M}$, $a, b \in M$ satisfy the assumptions of the lemma. The formula $A_n^k(S)$ can be interpreted in the model $\mathbb{M}$, where f_i run over functions in the sense of $\mathbb{M}$ (see Exercise 20.10), S, S_i runover sets function in the sense of $\mathbb{M}$ (see Exercise 20.8), and card S is the number of elements of S in the sense of $\mathbb{M}$ (see Exercise 20.14) [More precisely, the expression card $S_n > k$ should be replaced by the formula $\exists y\big(F(S_n, y) \wedge y > k\big)$, where F represents the function "the number of elements of S_n" described in Exercise 20.11].

If in the property $A_n^k(S)$ we restrict ourselves to sets $S, S_i \subseteq^M I^M(a,b)$ and to (partial) functions mapping $I^M(0,b)$ into $I^M(0,b)$ in $\mathbb{M}$, then S, S_i, f_i are elements of M not greater in $\mathbb{M}$ than the element 2^{4b^2+1} (see Exercise 20.14). Then the formula $A_n^k(S)$ is equivalent in $\mathbb{M}$ with the formula obtained from it by restricting all the quantifiers to 2^{4b^2+1}. Let $A_n^k(x,S)$ denote the formula

$$\exists u(u = 2^{4x^2+1} \wedge \forall f_1 \leq u \exists S_1 \leq u \forall f_2 \leq u \exists S_2 \leq u$$

$$\cdots \forall f_n \leq u \exists S_n \leq u(\bigwedge_{i=1}^{n} Fn(f_i) \wedge \bigwedge_{i=1}^{n-1} S_i \subseteq S_{i+1}$$

$$\wedge S_1 \subseteq S \wedge (\bigwedge_{i=1}^{n} S_i \text{ is an approximation of } f_i) \wedge \text{card } S_n > k)),$$

where "$u = 2^{4x^2+1}$" should be replaced by a Σ_1-formula representing the function 2^{4x^2+1} in PA and "$S_i \subseteq S_{i+1}$," "$S_1 \subseteq S$," and "S is an approximation of f_i" should be replaced by suitable formulas following natural definitions of the above properties expressed with the help of the formula representing the relation $e(k,a)$ from Exercise 20.8 (a), in particular following Definition 22.16 of the property "S is an approximation of f." Let $G(x,y,z)$ be a Σ_1-formula such that

$$\mathbb{N} \models G[n,k,m] \quad \text{if and only if} \quad m \text{ is the formula } A_n^k(x,S).$$

Let $\phi_2(x,y,z)$ be a universal Σ_1-formula for Σ_1-formulas of two variables (see Corollary 21.34).

Let $A(x,y,z,S)$ denote the formula

$$\exists w[G(x,y,w) \wedge \phi_2(w,z,S)].$$

Then we have for $n,k \in N$, $S \in M$:

$$(\mathbb{M} \models A[n,k,b,S]) \equiv (\mathbb{M} \models A_n^k[b,S]).$$

Thus, we have

$$\mathbb{M} \models A[n,k,b,I^M(a,b)] \quad \text{for} \quad n,k \in N.$$

In particular

$$\mathbb{M} \models A[n,n,b,I^M(a,b)] \quad \text{for} \quad n \in N.$$

By the overspill (see Exercise 13.13), we obtain

$$\mathbb{M} \models A[c,c,b,I^M(a,b)] \quad \text{for some} \quad c \in M \setminus N.$$

Let us fix an element $c \in M \setminus N$ with the above property.

Let $\hat{f}_1, \hat{f}_2, \ldots$ be an enumeration of all the functions mapping $\{x \in M\colon\ x \leq^M b\}$ into $\{x \in M\colon\ x \leq^M b\}$ definable in $\mathbb{M}$. To each such function there corresponds an element $f_i \in M$, its code (see Exercises 20.9 and 20.10) such that

$$\hat{f}_i = \{\langle k, l\rangle\colon\ e^M(J(k,l), f_i)\}.$$

We have, by Remark 22.18,

$$\mathbb{N} \models \forall x, y, z, S\{A(x+1, y, z, S) \rightarrow \forall f_1 (f_1 \text{ is a function}$$
$$\text{from I(0,b) into I(0,b)}) \rightarrow \exists S_1 \subseteq S[S_1 \text{ is an approximation of } f_1$$
$$\wedge \text{ card } S_1 > y \wedge A(x, y, z, S_1)]\},$$

where the verbal formulations are understood as the suitable formulas. Since $\mathbb{M} \equiv \mathbb{N}$, the same sentence is true in $\mathbb{M}$. Hence, it follows that if $x, y, S \in M$, $\mathbb{M} \models A[x+1, y, b, S]$ and $\hat{f}$ is a function from $\{x \in M\colon\ x \leq^M b\}$ into $\{x \in M\colon\ x \leq^M b\}$ which is definable in $\mathbb{M}$, then there is an $S_1 \in M$ such that

$S_1^* \subseteq S^*$ (where $a^* = \{x \in M\colon e^M(x, a)\}$),
S_1^* is an approximation of $\hat{F}$ in the sense of Definition 22.16 and $\mathbb{M} \models \text{card } S_1 > y$.

This follows from the form of the formula "S is an approximation of f."

Now we are ready to define the required strong initial segment.

We take $\hat{f}_1$. Let $S_1 \in M$ be such that $S_1^* \subseteq (I^M(a,b))^*$ and S_1 is an approximation of $\hat{f}_1$ satisfying $A[c-1, c, b, S_1]$ and card $S_1 > c$ in $\mathbb{M}$.

We take $\hat{f}_2$. Let $S_2 \in M$ be such that $S_2^* \subseteq S_1^*$ and S_2 is an approximation of $\hat{f}_2$ satisfying $A[c-2, c, b, S_2]$ and card $S_2 > c$ in $\mathbb{M}$.

Generally, given S_i, we take $S_{i+1} \in M$ such that $S_{i+1}^* \subseteq S_i^*$ and S_{i+1} is an approximation of $\hat{f}_{i+1}$ satisfying $A[c-(i+1), c, b, S_{i+1}]$ and card $S_{i+1} > c$ in $\mathbb{M}$.

In particular, we have card ${}^M S_i >^M n$ for every $i \in M$ and any $n \in N$. Therefore, every S_i^* is infinite (see Exercise 20.11).

Let d_i be the ith element of S_i^* in the ordering "$\leq^M$"; since S_i^* is parametrically definable (with the parameter S_i) it has a least, a second, and an ith element for $i \in N$. Let $I = \{x \in M\colon\ \exists\, i \in N\ x \leq^M d_i\}$. The set I so defined is an initial segment of M. We have $d_i + 1 \leq d_{i+1}$, since $S_{i+1}^* \subseteq S_i^*$. Hence I is closed under successor. Moreover, $a \in I$ and $b \notin I$ (since $S_i^* \subseteq \{x \in M\colon\ a \leq^M x \leq^M b\}$).

We shall show that I is strong in $\mathbb{M}$. Let $f\colon\ M \longrightarrow M$ be a definable function in $\mathbb{M}$. Then there is an $i \in N$ such that

$$f \cap (I^M(0,b))^* \times (I^M(0,b))^* = \hat{f}_i.$$

Let us take $e_i = \max S_i^*$. We shall show that for $x \in I$,

$$\hat{f}_i(x) \in I \quad \text{or} \quad \hat{f}_i(x) \geq^M e_i.$$

Let $x \in I$. Let $j \geq i$ be such that $x < d_j - 2$. Then $d_j \in S_i^*$ since $S_j^* \subseteq S_i^*$. Hence, S_i^* is an approximation of $\hat{f}_i, \hat{f}_i(x) < d_j^+$, where d_j^+ denotes the next after d_j element of S_i^* or $\hat{f}_i^*(x) \geq \max S_i^* = e_i$. Thus, it suffices to show that $d_j^+ \in I$. We know that d_j is the jth element of S_j^*. Since $S_{j+1}^* \subseteq S_j^*$ the jth element of S_{j+1}^* is $\geq^M d_j$. Hence, the $(j+1)$th element S_{j+1}^* is $\geq^M d_j^+$ (since $S_{j+1}^* \subseteq S_i^*$). Thus, $d_j^+ \leq^M d_{j+1}$, whence $d_j^+ \in I$.

Now, we shall show that for every $x \in I$,

$$f(x) \in I \quad \text{or} \quad f(x) \geq^M e_i.$$

Let us take $x \in I$. If $f(x) \leq^M b$, then $f(x) = \hat{f}_i(x)$ and we have $f(x) \in I$ or $f(x) \geq^M e_i$. If $f(x) >^M b$, then $f(x) >^M e_i$, since $e_i \leq^M b$ (since $S_i^* \subseteq (I^M(a,b))^*$). Thus, we have shown that I is strong in $\mathbb{M}$.

Now, we shall show that I is a universe of a model $\mathbb{I} \subseteq \mathbb{M}$ of PA. We have

$$e_1 \geq^M e_2 \geq^M e_3 \geq^M \cdots \geq^M I.$$

We shall show that for any formula $F(x_1, \ldots, x_n)$ there exists an $m \in N$, $e_{i_1}, \ldots, e_{i_m}$, and a bounded formula $F^*(x_1, \ldots, x_n, y_1, \ldots, y_m)$ such that for any $x_1, \ldots, x_n \in I$,

22.24 $$(\mathbb{I} \models F[x_1, \ldots, x_n]) \equiv (\mathbb{M} \models F^*[x_1, \ldots, x_n, e_{i_1}, \ldots, e_{i_m}]),$$

where $\mathbb{I} = \langle I;\ +^M \restriction I^2, \cdot^M \restriction I^2, 0^M, 1^M \rangle$.

First we shall show that the set I is closed under $+^M$ and $\cdot^M$. Suppose that for some $u, v \in I$, $u + v \notin I$. Let us consider the function $f(x) = u + x$. Then, f is definable in $\mathbb{M}$ and hence there is an $e \in M \setminus I$ such that, for $x \in I, f(x) \in I$ or $f(x) \geq^M e$. Hence, $f(v) \geq^M e$.

Let $x = e - 1 - u$. Then $x \in I$ since $x \leq^M v$. But $f(x) = e - 1$, whence $f(x) \notin I$ since $e \notin I$ and I is closed under successor. This is a contradiction with the choice of I, since $f(x) < e$.

Similarly we prove that I is closed under multiplication. Hence, I determines a subsystem $\mathbb{I} \subseteq \mathbb{M}$ with the universe I.

We infer that I is closed under the pairing function J^M. Since with the help of the pairing function J we may define an enumeration of sequences of length n for $n \in N$ $(\langle a_1, \ldots, a_n \rangle = J(a_1, j(a_2, J(\ldots J(a_{n-2}, J(a_{n-1}, a_n))\ldots))))$, it follows that I has the property from definition 22.23 for many argument functions.

Now, we shall show 22.24 by induction on F. If F is an open formula, then we define $F^* = F$, $m = 0$. Let

$$(\neg F)^* = \neg F^*, \quad (F \to G)^* = (F^* \to G^*).$$

Let us consider an F of the form $\exists y(x_1, \ldots, x_n, y)$ and let $m, e_{i_1}, \ldots, e_{i_m}$ correspond to the formula G. Let us consider the function

$$f(x_1, \ldots, x_n) = \min\{y\colon\ M \models G^*[x_1, \ldots, x_n, y, e_{i_1}, \ldots, e_{i_m}]\}.$$

The function f is definable in $\mathbb{M}$. Hence $f \cap \left(I^M(0, b)\right)^{*2} = \hat{f}_i$ for some $i \in N$ (after encoding the n-tuples in the domain of f as single numbers). Let $e_{i_{m+1}} = e_i$ and let $F^*(x_1, \ldots, x_n, y_1, \ldots, y_{m+1})$ be the formula

$$\exists y \leq y_{m+1}\ G^*(x_1, \ldots, x_n, y, y_1, \ldots, y_m).$$

It is easy to check that Definition 22.16 holds.

We shall show that $\mathbb{I} \models \mathrm{PA}$. Namely, we shall show that $\mathbb{I}$ satisfies the minimum principle (see Exercise 13.4). Let $F(a, x_1, \ldots, x_n)$ be an arbitrary formula and let $b_1, \ldots, b_n \in I$. Assume that $\mathbb{I} \models \exists x F[b_1, \ldots, b_n]$. Let $a \in I$ be such that $\mathbb{I} \models F[a, b_1, \ldots, b_n]$. By Definition 22.16 we have

$$\mathbb{M} \models F^*[a, b_1, \ldots, b_n, e_{i_1}, \ldots, e_{i_m}] \quad \text{for some} \quad e_{i_1}, \ldots, e_{i_m}.$$

We apply the minimum scheme to the formula F^* in $\mathbb{M}$ and we find a least element a_0 of M satisfying

$$\mathbb{M} \models F^*[a_0, b_1, \ldots, b_n, e_{i_1}, \ldots, e_{i_m}].$$

Then, $a_0 \in I$, since $a_0 \leq^M a$. Thus, by Definition 22.16, $\mathbb{I} \models F[a_0, b_1, \ldots, b_n]$.

Similarly we show that a_0 is the least element of I satisfying $F[b_1, \ldots, b_n]$ in $\mathbb{I}$. So we have finished the proof of lemma 3 and thus also of Wainer's theorem. □

22.25. Remark. Directly from the proof of Wainer's theorem we infer that the sentence $\forall x, y \exists z\ A\big(x, x, y, I(y, z)\big)$ is true in $\mathbb{N}$ and cannot be proved in PA. The first fact follows from Lemma 1 and the latter from Lemma 3. Thus, we have obtained another proof, different from that given in Chapters 20 and 21, of the incompleteness of the arithmetic PA.

The independence of PA of Goodstein's theorem is now shown from the following theorem.

22.26. Theorem (Cichon [C1]). *Let h: $N \times N \longrightarrow N$ be the function defined as follows:*

$$h(g, p) = min\{m\colon\ g_m(g, p) = 0\},$$

where $\{g_i(g, p)\colon\ i \in N\}$ is the Goodstein sequence in base p starting from g.

Let $\alpha = f_{g,p}(\omega)$ (note that then $g = G_\alpha(p)$ in view of 22.13). Then,

$$h(g, p) = H_\alpha(p + 1) - (p + 1).$$

Proof. We have by 22.9,

$$h(g, p) = \min\{n\colon\ P_{p+n}\big(P_{p+n-1}\big(\cdots P_{p+2}\big(P_{p+1}(\alpha)\big)\cdots\big)\big) = 0\}.$$

Let $y(p, \alpha)$ denote

$$\min\{y\colon\ P_{p+y}\big(P_{p+y-1}\big(\ldots P_{p+2}\big(P_{p+1}(\alpha)\big)\ldots\big)\big) = 0\}.$$

It suffices to show that the equality

22.27
$$y(p, \alpha) = H_\alpha(p+1) - (p+1)$$

is true for all p and all α.

We prove this by induction on α. We have

$$y(p, 0) = 0 = H_0(p+1) - (p+1).$$

Assume 22.27 for a number α and for all p. Then in particular

$$y(p+1, \alpha) = H_\alpha(p+2) - (p+2)$$

holds for any p. But $H_\alpha(p+2) = H_{\alpha+1}(p+1)$ and $y(p+1, \alpha) = y(p, \alpha+1) - 1$, since $P_{p+1}(\alpha+1) = \alpha$. Thus,

$$y(p, \alpha+1) - 1 = H_{\alpha+1}(p+1) - (p+2).$$

Hence,

$$y(p, \alpha+1) = H_{\alpha+1}(p+1) - (p+1).$$

Finally, assume 22.27 for all $\alpha < \lambda$ for a limit λ and for all p. We have

$$y(p, \lambda) = y\big(p, \{\lambda\}(p+1)\big)$$

by the definition of the function P_{p+1}. But

$$y\big(p, \{\lambda\}(p+1)\big) = H_{\{\lambda\}(p+1)}(p+1) - (p+1) = H_\lambda(p+1) - (p+1),$$

whence we get 22.27 for λ and the proof of Theorem 22.26 is completed. □

Now we shall show how the independence of Goodstein's theorem of the arithmetic PA follows from Theorem 22.26. To this end we shall prove some additional properties of the sequences H_α and G_α.

22.28. *Definition*. Let $\alpha \geq \beta$, $x \in N$. We shall write $\alpha \Rightarrow_x \beta$ if there is a finite sequence $\alpha = \alpha_0, \alpha_1, \ldots, \alpha_n = \beta$ such that $\alpha_{i+1} = \{\alpha_i\}(x)$, where for a nonlimit number $\gamma > 0$ we set $\{\gamma\}(x) = \gamma - 1$ and $\{0\}(x) = 0$.

22.29. Remark. *If $x, y, z \in N$, $z \geq y$, $\lambda < \varepsilon_0$, $x > 0$, then*

1. $\lambda \Rightarrow_x 0$,
2. $\{\lambda\}(z) \Rightarrow_x \{\lambda\}(y)$.

Proof. Property 1 can be proved easily by induction on λ. We prove property 2 again by induction on λ. The proof of the inductive step for a nonlimit λ and the proof for $\lambda = 0$ is immediate. Thus, assume that λ is a limit and assume Remark 22.29 for $\alpha < \lambda$. Let $\alpha = \beta + \omega^{\lambda_m}$ be the representation of λ in Cantor's normal form. Assume first that λ_m is a limit. Then we have $\{\lambda\}(z) = \beta + \omega^{\{\lambda_m\}(z)}$. By the inductive assumption $\{\lambda_m\}(z) < \{\lambda_m\}(y)$. Let

$$\{\lambda_m\}(z) = \alpha_0, \alpha_1, \ldots, \alpha_m = \{\lambda_m\}(y)$$

be the sequence existing by the definition of $\Rightarrow_x$. We shall show that

$$\beta + \omega^{\alpha_i} \Rightarrow_x \beta + \omega^{\alpha_{i+1}}.$$

If α_i is a limit, then

$$\{\beta + \omega^{\alpha_i}\}(x) = \beta + \omega^{\{\alpha_i\}}(x) = \beta + \omega^{\alpha_{i+1}}.$$

So assume that α_i is a successor. We have

$$\{\beta + \omega^{\alpha_i}\}(x) = \beta + \omega^{\alpha_{i+1}} x = \beta + \omega^{\alpha_{i+1}} + \omega^{\alpha_{i+1}}(x - 1) \Rightarrow_x \beta + \omega^{\alpha_{i+1}},$$

since $\omega^{\alpha_{i+1}} \cdot (x - 1) \Rightarrow_x 0$. Hence we have

$$\{\lambda\}(z) = \beta + \omega^{\alpha_0} \Rightarrow_x \beta + \omega^{\alpha_1} \Rightarrow_x \cdots \Rightarrow_x \beta + \omega^{\alpha_n} = \beta + \omega^{\{\lambda_m\}(y)} = \{\lambda\}(y).$$

Assume now that λ_m is a successor. We have

$$\{\lambda\}(z) = \beta + \omega^{\lambda_m - 1} \cdot z = \beta + \omega^{\lambda_m - 1} \cdot y + \omega^{\lambda_m - 1} \cdot (z - y) \Rightarrow_x \beta + \omega^{\lambda_m - 1} \cdot y$$
$$= \{\lambda\}(y),$$

since $\omega^{\lambda_m - 1} \cdot (z - y) \Rightarrow_x 0$. Thus Remark 22.29 has been proved. □

22.30. Remark.

1. *The functions H_α and G_α are increasing for $\alpha < \varepsilon_0$ and for positive arguments.*
2. *If $\alpha < \beta$, then $H_\alpha <_* H_\beta$, $G_\alpha <_* G_\beta$, where $f <_* g$ means that $f(x) < g(x)$ for almost all x.*
3. *Let us define $G_{\varepsilon_0}(x)$ as $G_{\omega_x}(x)$. Then $H_{\omega^3} \geq_* (G_{\varepsilon_0})^3$.*

Proof of 1. We show the following two properties by simultaneous induction on α:

(i) $\forall x > 0 \forall \beta[(\alpha \Rightarrow_x \beta \wedge \alpha > \beta) \rightarrow H_\alpha(x+1) > H_\beta(x+1)]$,
(ii) H_α is increasing for positive arguments.

The proof is easy. To prove the inductive step in (i) and (ii) for a limit α we use the property $\{\alpha\}(x+1) \Rightarrow_x \{\alpha\}(x)$ which follows from Remark 22.29.

We have two similar properties for the sequence G_α. In this case we may prove (i) and (ii) separately.

Proof of 2. Let us fix α. We prove point 2 by induction on $\beta < \alpha$. For a nonlimit β the proof of the inductive step is straightforward. So let us take a limit β and assume that, for $\alpha < \gamma < \beta$, point 2 holds, that is, $H_\alpha <_* H_\gamma$. Let us take x_0 such that $\{\beta\}(x_0) > \alpha$. By the inductive assumption $H_{\{\beta\}}(x_0) >_* H_\alpha$. On the other hand for $x \geq x_0$, $\beta \Rightarrow_x \{\beta\}(x_0)$ by Remark 22.29. Hence [and by (i) of the proof of point 1] we infer that $H_\beta >_* H_{\{\beta\}(x_0)}$ and thus $H_\beta >_* H_\alpha$. The reasoning for G_α and G_β is similar.

Proof of 3. By Remark 22.12 generalized to the case of $\alpha = \varepsilon_0$, we infer that

$$G_{\varepsilon_0}(x) = x^{x^{\cdot^{\cdot^{\cdot^{x}}}}},$$

where the number of levels is equal ot x. We shall show first that $H_{\omega^3}(x) \geq 2^{2^{\cdot^{\cdot^{\cdot^{2^x}}}}}$, where there are x levels, for $x \geq 1$. Let $h_n(x)$ be defined as $2^{2^{\cdot^{\cdot^{\cdot^{2^x}}}}}$ with n levels for $n \geq 1$ and let $h_0(x) = x$. We have $h_{n+1}(x) = 2^{h_n(x)}$. We shall show by induction on n that for all $x \geq 1$, $H_{\omega^2 \cdot n}(x) \geq h_n(x)$. We have

$$H_{\omega^2 \cdot 0}(x) = H_0(x) = x = h_0(x),$$

$$H_{\omega^2 \cdot (n+1)}(x) = H_{\omega^2 \cdot n + \omega^2}(x) = H_{\omega^2 \cdot n}(H_{\omega^2}(x))$$

by 22.14. Hence, and from the inductive assumption

$$H_{\omega^2 \cdot (n+1)}(x) \geq h_n(H_{\omega^2}(x)) = h_n(2^x \cdot x) \geq h_{n+1}(x).$$

Hence,

$$H_{\omega^3}(x) = H_{\omega^2 \cdot x}(x) \geq h_x(x) = 2^{2^{\cdot^{\cdot^{\cdot^{2^x}}}}}.$$

where the number of levels in $2^{2^{\cdot^{\cdot^{\cdot^{2^x}}}}}$ is equal x. Now let $x^{x^{\cdot^{\cdot^{\cdot^{x}}}}}$, where the number of levels is n, for $n \geq 1$, and let $g_0(x) = 1$. Then we have $g_{n+1}(x) = x^{g_n(x)}$ and $G_{\varepsilon_0}(x) = g_x(x)$ for $x \geq 1$. We shall show by induction on n that

$h_n(x) \geq x(g_n(x))^3$ for $x \geq 16$ and for all n. We have

$$h_0(x) = x = x(g_0(x))^3,$$
$$h_1(x) = 2^x \geq x \cdot x^3 = x \cdot (g_1(x))^3.$$

Assume the first inequality for $n \geq 1$. We have

$$h_{n+1}(x) = 2^{h_n(x)} \geq 2^{x(g_n(x))^3} = (2^x)^{(g_n(x))^3} \geq (x^2)^{(g_n(x))^3}$$
$$\geq x \cdot x^{(g_n(x))^3} \geq x \cdot x^{3g_n(x)} = x \cdot (x^{g_n(x)})^3 = x \cdot (g_{n+1}(x))^3,$$

since for $n \geq 1$, $x \geq 16$ we have

$$(g_n(x))^3 \geq 3g_n(x), \quad 2^x \geq x^2.$$

Hence, for $x \geq 16$,

$$H_{\omega^3}(x) \geq h_x(x) \geq (g_x(x))^3 = (G_{\varepsilon_0}(x))^3.$$

So the proof of Remark 22.30 has been completed. □

Now, we can prove the independence of Goodstein's theorem from PA. Suppose that Goodstein's theorem has a proof in PA. Let us extend the function $h(g, p)$ onto the pairs $\langle g, 0\rangle$ and $\langle g, 1\rangle$ defining $h(g, 0) = h(g, 1) = 0$. Then, so extended the function h is total in PA. Let us consider the recursive function h' defined as follows:

$$h'(x) = h(K(x), L(x)) + L(x) + 1,$$

that is,

$$h'(J(g, p)) = h(g, p) + p + 1 \quad \text{for all } g, p.$$

Then h' is also total in PA. By Wainer's theorem there is a $\beta < \varepsilon_0$ such that $h' <_* H_\beta$. By Remark 22.30, point 2 we may assume that β is of the form ω^γ and that $\gamma \geq 3$. Let x_0 be such that for $x \geq x_0$, $h'(x) \leq H_\beta(x)$. Let us fix a number $\alpha < \varepsilon_0$ such that $\alpha \geq \omega^\gamma + \omega^3$. We shall show that for infinitely many p $h'(J(G_\alpha(p), p)) \leq H_{\omega^\gamma + \omega^3}(p)$. We have

$$J(G_\alpha(p), p) \leq (G_\alpha(p))^3 \quad \text{for } p \geq 3.$$

Let us take a p such that $p \geq 3$ and $J(G_\alpha(p), p) \geq x_0$. Then,

$$h'(J(G_\alpha(p), p)) \leq H_{\omega^\gamma}(J(G_\alpha(p), p)) \leq H_{\omega^\gamma}((G_\alpha(p))^3)$$

by Remark 22.30, point 1. Let x_1 be such that for $x \geq x_1$, $G_\alpha(x) < G_{\varepsilon_0}(x)$ and $(G_{\varepsilon_0}(x))^3 \leq H_{\omega^3}(x)$. Assume that p satisfies the additional condition $p \geq x_1$. Then, we have

$$\begin{aligned} h'(J(G_\alpha(p),p)) &\leq H_{\omega^\gamma}((G_\alpha(p))^3) \leq H_{\omega^\gamma}((G_{\varepsilon_0}(p))^3) \\ &\leq H_{\omega^\gamma}(H_{\omega^3}(p)) = H_{\omega^\gamma+\omega^3}(p), \end{aligned}$$

where the last equality can be proved similarly as to Remark 22.14.

We have shown that for infinitely many p, $h'(J(G_\alpha(p),p)) \leq H_{\beta+\omega^3}(p)$.

Now, we may take p such that for $g = G_\alpha(p)$, $\alpha = f_{g,p}(\omega)$—p has to be larger than the coefficients in the representation of α in pure basis ω (see Remark 22.12). In this case we have, in view of Theorem 22.26 and Remark 22.13,

$$h'(J(G_\alpha(p),p)) = h'(J(g,p)) = h(g,p) + p + 1 = H_\alpha(p+1) = H_{\alpha+1}(p).$$

But, by Remark 22.30, point 2, $H_{\alpha+1} >_* H_{\beta+\omega^3}$, a contradiction. Thus, Goodstein's theorem has no proof in PA.

EXERCISES

22.1. Show that if a set A is $\alpha + 1$-large, $a \in A$, then $A \setminus \{a\}$ contains an α-large subset.

22.2. Let $F_0(x) = x + 1$, $F_{\alpha+1}(x) = F_\alpha^{(x)}(x)$, $F_\alpha(x) = F_{\{\lambda\}(x)}(x)$. Show that for $\alpha < \varepsilon_0$, $F_\alpha = H_{\omega^\alpha}$.

22.3. Using Exercise 22.2, show that the functions H_α for $\alpha < \omega^\omega$ are total in PA.

22.4. Prove that $H_\alpha \in \mathrm{PR}$ for $\alpha < \varepsilon_0$.

22.5. Find the length of the Goodstein sequence in base 2 starting from 2.

23
TARSKI'S THEOREM

In this chapter we prove a famous theorem of Tarski (cf. [T3])—the theory T of ordered real closed fields admits the elimination of quantifiers. It follows that theory T is complete and, in addition, effectively decidable; that is, the set of the theorems of T is recursive, (see the remarks at the end of the chapter). Let us note also that from the Tarski theorem a solution of Hilbert's Seventeenth problem follows easily; see the Exercises.

The axiom system of the theory of ordered real closed fields is given in Exercises 17.7 and 17.8. In particular, we consider fields with subtraction, so we may identify terms with polynomials of several variables with integer coefficients.

Let K be an arbitrary real closed field. Thus, Tarski's theorem states that any formula (of the language of ordered fields) is equivalent in K with an open formula (with the same free variables). In the proof we assume for definiteness $\mathrm{K} = \mathbb{R}$ (the field of the reals), but we use only those properties of $\mathbb{R}$ that hold in any real closed field. In addition to the ordinary properties of ordered fields we use the following two properties:

1. If a polynomial f changes a sign in the interval (a, b), then f has a root between a and b,
2. between two roots of a polynomial f there lies a root of the derivative f',

for polynomials f of one variable and with coefficients in the given field.

Examples

1. Consider the formula

$$\varphi(a, b):\ \exists y(ay + b = 0 \wedge y \geq 0).$$

This formula is equivalent with the following open formula:

$$\psi(a, b):\ (a < b \wedge b \geq 0) \vee (a > 0 \wedge b \leq 0) \vee (a = 0 \wedge b = 0).$$

2. Consider the formula

$$\varphi(a, b, c):\ \exists y(ay^2 + by + c = 0).$$

It is equivalent with the following open formula:

$$\psi(a,b,c):\ b^2-4ac\geq 0.$$

3. Consider the formula

$$\varphi(b,c,d):\ \exists y_1,y_2(y_1^3+by_1^2+cy_1+d=0\wedge y_2^3+by_2^2+cy_2+d=0\wedge y_1\neq y_2).$$

It is equivalent with the open formula

$$\psi(b,c,d): 3c-b^2<0\wedge 4c^3-b^2c^2-18bcd+4b^3d+27d^2\leq 0.$$

Proof of Tarski's Theorem (from Cohen [C2]; see also Hörmander [H2]).

Lemma 1. *Assume that we are given a formula $\varphi(x_1,\dots,x_n)$ of the form*

$$\exists y(\bigwedge_i p_i(x_1,\dots,x_n,y)=0\wedge\bigwedge_j q_j(x_1,\dots,x_n,y)>0),$$

where p_i and q_j are many variable polynomials with integer coefficients. Then, φ is equivalent with an open formula.

23.1. *Definition*. Let $p_1(x_1,\dots,x_n,y),\dots,p_m(x_1,\dots,x_n,y)$ be a finite set of polynomials with integer coefficients. We define the function $\mathrm{SGN}(x_1,\dots,x_n)$as follows. If $x_1,\dots,x_n\in R$ are given, then let k be the number of all real roots of those polynomials $p_1(x_1,\dots,x_n,y),\dots,$ $p_m(x_1,\dots,x_n,y)$ treated as polynomials of the variable y which are of nonzero degree with respect to y. Let $\alpha_1,\dots,\alpha_k$ be all the roots of those polynomials in increasing order.

Let (α_i,α_j) denote the open interval with the end points α_i,α_j, and similarly, $(-\infty,\alpha_i)$ or (α_i,∞). Instead of writing $x_1,\dots,x_n$ we shall write $\bar{x}$ for short. By $\mathrm{sg}\ p_i(\bar{x},\cdot)\restriction(\alpha_i,\alpha_j)$ we shall denote the sign of the polynomial $p_i(\bar{x},y)$, treated as a polynomial of the variable y in the interval (α_i,α_j), where the sign of z is 1, if $z>0$; 0, if $z=0$; and -1, if $z<0$.

We define $\mathrm{SGN}(\bar{x})$ as the sequence:

$$\begin{aligned}\langle k,\ &\mathrm{sg}\ p_1(\bar{x},\cdot)\restriction(-\infty,\alpha_1),\mathrm{sg}\ p_1(\bar{x},\alpha_1),\mathrm{sg}\ p_1(\bar{x},\cdot)\restriction(\alpha_1,\alpha_2),\mathrm{sg}\ p_1(\bar{x},\alpha_2),\dots,\\ &\mathrm{sg}\ p_1(\bar{x},\cdot)\restriction(\alpha_{k-1},\alpha_k),\mathrm{sg}\ p_1(\bar{x},\alpha_k),\mathrm{sg}\ p_1(\bar{x},\cdot)\restriction(\alpha_k,\infty),\\ &\mathrm{sg}\ p_2(\bar{x},\cdot)\restriction(-\infty,\alpha_1),\mathrm{sg}\ p_2(\bar{x},\alpha_1),\mathrm{sg}\ p_2(\bar{x},\cdot)\restriction(\alpha_1,\alpha_2),\mathrm{sg}\ p_2(\bar{x},\alpha_2),\dots,\\ &\mathrm{sg}\ p_2(\bar{x},\cdot)\restriction(\alpha_{k-1},\alpha_k),\mathrm{sg}\ p_2(\bar{x},\alpha_k),\mathrm{sg}\ p_2(\bar{x},\cdot)\restriction(\alpha_k,\infty),\dots,\\ &\mathrm{sg}\ p_m(\bar{x},\cdot)\restriction(-\infty,\alpha_1),\mathrm{sg}\ p_m(\bar{x},\alpha_1),\mathrm{sg}\ p_m(\bar{x},\cdot)\restriction(\alpha_1,\alpha_2),\mathrm{sg}\ p_m(\bar{x},\alpha_2),\dots,\\ &\mathrm{sg}\ p_m(\bar{x},\cdot)\restriction(\alpha_{k-1},\alpha_k),\mathrm{sg}\ p_m(\bar{x},\alpha_k),\mathrm{sg}\ p_m(\bar{x},\cdot)\restriction(\alpha_k,\infty)\rangle.\end{aligned}$$

Roughly, $\mathrm{SGN}(\bar{x})$ consists of the signs of the p's in the intervals determined by the α's and at the α's themselves.

Lemma 2. *Assume that $p_1, \ldots, p_m$ are given. Let $k \in N$ and let s be a sequence of zeros, ones and minus ones of length $(2k+1) \cdot m$. Then there is an open formula $\psi(\bar{x})$ defining the relation* $\mathrm{SGN}(\bar{x}) = \langle k \rangle * s$.

Proof of lemma 2. Let

23.2
$$d = (\text{maximum of the degrees of } p_1, \ldots, p_m \text{ with respect to y}),$$
$$l = (\text{the number of the polynomials of degree } d \text{ w.r.t. } y \text{ among } p_1, \ldots, p_m).$$

We prove the lemma by induction with respect to the pairs $\langle d, l \rangle$ ordered lexicographically. Consider first the case: $d = 0$, l arbitrary, as the initial step. In this case we have $p_j(\bar{x}, y) = p_j(\bar{x})$.

Let a sequence $\langle k \rangle * s$ be given. If $k \neq 0$, then we define $\psi(\bar{x})$ as a fixed false sentence, for example, $0 = 1$. If $k = 0$, that is, $\langle k \rangle * s$ is of the form $\langle 0 \rangle * s$, where s is a sequence of length m, then let $\psi_j(\bar{x})$ be the formula

$$p_j(\bar{x}) > 0, \quad \text{if } s(j) = 1,$$
$$p_j(\bar{x}) = 0, \quad \text{if } s(j) = 0,$$
$$p_j(\bar{x}) < 0, \quad \text{if } s(j) = -1.$$

Clearly, the formula $\psi(\bar{x})$ defined as

$$\bigwedge_{j=1}^{m} \psi_j(\bar{x})$$

satisfies the conclusion.

The inductive step. Assume that d, l are given, $d \geq 1$, $l \geq 1$. We assume inductively that for all the systems $q_1, \ldots, q_{m'}$ for which d' is less than d and l' is arbitrary or $d' = d$ and $l' < l$ the lemma holds.

Let a sequence $\langle k \rangle * s$ be given. Without loss of generality we may assume that p_m is of degree d with respect to y. Let us denote by d_j the degree of p_j with respect to y (we have $d_m = d$). Assume that $p_j(\bar{x}, y)$ is of the form

$$a_j(\bar{x}) \cdot y^{d_j} + \bar{p}_j(\bar{x}, y).$$

First we shall show that there is a sequence $q_1, \ldots, q_{m'}$ for which $d' = d(q_1, \ldots, q_{m'}) < d$ or $d' = d$ and $l' = l(q_1, \ldots, q_{m'}) < l$ and there is a finite set A of sequences $\langle \varepsilon \rangle * \langle k' \rangle * s'$, $\varepsilon \in \{-1, 1\}$ such that for all $\bar{x} \in R$

satisfying $a_m(\bar{x}) \neq 0$ we have

$$\text{SGN}(\bar{x}) = \langle k \rangle * s$$

23.3 $$\text{if and only if} \quad \bigvee_{\langle \varepsilon \rangle * \langle k' \rangle * s' \in A} \{\text{SGN}'(\bar{x}) = \langle k' \rangle * s' \wedge \text{sg } a_m(\bar{x}) = \varepsilon\},$$

where SGN′ denotes the function from definition 23.1 for the system $q_1, \ldots, q_{m'}$.

Let $r_{p_m, p_j}(\bar{x}, y)$ denote the remainder of the division of p_m by p_j (as polynomials of the variable y) multiplied by a certain even power of $a_j(\bar{x})$, such that this should be a polynomial of variables $\bar{x}, y$ and let $r_{p_m, p'_m}(\bar{x}, y)$ denote the remainder of the division of p_m by the derivative p'_m with respect to y (as polynomials of the variable y) multiplied by a certain even power of $a_m(\bar{x})$, such that it should be a polynomial of the variables $\bar{x}, y$.

Let $q_1, \ldots, q_{m'}$ be the system

23.4 $$p_1, \ldots, p_{m-1}, p'_m, r_{p_m, p_1}, \ldots, r_{p_m, p_{m-1}}, r_{p_m, p'_m}.$$

If by d' and l' we denote the suitable numbers satisfying 23.2 for this system, then $\langle d', l' \rangle$ is lexicographically earlier than $\langle d, l \rangle$ (since $d \geq 1$).

We shall show that for $\bar{x} \in R$ satisfying

23.5 $$a_m(\bar{x}) \neq 0$$

the value $\text{SGN}'(\bar{x})$ and sg $a_m(\bar{x})$, where SGN′ satisfies Definition 23.1 for the system 23.4, uniquely determine $\text{SGN}(\bar{x})$ (where SGN satisfies Definition 23.1 for $p_1, \ldots, p_m$).

For the proof, let $\bar{x} \in R$ satisfying 23.5 be given. Assume that $\text{SGN}'(\bar{x}) = \langle k' \rangle * s'$. Let $\beta_1, \ldots, \beta_l$, $l \leq k'$, be all the real roots of $p_1, \ldots, p_{m-1}, p'_m$ as functions of the variable y under our fixed $\bar{x}$ and assume $\beta_1 < \beta_2 < \cdots < \beta_l$. Assume $-\infty = \beta_0$ and $\infty = \beta_{l+1}$ and let the pairs $\langle q, (\beta_i, \beta_{i+1}) \rangle$ and pairs $\langle q, \{\beta_i\} \rangle$ for $q \in \{q_1, \ldots, q_{m'}\}$ be enumerated by some integers $j \leq (2k' + 1) \cdot m'$ so that

$$\text{sg } q(\bar{x}, \cdot) \restriction (\beta_i, \beta_{i+1}) = s'(\langle q, (\beta_i, \beta_{i+1}) \rangle),$$
$$\text{sg } q(\bar{x}, \beta_i) = s'(\langle q, \{\beta_i\} \rangle),$$

where pairs are identified with their numbers.

Now we shall show how $\langle k' \rangle * s'$ determines the number k. Let $\alpha_1, \ldots, \alpha_{\underline{k}}$ be all the roots of $p_m(\bar{x}, \cdot)$, $\alpha_1 < \alpha_2 < \cdots < \alpha_{\underline{k}}$. By 23.4, $q_1, \ldots, q_{m-1}, q_m$ are equal to $p_1, \ldots, p_{m-1}, p'_m$, respectively. Let $j(i) \in \{1, \ldots, m\}$ be chosen in such a way that β_i is a root of $q_{j(i)}$ for $i = 1, \ldots, l$.

We have

$$\alpha_1 \in (-\infty, \beta_1) \quad \text{if and only if} \quad (\text{sg } a_m(\bar{x})) \cdot (-1)^d = -s'(\langle r_{p_m, q_{j(i)}}, \{\beta_1\} \rangle).$$

Let us explain this as follows. Since in the interval $(-\infty, \beta_1)$ there is no root of $p'_m(\bar{x}, \cdot)$, $p_m(\bar{x}, y)$ changes monotonically in this interval (as a function of y), and thus it has a root in this interval provided that it changes sign. The sign of $p_m(\bar{x}, \cdot)$ at $-\infty$ (for sufficiently small y) is equal to $(\text{sg } a_m(\bar{x})) \cdot (-1)^d$ and the sign at β_1 is equal to sg $r_{p_m, q_{j(1)}}(\bar{x}, \beta_1)$, and thus to $s'(\langle r_{p_m, q_{j(1)}}, \{\beta_1\}\rangle)$.

We have $\alpha_1 = \beta_1$ if $s'(\langle r_{p_m, q_{j(1)}}, \{\beta_1\}\rangle) = 0$. Similarly,

$$\alpha_{\underline{k}} \in (\beta_l, \infty) \quad \text{if and only if} \quad \text{sg } a_m(\bar{x}) = -s'(\langle r_{p_m, q_{j(l)}}, \{\beta_l\}\rangle)$$

and

$$\exists j(\alpha_j \in (\beta_i, \beta_{i+1})) \quad \text{for } i = 1, \ldots, l$$
$$\text{if and only if} \quad s'(\langle r_{p_m, q_{j(i)}}, \{\beta_i\}\rangle) = -s'(\langle r_{p_m, q_{j(i+1)}}, \{\beta_l\}\rangle),$$

and

$$\exists j(\alpha_j = \beta_i) \quad \text{if and only if} \quad s'(\langle r_{p_m, q_{j(i)}}, \{\beta_i\}\rangle) = 0.$$

Thus, we have

$$\begin{aligned} k = &\big(\text{the number of roots of } p_1(\bar{x}, \cdot), \ldots, p_{m-1}(\bar{x}, \cdot)\big) \\ &+ \big(\text{the number of intervals } (\beta_i, \beta_{i+1}),\ i = 0, \ldots, l+1, \text{ in which} \\ &\quad \text{there is a root of } p_m(\bar{x}, \cdot) \text{ which is not a root of } p_1, \ldots, p_{m-1}\big) \\ &+ \big(\text{the number of points } \{\beta_i\} \text{ which are a root of } p_m(\bar{x}, \cdot) \\ &\quad \text{which is not a root of } p_1, \ldots, p_{m-1}\big). \end{aligned}$$

So, we can find k if s' is known. Also, we may reconstruct the mutual position of $\beta_1, \ldots, \beta_l$ and $\alpha_1, \ldots, \alpha_{\underline{k}}$ in the ordering on R.

Now, we shall find the values

$$\text{sg } (p_m \restriction (\beta_i, \beta_{i+1})), \quad \text{sg } (p_m \restriction \{\beta_i\}), \quad \text{sg } (p_m \restriction (\beta_i, \alpha_{\underline{l}}),$$
$$\text{sg } (p_m \restriction (\alpha_{\underline{l}}, \beta_{i+1})), \quad \text{sg } (q_j \restriction (\beta_i, \alpha_{\underline{l}})), \quad \text{sg } (q_j \restriction \{\alpha_{\underline{l}}),$$
$$\text{sg } (q_j \restriction (\alpha_{\underline{l}}, \beta_{i+1})),$$

where no $\alpha \underline{j}$ or β_j lies in any of the above intervals [for simplicity we write q instead of $q(\bar{x}, \cdot)$]. Notice that if the interval $(\alpha_{\underline{l}}, \beta_{i+1})$ or $(\beta_i, \alpha_{\underline{l}})$ has the above property (is taken under consideration), then β_{i+1} or β_i, respectively, is not a root of p_m; otherwise another root of p'_m would lie between two roots of p_m.

For $i = 0, \ldots, l$, $\underline{l} = 1, \ldots, \underline{k}$ we have $i \neq l$ provided that β_{i+1} occurs in the formula below and $i \neq 0$ provided β_i occurs below.

$$\text{sg}\,\big(p_m \upharpoonright (\beta_i, \beta_{i+1})\big) = \begin{cases} s'(\langle r_{p_m, q_{j(i)}}, \{\beta_i\}\rangle), & \text{provided this is} \neq 0, \\ s'(\langle r_{p_m, q_{j(i+1)}} \{\beta_{i+1}\rangle), & \text{provided this is} \neq 0, \end{cases}$$

$$\text{sg}\,(p_m \upharpoonright \{\beta_i\}) = s'(r_{p_m, q_{j(i)}}, \{\beta_i\}\rangle),$$

$$\text{sg}\,\big(p_m \upharpoonright (\beta_i, \alpha_{\underline{l}})\big) = s'(\langle r_{p_m, q_{j(i)}}, \{\beta_i\}\rangle),$$

$$\text{sg}\,\big(p_m \upharpoonright (\alpha_{\underline{l}}, \beta_{i+1})\big) = s'(\langle r_{p_m, q_{j(i+1)}}, \{\beta_{i+1}\}\rangle),$$

$$\text{sg}\,\big(q_j \upharpoonright (\beta_i, \alpha_{\underline{l}})\big) = s'\big(\langle q_j, (\beta_i, \beta_{i+1})\rangle\big),$$

$$\text{sg}\,\big(q_j \upharpoonright \{\alpha_{\underline{l}}\}\big) = s'\big(\langle q_j, (\beta_i, \beta_{i+1})\rangle\big),$$

$$\text{sg}\,\big(q_j \upharpoonright (\alpha_{\underline{l}}, \beta_{i+1})\big) = s'\big(\langle q_j, (\beta_i, \beta_{i+1})\rangle\big),$$

where $\beta_i \leq \alpha_{\underline{l}} < \beta_{i+1}$ or $\beta_i < \alpha_{\underline{l}} \leq \beta_{i+1}$ are consecutive roots of $p_1, \ldots, p_{m-1}, p'_m, p_m$.

Let now $\gamma_1, \ldots, \gamma_k$ be those of $\beta_1, \ldots, \beta_l, \alpha_1, \ldots, \alpha_{\underline{k}}$ which are roots of $p_1, \ldots, p_{m-1}, p_m$. Let

$$\gamma_0 = -\infty, \quad \gamma_{k+1} = \infty, \quad \gamma_0 < \gamma_1 < \ldots < \gamma_k < \gamma_{k+1}.$$

If we know all the signs of all p_i in all intervals determined by β_i, $\alpha_{\underline{l}}$ for $i = 0, \ldots, l+1$, $\underline{l} = 1, \ldots, \underline{k}$, we can find $s\big(\langle p_j, (\gamma_i, \gamma_{i+1})\rangle\big)$ for all $j = 1, \ldots, m$, $i = 0, \ldots, k$. The signs of p_j in the finite intervals determined by β_i, $\alpha_{\underline{l}}$ have been fixed according to the values of s' and the signs in the infinite intervals are determined by the sign of $a_m(\bar{x})$ for $j = m$ and by suitable values of s' for $j < m$.

For a given k, s, $s \in (\{-1, 0, 1\})^{(2k+1)\cdot m}$, let us take as A the set of such sequences $\langle \varepsilon \rangle * \langle k' \rangle * s'$ that for arbitrary $\bar{x} \in R$, if $\bar{x}$ satisfy

$$\big(a_m(\bar{x}) \neq 0\big) \wedge \big(\text{sg}\; a_m(\bar{x}) = \varepsilon\big) \wedge \big(\text{SGN}'(\bar{x}) = \langle k' \rangle * s'\big),$$

then $\text{SGN}(\bar{x}) = \langle k \rangle * s$. Then 23.3 holds for $\bar{x} \in R$ such that $a_m(\bar{x}) \neq 0$.

Now we shall modify the equivalence 23.3 so that it should hold for arbitrary $\bar{x} \in R$. Consider the system $p_1, \ldots, p_{m-1}, \bar{p}_m$, where $\bar{p}_m$ is a polynomial of degree $\leq d-1$ with respect to y which results from p_m by rejecting the term $a_m(\bar{x}) \cdot y^d$. Let SGN″ be the function satisfying Definition 23.1 for this system. Then for $\bar{x}$ satisfying $a_m(\bar{x}) = 0$, $\text{SGN}(\bar{x}) = \text{SGN}''(x)$.

Finally, we have

$$\mathrm{SGN}(\bar{x}) = \langle k \rangle * s \quad \text{if and only if} \quad \Big[a_m(\bar{x}) \neq 0 \bigvee_{\langle \varepsilon \rangle * \langle k' \rangle * s' \in A} \big\{\mathrm{SGN}'(\bar{x}) = \langle k' \rangle * s' \wedge \mathrm{sg}\ a_m(\bar{x}) = \varepsilon\big\}\Big] \vee [a_m(\bar{x}) = 0 \wedge \mathrm{SGN}''(\bar{x}) = \langle k \rangle * s]$$

for arbitrary $\bar{x} \in R$.

Let $\psi'_{k',s'}(\bar{x})$ be an open formula (existing by the inductive assumption) such that

$$\mathrm{SGN}'(\bar{x}) = \langle k' \rangle * s' \quad \text{if and only if} \quad \psi'_{k',s'}(\bar{x})$$

for all $\bar{x} \in R$ and let $\psi''_{k,s}(\bar{x})$ be the open formula existing by the inductive assumption such that

$$\mathrm{SGN}''(\bar{x}) = \langle k \rangle * s \quad \text{if and only if} \quad \psi''_{k,s}(\bar{x})$$

for $\bar{x} \in R$. We have

$$\mathrm{SGN}(\bar{x}) = \langle k \rangle * s \quad \text{if and only if} \quad \Big[a_m(\bar{x}) \neq 0 \bigvee_{\langle \varepsilon \rangle * \langle k' \rangle * s' \in A} \big\{\psi'_{k',s'}(\bar{x}) \wedge \mathrm{sg}\ a_m(\bar{x}) = \varepsilon\big\}\Big] \vee [a_m(\bar{x}) = 0 \wedge \psi''_{k,s}(\bar{x})].$$

Evidently, the relation sg $a_m(\bar{x}) = \varepsilon$ is definable by an open formula. Thus we have proved Lemma 2.

Proof of Lemma 1. Consider the function $\mathrm{SGN}(\bar{x})$ for the system p_j, q_j. There is a finite set of sequences $\langle k \rangle * s \in A$ such that $\varphi(\bar{x})$ is satisfied if and only if $\mathrm{SGN}(\bar{x}) \in A$. For every sequence $\langle k \rangle * s \in A$ there exists, by lemma 2, an open formula $\psi_{k,s}(\bar{x})$ defining the relation $\mathrm{SGN}(\bar{x}) = \langle k \rangle * s$.

Hence, we have

$$\varphi(\bar{x}) \quad \text{if and only if} \quad \bigvee_{\langle k \rangle * s \in A} \psi_{k,s}(\bar{x}).$$

This completes the proof of Lemma 1.

The proof of the theorem is now straightforward. Applying the remarks preceding Theorem 17.2, we infer that to prove the theorem it suffices to eliminate the quantifier $\exists$ from formulas of the form $\exists xF$, where F is an open formula. The atomic formulas and their negations have, in our case, the form $f = g$, $f \neq g$, $f > g$, $f \leq g$, where f and g are terms, that is, in this case polynomials with integer coefficients. The formula $f = g$ is equivalent with the formula $f - g = 0$, the formula $f \neq g$ is equivalent with the formula $f - g > 0 \vee g - f > 0$, the formula $f > g$ is equivalent with $f - g > 0$, and finally the formula $f \leq g$ with $g - f > 0 \vee f = g$. Hence every open formula is equivalent with a disjunction $F_1 \vee \cdots \vee F_n$, where every F_i has the form $G_1 \wedge \cdots \wedge G_m$ and G_j is a formula of the form $f = 0$ or $f > 0$ for a polynomial f with integer coefficients. As in the proof of Theorem 17.2, we show that it is enough to eliminate the quantifier from formulas of the form $\exists x(G_1 \wedge \cdots \wedge G_m)$, where G_i has the form $f = 0$ or $f > 0$. Therefore, the proof of the theorem is now reduced to the proof of Lemma 1, and thus Tarski's theorem has been proved. □

Let us note that a similar result holds also for p-adic fields; see Cohen [C2].

Let us also note a few other results and problems connected with Tarski's theorem. In the beginning of Chapter 17 we introduced the notion of model completeness of a theory. From Tarski's theorem it follows that the theory of real closed fields is model complete—see the remark following Section 17.1. As an illustration let us mention a corollary of the Hilbert "Nullstellensatz":

> If a system of polynomial equations and inequalities F with real algebraic coefficients has a real "zero" then it has a real algebraic "zero." In this case K is the field of real algebraic numbers and L is $\mathbb{R}$.

From the model completeness of the theory T of real closed fields follows its completeness. To prove this, let us show that any two real closed fields are equivalent. So let K, L be given. Let $\bar{Q}$ denote the field of real algebraic numbers. We have $\bar{Q} \subseteq K$ and $\bar{Q} \subseteq L$—in fact $\bar{Q}$ is a prime model of T (see Section 16.8). By model completeness, $\bar{Q} \leq K$ and $\bar{Q} \leq L$. In particular $\bar{Q} \equiv K$ and $\bar{Q} \equiv L$; thus $K \equiv L$. In fact, we have just proved that any model complete theory that has a prime model is complete.

From the completeness of T its decidability follows. Consider the set of theorems of T. Clearly, this set is $\Sigma_1(\mathbb{N})$. However, its complement is also $\Sigma_1(\mathbb{N})$ since a sentence ϕ is not a theorem if and only if the negation of ϕ is a theorem. Hence it follows that the set of theorems of T is $\Delta_1(\mathbb{N})$, and thus recursive.

Again we have proved a more general fact—a recursive complete theory is decidable.

Another feature of a real closed field that follows from Tarski's theorem is the form of its definable subsets. Let K be a real closed field. Let $A \subseteq K^n$ be called *basic* if

$$A = \{\langle a_1, \ldots, a_n \rangle \in K \colon F(a_1, \ldots, a_n)\}$$

where F is a system of equations and inequalities (sharp inequalities).

Thus Tarski's theorem can be reformulated as:

> Every definable (in K by means of $+$ and $\cdot$) subset of K^n is a finite union of basic sets, the so called *semialgebraic* set. For $n = 1$, it is a union of finitely many intervals. In particular, a projection of a semialgebraic set in K^n onto K^m for $m \leq n$ is again semialgebraic.

This feature of a real closed field has been generalized to the notion of *o-minimality* ([PS]).

A structure $\mathbb{M} = \langle M, <, \mathcal{R}, \mathcal{F}, \mathcal{C} \rangle$ is said to be o-minimal if $<$ is a total ordering on M and every definable (with parameters) subset of M is a finite union of points in M and intervals (a, b), where $a \in M$ or $a = -\infty$ and $b \in M$ or $b = +\infty$. A structure $\mathbb{M}$ is said to be *strongly o-minimal* if every $\mathbb{M}'$ which is equivalent to $\mathbb{M}$ is o-minimal.

The following two theorems are basic theorems about the notion of o-minimality ([KPS]).

Theorem 1. Let $\mathbb{M}$ be o-minimal. Then every definable set $X \subseteq M$ is a disjoint union of finitely many definably connected sets X, where X is called definably connected if there do not exist disjoint nonempty definable sets Y_1, Y_2, both open in X such that $X = Y_1 \cup Y_2$.

Theorem 2. *If* $\mathbb{M}$ *is o-minimal, then it is strongly o-minimal.*

The second theorem states that o-minimality is the feature of $Th(\mathbb{M})$ rather than of $\mathbb{M}$ itself.

The o-minimality of a real closed field follows from the elimination of quantifiers for the theory of real closed fields. However, we can prove the o-minimality of certain structures without proving the elimination of quantifiers for their theories.

Tarski posed the following problem: Let T be the theory of all the sentences true about the reals expressible by means of $+$ and $\cdot$ and exponentiation. More precisely, we let exp: $R \longrightarrow R$ be the exponential function on R and we let T be the theory of $\langle R, <, +, \cdot, \exp, 0, 1 \rangle$ in the language with an additional function symbol for exp. Is a corresponding version of Tarski's elimination of quantifiers true for this theory? In particular, is this theory model complete? Recently Wilkie [W4] proved that this theory *is* model complete. Also he proved (using results of Khovanskii [K5]) that it is o-minimal although it does not admit elimination of quantifiers (see also [vdD].

EXERCISES

23.1. Find an open formula equivalent with the formula $\exists x(f(x) = 0)$, where f is of degree 3.

23.2. Show that every basic set in R^n is a union of finitely many connected algebraic surfaces. For $n = 1$ it is a union of finitely many intervals.

23.3. Show that a projection of an algebraic surface in R^n onto a subspace R^m ($m \leq n$) is a finite union of algebraic surfaces.

For a field K let ΣK^2 denote the set of sums of squares $a_1^2 + \cdots + a_n^2$, with $a_1, \ldots, a_n \in K$.

23.4. For any field K the set ΣK^2 is closed under addition, multiplication, and division.

A field K is said to be formally real if $\Sigma K^2 \cap -\Sigma K^2 = \{0\}$.

23.5. If K can be ordered (i.e., if there is a $\leq$ such that $\langle K, \leq \rangle$ is an ordered field), then K is formally real.

23.6. Let K be formally real and let $x \in K$ be such that $-x \notin \Sigma K^2$.

(a) The set $S = \{a + bx\colon\ a, b \in \Sigma K^2\}$ is closed under addition, multiplication, and also $S \cap -S = \{0\}$ (use Exercise 23.5).

(b) If an $S \subseteq K$ is maximal with respect to the properties listed in (a), then also $S \cup -S = K$.

(c) For an S as in (b) show that the relation $a \leq_S b$ if and only if $b - a \in S$, is a field ordering on K.

(d) There is a maximal S as in (a) such that $x \in S$.

Hence, if K is formally real and x is not a sum of squares, then there is a field ordering $\leq$ on K in which $x < 0$. In particular, a field K expands to an ordered field if and only if K is formally real.

A theorem of Artin and Schreier says that every ordered field K has an ordered real closed algebraic extension $\bar{K}$. $\bar{K}$ is uniquely determined and called the real closure of K.

23.7. Let $f \in \mathbb{Q}(\xi_1, \ldots, \xi_n)$ be a rational function of n indeterminates. Assume that f is not a sum of squares and let φ be the sentence $\exists x_1, \ldots, x_n(f(x_1, \ldots, x_n) < 0)$.

(a) The sentence φ holds in the real closure of $\mathbb{Q}(\xi_1, \ldots, \xi_n)$.

(b) φ holds in the real closure of $\mathbb{Q}$ (apply the Tarski theorem).

(c) φ holds in $\mathbb{Q}$ ($\mathbb{Q}$ is dense in its real closure).

Hence, if $f \in \mathbb{Q}(\xi_1, \ldots, \xi_n)$ is positive definite (i.e., assumes nonnegative values only), then f is a sum of squares. This is the Seventeenth Hilbert problem. Moreover, $\mathbb{Q}$ can be replaced by $\mathbb{R}$ or any K such that K has a unique ordering and is dense in its real closure $\bar{K}$; see [R2].

24

MATIYASEVICH'S THEOREM

The tenth Hilbert problem concerns the existence of an algorithm deciding whether any arithmetical polynomial $f(x_1, \ldots, x_n)$ has a zero in N or not. The problem was ultimately solved (negatively) in the late sixties by Matiyasevich [M3]. It should be noted here that an important contribution to the solution had been done in 1961 by Davis, Putnam and Robinson [DPR], who proved, for PA with the additional primitive operation x^y, the so-called Davis conjecture saying that every Σ_1-definable set is diophantine. Actually, Matiyasevich proved the conjecture for PA itself. This fact implies easily the negative solution of Hilbert's tenth problem.

Arithmetical formulas of the form

$$\exists y_1, \ldots, y_k[t = s], \quad \text{where } t, s \in \mathrm{Tm}$$

are called *diophantine*. Every existential formula (see Exercise 5.7) is equivalent in PA with a diophantine formula (see Exercise 13.10). Also, we shall call diophantine the formulas equivalent in $\mathbb{N}$ with formulas of the above form.

Sets (relations) $A \subseteq N^n$ are called diophantine if they are definable in $\mathbb{N}$ by diophantine formulas:

$$\langle a_1, \ldots, a_n \rangle \in A \quad \text{if and only if} \quad \mathbb{N} \models \exists y_1, \ldots, y_k(t = s)[a_1, \ldots, a_n].$$

To arithmetical terms t, s there correspond polynomials f_t, f_s (cf. Section 18.4). Setting $f = f_t - f_s$ (the algebraic difference) we obtain

$$A = \{\langle a_1, \ldots, a_n \rangle \in N^n \colon \exists y_1, \ldots, y_k \in N[f(a_1, \ldots, a_n, y_1, \ldots, y_k) = 0]\}.$$

Thus, diophantine sets can be characterized as sets of sequences $\langle a_1, \ldots, a_n \rangle \in N^n$ of parameters for which a certain polynomial f with integer coefficients has a zero in N. Moreover, the condition "zero in N" can be replaced by "zero in $\mathbb{Z}$," see Example 24.6.

In connection with the tenth Hilbert problem Matiyasevich proved the following theorem.

24.1. Theorem. *Every set $A \subseteq N^n$ of class $\Sigma_1(\mathbb{N})$ is diophantine.*

Hence a negative solution of the Hilbert problem follows immediately—is there a method (an algorithm) enabling us to decide whether a polynomial f with integer coefficients has a zero in $\mathbb{Z}$? To see this let us identify polynomials f with terms of the language of the theory of rings with subtraction defined in $\mathbb{N}$ as in Chapter 19. With such an identification, the set $W \subseteq N$ of polynomials f with integer coefficients is of class PR. Let

$$h(f) = \begin{cases} 0, & \text{if } f \in W \text{ and } f \text{ has a zero in } \mathbb{Z}, \\ 1, & \text{otherwise.} \end{cases}$$

The existence of a method required in Hilbert's problem means, in a precise language, that the function h is recursive, that is, $h \in \Sigma_1(\mathbb{N})$. However, in view of the Matiyasevich theorem, it is easy to see that $h \notin \Sigma_1(\mathbb{N})$. To explain this we let $A \subseteq N$ be a set satisfying the condition $A \in \Sigma_1(\mathbb{N}) \setminus \Pi_1(\mathbb{N})$, for example, we may take as A the set of theorems of the arithmetic PA or the universal set u_1—see Exercises 20.7 and 21.9. By the Matiyasevich theorem there is a polynomial $f(x, y_1, \ldots, y_k)$ such that

$$A = \{a \in N\colon\ \exists y_1, \ldots, y_k(f(a, y_1, \ldots, y_k) = 0)\}.$$

If h were a recursive function, then the set

$$A = \{a \in N\colon\ h(f(\Lambda_a/x)) = 0\}$$

would be of class $\Sigma_1(\mathbb{N}) \cap \Pi_1(\mathbb{N})$, as the zero set for a certain recursive function, contrary to the choice of A.

Before the proof of Theorem 24.1 (which will be based on Manin [M2]) let us give a few examples of diophantine relations and formulas.

24.2. *Example.* Directly from the definition it follows that the relations $=$, $<$ and $\leq$ are diophantine.

24.3. *Example.* The congruence $x \equiv y \bmod m$ is a diophantine relation since it is definable in $\mathbb{N}$ by the formula $\exists z[x = y + m \cdot z \vee y = x + m \cdot z]$.

24.4. *Example.* If $F_1, \ldots, F_m$ are diophantine formulas then also the conjunction $F_1 \wedge \cdots \wedge F_m$ and the disjunction $F_1 \vee \cdots \vee F_m$ are diophantine.

To prove this we let the polynomials f_i be chosen so that

$$\mathbb{N} \models F_i(a_1, \ldots, a_n) \quad \text{if and only if} \quad f_i(a_1, \ldots, a_n, y_1, \ldots, y_k) \text{ has a zero in } N,$$

$i = 1, \ldots, m$. Then, the polynomial $f_1^2 + \cdots + f_m^2$ is as required for the conjunction and $f_1 \cdot \cdots \cdot f_n$ was required for the disjunction.

24.5. *Example.* The graphs of the partial functions $z = x/y$ and $z = [x/y]$ (the integer part of the number x/y) are diophantine. This follows immediately from the equivalence

$$z = \frac{x}{y} \quad \text{if and only if} \quad z \cdot x = y,$$

$$z = \left[\frac{x}{y}\right] \quad \text{if and only if} \quad z \cdot y \leq x < (z+1) \cdot y.$$

24.6. *Example.* For every polynomial f there is a polynomial g such that for any system of parameters $a_1, \ldots, a_n$ we have

$$(*) \quad \begin{aligned} &f(a_1, \ldots, a_n, y_1, \ldots, y_k) \text{ has a zero in } N \\ &\qquad \text{iff} \quad g(a_1, \ldots, a_n, z_1, \ldots, z_l) \text{ has a zero in } Z. \end{aligned}$$

Conversely, for every g there is an f such that $(*)$ holds.

We explain this as follows. By the Lagrange theorem every natural number is a sum of four squares of integers. Hence, the substitution of $y_i = z_{i1}^2 + \cdots + z_{i4}^2$ in f, for $i = 1, \ldots, k$ yields a polynomial g satisfying $(*)$. Conversely, if a polynomial g is given then let $\varepsilon = \langle \varepsilon_1, \ldots, \varepsilon_l \rangle$ run over all the sequences of length l consisting of the numbers 1 and -1. If

$$g_\varepsilon(x_1, \ldots, x_n, z_1, \ldots, z_l) = g(x_1, \ldots, x_n, \varepsilon_1 z_1, \ldots, \varepsilon_l z_l),$$

then the polynomial $f = \prod_\varepsilon g_\varepsilon$ satisfies the condition $(*)$.

24.7. *Example.* The set of composite numbers is diophantine since it is definable by the formula $F(x)$,

$$\exists y, z[y > 1 \wedge z > 1 \wedge y \cdot z = x].$$

24.8. *Example.* The relation $(a, b) = 1$ (a, b are relatively prime) is diophantine since

$$(a, b) = 1 \quad \text{if and only if} \quad \exists x, y[(ax - by)^2 = 1].$$

The class D of diophantine formulas contains the atomic formulas $t = s$ and their negations [since $t \neq s$ if and only if $(t < s \vee s < t)$] and is closed under taking conjunctions and disjunctions. Hence this class contains all the open formulas. Moreover, if $F \in D$, then also $\exists y_1, \ldots, y_l F \in D$. Hence the class D

contains all the existential formulas. Matiyasevich's theorem states that $\Sigma_1 \subseteq D$. Thus, to prove Matiyasevich's theorem it remains to show that D is closed under the operation of adding a bounded quantifier. Thus, we have to prove the following lemma:

24.9. Lemma. *If $A \subseteq N^{n+2}$ is a diophantine set, then also $\forall z < xA$ is diophantine.*

Proof. The proof of this lemma will occupy the rest of this chapter. First we shall give the general idea of the proof. For the set A there is a polynomial f such that

$$\langle z, x, a_1, \ldots, a_n \rangle \in A \quad \text{iff} \quad f(z, x, a_1, \ldots, a_n, y_0, \ldots, y_k) \text{ has a zero in } N.$$

Hence

$$\langle x, a_1, \ldots, a_n \rangle \in \forall z < x\, A \quad \text{if and only if} \quad \forall z < x \exists y_0, \ldots, y_k$$
$$[f(z, x, a_1, \ldots, a_n, y_0, \ldots, y_k) = 0].$$

The condition at the right hand side of this equivalence is equivalent to the existence of a matrix

$$\begin{bmatrix} y_{0,0} & \cdots & y_{0,k} \\ & \cdots & \\ y_{x-1,0} & \cdots & y_{x-1,k} \end{bmatrix}$$

whose every row $y_{z,0}, \ldots, y_{z,k}$, $z = 0, \ldots, x-1$, satisfies the equation

$$f(z, x, a_1, \ldots, a_n, y_{z,0}, \ldots, y_{z,k}) = 0.$$

This, in turn, is equivalent with the validity in $\mathbb{N}$ of the formula

24.10 $$\exists Y_0, \ldots, Y_k \Big(\bigwedge_{i=0}^{k} (Y_i = \langle y_{0,i}, \ldots, y_{x-1,i} \rangle$$
is the code of a sequence of length x)
$$\wedge\ \forall z < x \big[f\big(z, x, a_1, \ldots, a_n, Y_0(z), \ldots, Y_k(z)\big) = 0 \big] \Big).$$

Thus, we have to show that this formula is equivalent with a diophantine one. The coding of sequences is by means of the β function, that is Y_i is the code of the sequence $\langle y_{0,i}, \ldots, y_{x-1,i} \rangle$ if

24.11 $$Y_i \equiv y_{z,i} \bmod \big((z+1)K! + 1\big)$$

for $z = 0, \ldots, x-1$, where K is large enough. Thus, it suffices to prove that the relation $R(x, a_1, \ldots, a_n, Y_0, \ldots, Y_k, K)$

$$\forall z < x \big[f\big(z, x, a_1, \ldots, a_n, r\big(Y_0, (z+1)K! + 1\big), \ldots, r\big(Y_k, (z+1)K! + 1\big)\big) = 0 \big]$$

is diophantine. In particular we have to prove that the graph of the function $y = x!$ is diophantine. It is more convenient to prove first that the graph of the exponential function is diophantine. We shall show this by considering the sequence of solutions of Pell's equation. In particular, the graph of the function $y = y(a, z)$, where y is the second coordinate of the zth solution of Pell's equation is diophantine.

STEP 1. Here, we shall deal with the function $y(a, z)$. The diophantine equation

$$x^2 - dy^2 = 1,$$

where d is not a square, is called *Pell's equation*. It has been widely studied in number theory; it has many elegant properties. A particular case of Pell's equation is the equation

$$x^2 - (a^2 - 1)y^2 = 1,$$

where $a > 1$. In the sequel we shall call such an equation "the *ath Pell's equation*." It is easy to see that the pairs

$$\langle x, y \rangle = \langle 1, 0 \rangle, \qquad \langle x, y \rangle = \langle a, 1 \rangle$$

are solutions of that equation. In general, if a pair $\langle x, y \rangle$ is a solution, then the pair $\langle x', y' \rangle$, where

$$x' = ax + (a^2 - 1)y, \qquad y' = x + ay,$$

is also a solution.

Consider the linear mapping from $N \times N$ into $N \times N$ given by the matrix

24.12
$$\begin{bmatrix} a & a^2 - 1 \\ 1 & a \end{bmatrix}.$$

By the nth solution of the ath Pell's equation we shall mean the pair $\langle x, y \rangle$ which is the value of the nth iteration of the above mapping applied to the pair $\langle 1, 0 \rangle$. Thus the pair $\langle 1, 0 \rangle$ is the zeroth solution, the pair $\langle a, 1 \rangle$ the first solution, the pair $\langle 2a^2 - 1, 2a \rangle$ the second solution, the pair $\langle 4a^3 - 3a, 4a^2 - 1 \rangle$ the third solution, and so on.

Property 1. *The above sequence exhausts all the solutions of the ath Pell's equation with both coordinates positive. Moreover, we have $y(a, z) \geq z$ for every z.*

All the properties of solutions of Pell's equations of a number theoretic character will be proved later.

For a given z there is a polynomial f_z such that $y(a, z) = f_z(a)$, namely,

$$y(a, z) = \sum_{i=1}^{[(z+1)/2]} \binom{z}{2l-1} a^{z-2l+1}(a^2 - 1)^{l-1}.$$

(The proof will be given later.) Our aim is to find a polynomial f independent of z such that for any $a, z, a > 1$ we have

$$y = y(a, z) \quad \text{if and only if} \quad \exists w_1, \ldots, w_k \big(f(a, z, y, w_1, \ldots, w_k) = 0\big).$$

We shall also write $y_z(a)$ instead of $y(a, z)$ and $\langle x_z(a), y_z(a)\rangle$ will denote the zth solution.

The main property of the solutions of Pell's equations needed for the construction of the polynomial f is the following.

Property 2. $y(a, z) \equiv z \bmod (a - 1)$.

Observe that basing on this property we can define diophantically the family of solutions of the ath equation with indices congruent to z modulo $a - 1$. Namely, we have (by Property 1)

$$\exists u\big(y = y(a, u(a-1) + z) \wedge a > 1\big)$$
$$\text{iff} \quad \exists x\big(x^2 - (a^2 - 1)y^2 = 1 \wedge y \equiv z \bmod(a - 1) \wedge a > 1\big).$$

By Examples 24.3 and 24.4, we infer that the right-hand side of the equivalence is diophantine. Now we are aiming to strengthen the right-hand side of the equivalence so that $y = y(a, z)$ should follow from it, that is, we have to distinguish in a diophantine way the zth solution from among the $\big(u(a-1) + z\big)$th.

Property 3. *The congruence* $\big(y_v(a) \equiv y_w(a) \bmod x_w(a)\big)$ *implies* ($v \equiv w \bmod 2u$ or $v \equiv -w \bmod 2u$) *for arbitrary* v, w, u.

Assume now that y satisfies

(1) $x^2 - (a^2 - 1)y^2 = 1$, for some $x, a > 1$,
(2) $y \equiv z \bmod(a - 1)$.

By Property 1 there is a w such that $y = y_w(a)$. We want to add to the above two conditions some further ones so that $w = z$ should follow from them. We shall use Property 3. Let us consider an auxiliary solution of the ath equation $\langle x_u(a), y_u(a)\rangle$. We introduce new variables x_1, y_1 to denote $x_u(a)$, $y_u(a)$, respectively. Consider the following system of conditions:

(1) $x^2 - (a^2 - 1)y^2 = 1 \wedge a > 1$,
(2) $y \equiv z \bmod (a - 1)$,
(3) $x_1^2 - (a^2 - 1)y_1^2 = 1$,
(4) $y \geq z$.

We look next for conditions such that jointly with (1) to (4) they should imply $w = z$. We aim to formulate such diophantine conditions that jointly with (1) to (4) imply

(i) $y \equiv y_v(a) \bmod x_u(a)$ for some v,
(ii) $y|u$,
(iii) $v \equiv z \bmod 2y$.

First, we shall show that (1) to (4) together with (i) to (iii) imply $w = z$, that is, $y = y_z(a)$. Next, we shall find diophantine conditions (5) to (10) such that (i) and (ii) will follow from (1) to (10).

From (i) and from the definition of w we obtain

$$y_w(a) \equiv y_v(a) \bmod x_u(a).$$

Now, from Property 3 we infer

$$v \equiv w \bmod 2u \quad \text{or} \quad v \equiv -w \bmod 2u.$$

So, from (ii) and (iii) we get

$$z \equiv w \bmod 2y \quad \text{or} \quad z \equiv -w \bmod 2y.$$

But from the definition of w and from Property 1 we infer that $y \geq w$. This together with (4) gives $z + w \leq 2y$ and $|z - w| < 2y$. Hence, and from the above congruences, $z = w = y$ or $z - w = 0$ and thus $z = w$.

Now, we shall formulate some of the announced conditions (5) to (10). To this end let us introduce auxiliary variables x_2, y_2, A. Let

(5) $x_2^2 - (A^2 - 1)y_2^2 = 1$,
(6) $y_2 \equiv y \bmod x_1$,
(7) $A \equiv a \bmod x_1$.

We shall show that (1) to (7) imply (i).

Property 4. $a \equiv b \bmod c \rightarrow y_v(a) \equiv y_v(b) \bmod c$ *for any* v.

By (5), $\langle x_2, y_2 \rangle$ is a solution of the Ath equation. Let v be the index of this solution in the sequence of the solutions of the Ath equation. By conditions (4) and (7) we have

$$y_v(A) \equiv y_v(a) \text{mod } x_u(a),$$

whence

$$y_2 \equiv y_v(a) \text{mod } x_u(a).$$

But by (6), $y_2 \equiv y \text{ mod } x_u(a)$, whence $y_v(a) \equiv y \text{ mod } x_u(a)$, that is (i).

How to ensure that (ii) and (iii) will hold by adding diophantine conditions to (1) to (7)? Let

(8) $y_2 \equiv z \text{ mod } 2y$,
(9) $A \equiv 1 \text{ mod } 2y$.

Assume (1) to (9). We have $y_2 = y_v(A)$ by the definition of v, $y_v(A) \equiv v \text{ mod}(A-1)$ by Property 2. Thus, $y_v(A) \equiv v \text{ mod } 2y$ by (9), that is, $y_2 \equiv v \text{ mod } 2y$ and by (8) we have $v \equiv z \text{ mod } 2y$, that is, (iii).

Thus, it remains to ensure (ii). We add to (1) to (9) the following condition:

(10) $y_1 \equiv 0 \text{ mod } y^2$.

From (1) to (10) we infer $y^2 | y_1$, but $y_1 = y_w(a)$ by the definition of u, hence $y^2 | y_w(a)$.

Consider the next property.

Property 5. *The condition $y_w^2(a) | y_u(a)$ implies $y_w(a) | u$, for any u, w.*

Hence, we infer $y|u$, that is, (ii). Finally, we have the following system of conditions:

(1) $x^2 - (a^2-1)y^2 = 1 \wedge a > 1$,
(2) $y \equiv z \text{ mod } (a-1)$,
(3) $x_1^2 - (a^2-1)y_1^2 = 1$,
(4) $y \geq z$,
(5) $x_2^2 - (A^2-1)y_2^2 = 1$,
(6) $y_2 \equiv y \text{ mod } x_1$,
(7) $A \equiv a \text{ mod } x_1$,
(8) $y_2 \equiv z \text{ mod } 2y$,
(9) $A \equiv 1 \text{ mod } 2y$,
(10) $y_1 \equiv 0 \text{ mod } y^2$.

We have shown that

$$\exists x, x_1, y_1, A, x_2, y_2 \text{ (1) to (10) hold implies } y = y_z(a).$$

By Example 24.4, the left-hand side of the implication is diophantine. Thus, it remains to show the converse implication and to prove the properties. So we shall show that if $y = y_z(a)$, then there exists x, x_1, y_1, A, x_2, y_2 satisfying (1) to (10). Assume that $y = y_z(a)$. Let $x = x_z(a)$. Then (1), (2) and (4) hold. Consider the equation

$$X^2 - (a^2 - 1) \cdot (2y^2) Y^2 = 1.$$

Since $(a^2 - 1)(2y^2)^2$ is not a square, the above equation has a solution. Let $x_1, \bar{y}_1$ be any solution of the above equation and let $y_1 = 2y^2 \bar{y}_1$. Then $\langle x_1, y_1 \rangle$ is a solution of the ath equation such that $2y^2 | y_1$. Hence, we have (3) and (10).

Let $A = a + x_1^2(x_1^2 - a)$. Then, we have (7). Moreover, since $x_1^2 = 1 + (a^2 - 1)y_1^2$, we have

$$A = a + \left(1 + (a^2 - 1)y_1^2\right) \cdot \left(1 + (a^2 - 1)y_1^2 - a\right).$$

Hence $A \equiv 1 \bmod y_1^2$, but $2y | y_1^2$. Thus, $A \equiv 1 \bmod 2y$, that is, (9).

Let $\langle x_2, y_2 \rangle$ be the zth solution of the Ath equation. Then, we have (5). Since $A \equiv a \bmod x_1$, therefore $y_z(A) \equiv y_z(a) \bmod x_1$, by Property 4. Thus, $y_2 \equiv y \bmod x_1$, that is, (6). We have $y_z(A) \equiv z \bmod (A - 1)$ by property 2, but, $2y | A - 1$. Thus, $y_z(A) \equiv z \bmod 2y$, that is, $y_2 \equiv z \bmod 2y$ and hence (8).

To finish step 1 it remains to prove all the properties.

Proof of Property 1. First, notice that if a pair $\langle x, y \rangle$ satisfies the inequality

24.13 $$1 < x + y\sqrt{d} < a + \sqrt{d}, \quad \text{where } d = a^2 - 1,$$

then $\langle x, y \rangle$ is not a solution of the ath equation.

Suppose the converse; suppose that $\langle x, y \rangle$ satisfies the ath equation and 24.13. Then we have

$$1 = (a + \sqrt{d})(a - \sqrt{d}) = (x + y\sqrt{d})(x - y\sqrt{d}).$$

Hence, by 24.13, $a - \sqrt{d} < x - y\sqrt{d} < 1$, that is, $-1 < -x + y\sqrt{d} < -a + \sqrt{d}$. Adding this inequality side by side to 24.13 we obtain $0 < 2y\sqrt{d} < 2\sqrt{d}$, hence $0 < y < 1$, a contradiction.

Now, we shall show that every solution $\langle x, y \rangle$ of the ath equation is of the form $\langle x_n(a), y_n(a) \rangle$ for a certain n. We can easily check that

24.14 $$x_n(a) + y_n(a)\sqrt{d} = (a + \sqrt{d})^n.$$

Since the sequence $(a+\sqrt{d})^n$ diverges to infinity, there is such an n for which

$$(a+\sqrt{d})^n \leq x+y\sqrt{d} < (a+\sqrt{d})^{n+1}.$$

Suppose that

$$(a+\sqrt{d})^n < x+y\sqrt{d} \leq a+\sqrt{d})^{n+1}.$$

That is,

24.15 $$x_n(a)+y_n(a)\sqrt{d} < x+y\sqrt{d} < \left(x_n(a)+y_n(a)\sqrt{d}\right)(a+\sqrt{d}).$$

Since

$$\left(x_n(a)+y_n(a)\sqrt{d}\right)\left(x_n(a)-y_n(a)\sqrt{d}\right) = 1,$$

we have $x_n(a)-y_n(a)\sqrt{d} > 0$.

Multiplying 24.15 by $x_n(a)-y_n(a)\sqrt{d}$ we obtain $1 < (x+y\sqrt{d})\left(x_n(a)-y_n(a)\sqrt{d}\right) < a+\sqrt{d}$,whence 24.13, a contradiction.

It is easy to show that the sequence $y_n(a)$ is increasing, whence $y_n(a) \geq n$ for every n which completes the proof.

Proof of Property 2. We show by induction on n that

24.16
$$y_n(a) = \sum_{i=1}^{[(n+1)/2]} \binom{n}{2l-1} a^{n-2l+1}(a^2-1)^{l-1}$$

$$x_n(a) = a^n + \sum_{i=1}^{[n/2]} \binom{n}{2l} a^{n-2l}(a^2-1)^l,$$

whence we immediately infer $y_n(a) \equiv n \bmod (a-1)$, which completes the proof.

Proof of Property 4. By 24.16 $\left(y_n(a)-y_n(b)\right) \equiv 0 \bmod (a-b)$, whence we obtain Property 4.

Proof of Property 3. By 24.14 for an arbitrary n we obtain

$$x_{n\pm m}(a)+y_{n\pm m}(a)\sqrt{d} = \left(x_n(a)+y_n(a)\sqrt{d}\right)\left(x_m(a)\pm y_m(a)\sqrt{d}\right),$$

whence

24.17
$$x_{n\pm m}(a) = x_n(a)x_m(a) \pm dy_n(a)y_m(a)$$

$$y_{n\pm m}(a) = \pm x_n(a)y_m(a) + x_m(a)y_n(a).$$

Hence,

$$y_{2n\pm m}(a) = y_{n+(n\pm m)}(a) \equiv x_{n\pm m}(a)y_n(a) \bmod x_n(a)$$
$$\equiv \pm dy_n^2(a)y_m(a) \bmod x_n(a) \equiv \mp y_m(a) \bmod x_n(a).$$

Similarly

$$y_{4n\pm m}(a) = y_{2n+(2n\pm m)}(a) \equiv -y_{2n\pm m}(a) \bmod x_n(a) \equiv y_{\pm m}(a) \bmod x_n(a),$$
$$y_{-m}(a) \equiv -y_{2n-m}(a) \equiv -y_m(a) \bmod x_n(a),$$

where by $\langle x_{-m}(a), y_{-m}(a)\rangle$ we denote the result of the mth iteration applied to the pair $\langle 1, 0\rangle$ of the mapping that is converse to the mapping given by the matrix 24.12.

Let a number $r \in Z$ be called the *strict remainder* of the division of $y_k(a)$ by $x_n(a)$ if $y_k(a) \equiv r \bmod x_n(a)$ and $|r| \leq \frac{1}{2}x_n(a)$. We infer that the function $f(k) =$ [strict remainder of the division of $y_k(a)$ by $x_n(a)$] has the period $4n$ and it changes sign every $2n$. Moreover, all its values are determined by the values in the interval $[0, n]$ since for $m \in [0, n]$

$$y_{n+m}(a) \equiv -y_{-(n+m)}(a) \equiv -y_{-n-m}(a) \equiv y_{n-m}(a) \bmod x_n(a).$$

To conclude the proof of Property 3 it is enough to show that for $m \in [0, n]$, the $y_m(a)$s all have different and less than $\frac{1}{2}x_n(a)$ remainders of the division by $x_n(a)$. But for $a \geq 3$ we have $y_m(a) < \frac{1}{2}x_n(a)$ for $m \in [0, n]$, since

$$4y_m^2(a) < (a^2 - 1)y_n^2(a) + 1 = x_n^2(a).$$

For $a = 2$, we have $y_m(a) < \frac{1}{2}x_n(a)$ for $m \leq n - 1$, and if $m = n$, then $y_m(a) < 1/\sqrt{3}x_n(a)$. Thus, the proof has been completed.

Proof of Property 5. From the equality

$$x_{nk}(a) + y_{nk}(a)\sqrt{a^2 - 1} = \left(x_n(a) + y_n(a)\sqrt{a^2 - 1}\right)^k$$

we obtain

$$y_{nk}(a) = \sum_{j \leq k,\, j \equiv 1 \bmod 2} \binom{k}{j} \left(x_n(a)\right)^{k-j} \left(y_n(a)\right)^j (a^2 - 1)^{(j-1)/2}.$$

In particular,

24.18 $$y_{nk}(a) \equiv k\left(x_n(a)\right)^{k-1} y_n(a) \bmod y_n^2(a).$$

Now we shall show that from the assumption $y_n(a) \equiv 0 \bmod y_k(a)$ follows $k|n$. Let us represent n as $n = kq + r$, $0 \le r < k$. We have, by 24.17,

$$y_n(a) = x_r(a)y_{kq}(a) + x_{kq}(a)y_r(a).$$

Also we have

24.19 $$y_k(a)|y_{kq}(a) \quad \text{for arbitrary} \quad k, q,$$

which can be easily proved by induction on q using the equality

$$y_{k(q+1)}(a) = x_k(a)y_{kq}(a) + x_{kq}(a)y_k(a),$$

which follows from 24.17. Thus, $y_k(a)|x_{kq}(a)y_r(a)$, but $(y_k(a), x_{kq}(a)) = 1$ [if $d|y_k(a)$ and $d|x_{kq}(a)$, then $d|y_{kq}(a)$ by 24.19, whence $d = 1$ since $(x_{kq}(a), y_{kq}(a)) = 1$]. Hence, $y_k(a)|y_r(a)$, which, together with $r < k$, gives $r = 0$.

Assume $y_n(a) \equiv 0 \bmod y_k(a)$. Then $k|n$. Replacing nk by n and n by k in 24.18 we obtain

$$y_n(a) \equiv \frac{n}{k}\left(x_k(a)\right)^{(n/k)-1} \cdot y_k(a) \bmod y_k^2(a).$$

Since $x_k(a)$ and $y_k(a)$ are relatively prime we obtain the implication

$$y_n(a) \equiv 0 \bmod y_k^2(a) \to \frac{n}{k} \equiv 0 \bmod y_k(a) \to n \equiv 0 \bmod y_k(a).$$

This completes the proof of Property 5.

STEP 2. Now we shall show that the graph of the exponential function $y = x^z$ is diophantine. First we shall show that for a sufficiently large number K the closest integer to the number $(y_{z+1}(Kx))/y_{z+1}(K)$ is x^z.

It is easy to show by induction on z that for an arbitrary z and $a > 1$,

$$(2a-1)^z \le y_{z+1}(a) \le (2a)^z.$$

Hence, we infer

24.20 $$x^z\left(1 - \frac{1}{2Kx}\right)^z = \frac{(2Kx-1)^z}{(2K)^z} \le \frac{y_{z+1}(Kx)}{y_{z+1}(K)} \le \frac{(2Kx)^z}{(2K-1)^z} = x^z \frac{1}{(1-(1/2K))^z}.$$

Notice that for any m, n we have

$$\left(1 - \frac{1}{m}\right)^n \ge 1 - \frac{n}{m}, \quad 2n \le m \to \left(1 - \frac{n}{m}\right)^{-1} \le 1 + \frac{2n}{m}.$$

Thus, for $K \geq 2z$ we have

$$\left(1 - \frac{1}{2Kx}\right)^z \geq 1 - \frac{z}{2Kx},$$

and

$$\frac{1}{(1-(1/2K))^z} \leq \frac{1}{1-(z/2K)} = \left(1 - \frac{z}{2K}\right)^{-1} \leq 1 + \frac{z}{K}.$$

Hence, from 24.20 we obtain

$$x^z \cdot \left(1 - \frac{z}{2Kx}\right) \leq \frac{y_{z+1}(Kx)}{y_{z+1}(K)} \leq x^z\left(1 + \frac{z}{K}\right),$$

$$x^z - \frac{x^z \cdot z}{2Kx} \leq \frac{y_{z+1}(Kx)}{y_{z+1}(K)} \leq x^z + \frac{x^z \cdot z}{K}.$$

Now, if $K > 2z \cdot x^z$, $z, x \neq 0$, then we obtain

$$x^z - \frac{1}{2} < \frac{y_{z+1}(Kx)}{y_{z+1}(K)} < x^z + \frac{1}{2}.$$

Hence, we have

$$\begin{aligned} y = x^z \quad \text{iff} \quad & (x = 0 \wedge y = 0 \wedge z \neq 0) \vee (z = 0 \wedge y = 1) \\ & \vee\big[x > 0 \wedge z > 0 \wedge \exists y_1, y_2, y_3, K\big(y_1 = y_{z+1}(x) \\ & \wedge K > 2z \cdot y_1 \wedge y_2 = y_{z+1}(K) \wedge y_3 = y_{z+1}(Kx) \\ & \wedge[0 \leq y - \frac{y_3}{y_2} < \frac{1}{2} \vee 0 \leq \frac{y_3}{y_2} - y < \frac{1}{2}]\big)\big]. \end{aligned}$$

By 24.4 and 24.5, the graph of the exponential function $y = x^z$ is diophantine.

STEP 3. The graph of the function $y = \binom{x}{z}$ for $x \geq z$ is diophantine. Consider the following remark:

Remark. *If* $u > n^k$ *and* $n \geq k$, *then* $\binom{n}{k}$ *is the remainder of the division of* $[(u+1)^n/u^k]$ *by* u.

Proof of the remark. We have

$$\frac{(u+1)^n}{u^k} = \sum_{i=k+1}^{n} \binom{n}{i} \cdot u^{i-k} + \binom{n}{k} + \sum_{i=0}^{k-1} \binom{n}{i} \cdot u^{i-k}.$$

If $u > n^k$, then the last sum is less than 1 and the first sum is divisible by u. This completes the proof of the remark.

We have

$$x \geq z \wedge y = \binom{x}{z}$$

$$\text{iff} \quad \exists u, v\left(u > x^z \wedge v = \left[\frac{(u+1)^x}{u^z}\right] \wedge y \equiv v \bmod u \wedge y < u \wedge x \geq z\right)$$

$$\text{iff} \quad \exists u, v, y_1, y_2, y_3\big(y_1 = x^z \wedge y_2 = (u+1)^x \wedge y_3 = u^z$$

$$\wedge\, U > y_1 \wedge v = \left[\frac{y_2}{y_3}\right] \wedge y \equiv v \bmod u \wedge y < u \wedge x \geq z\big).$$

Hence we infer that the graph of the function $y = \binom{x}{z}$ for $x \geq z$ is diophantine.

STEP 4. We shall show that the graph of the function $y = x!$ is diophantine.

Remark. *If $k > 0$ and $n > (2k)^{k+1}$, then $k! = [n^k/\binom{n}{k}]$.*

Proof of the remark.

$$\frac{n^k}{\binom{n}{k}} = \frac{n^k \cdot k!}{n(n-1)\ldots(n-k+1)} = k! \cdot \frac{1}{(1-\frac{1}{n})\ldots(1-(k-1)/n)}$$

$$< k! \cdot \frac{1}{(1-k/n)^k}.$$

But

$$\frac{1}{(1-k/n)} = 1 + \frac{k}{n} + \frac{k^2}{n} + \cdots = 1 + \frac{k}{n}\left(1 + \frac{k}{n} + \left(\frac{k}{n}\right)^2 + \cdots\right)$$

$$< 1 + \frac{k}{n}\left(1 + \frac{1}{2} + \frac{1}{4} + \cdots\right) = 1 + \frac{2k}{n}$$

and

$$\left(1 + \frac{2k}{n}\right)^k = \sum_{j=0}^{k} \binom{k}{j} \cdot \left(\frac{2k}{n}\right)^j < 1 + \frac{2k}{n} \sum_{j=1}^{k} \binom{k}{j} < 1 + \frac{2k}{n} \cdot 2^k,$$

therefore

$$\frac{n^k}{\binom{n}{k}} < k! + \frac{2k}{n} \cdot k! \cdot 2^k < k! + \frac{2^{k+1} \cdot k^{k+1}}{n} < k! + 1,$$

which proves the remark.

Thus, we have

$$y = x! \quad \text{iff} \quad \exists n\left(n > (2x)^{x+1} \wedge y = \left[\frac{n^x}{\binom{n}{x}}\right]\right)$$

$$\text{iff} \quad \exists n, y_1, y_2, y_3\left(y_1 = (2x)^{x+1} \wedge y_2 = n^x \wedge y_3 = \binom{n}{x} \wedge n > y_1 \wedge y = \left[\frac{y_2}{y_3}\right]\right).$$

STEP 5. The graph of the function $y = \prod_{0 \leq z < x} (1 + (z+1)x_1)$ as a function of x, x_1 is diophantine.

Remark. *Assume* $x_1 \cdot w \equiv 1 \bmod u$. *Then*

$$\prod_{z=0}^{x-1}(1 + (z+1)x_1) \equiv x_1^x \cdot x!\binom{w+x}{x} \bmod u.$$

Proof of the remark.

$$\begin{aligned} x_1^x \cdot x! \cdot \binom{w+x}{x} &= x_1^x \cdot (w+x) \cdot (w+x-1) \cdot \cdots \cdot (w+1) \\ &= (x_1 w + x x_1) \cdot (x_1 w + (x-1)x_1) \cdot \cdots \cdot (x_1 w + x_1) \\ &\equiv (1 + x x_1) \cdot (1 + (x-1)x_1) \cdot \cdots \cdot (1 + x_1) \bmod u. \end{aligned}$$

This completes the proof of the remark.

Let $u = x_1 \cdot (1 + x_1 x)^x + 1$. Then $(u, x_1) = 1$ and $u > \prod_{z=0}^{x-1}(1 + (z+1)x_1)$. Hence there is such a w that $x_1 w \equiv 1 \bmod u$. Let us take such a w. Then $\prod_{z=0}^{x-1}(1 + (z+1)x_1)$ is the remainder of the division of $x_1^x \cdot x! \cdot \binom{w+x}{x}$ by u. Thus, we have

$$y = \prod_{z=0}^{x-1}(1 + (z+1)x_1) \quad \text{iff} \quad \exists u, w(u > y \wedge x_1 w \equiv 1 \bmod u$$

$$\wedge\, x_1^x \cdot x!\binom{w+x}{x} \equiv y \bmod u \wedge u > x_1 \cdot (1 + x_1 x)x)$$

$$\text{iff} \quad \exists u, w, y_1, y_2, y_3, y_4, y_5(u > y \wedge x_1 w \equiv 1 \bmod u \wedge y_1 = x_1^2$$

$$\wedge\, y_2 = x! \wedge y_3 = \binom{w+x}{x} \wedge y_1 y_2 y_3 \equiv y \bmod u$$

$$\wedge\, y_4 = 1 + x_1 x \wedge y_5 = y_4^x \wedge u > y_5).$$

STEP 6. The graph of the function $y = \prod_{0 \le j < x_1} (x_2 - j)$ for $x_2 > x_1$ is diophantine. We have

$$y = \prod_{0 \le j \le x_1} (x_2 - j) \wedge x_2 > x_1 \quad \text{if and only if} \quad y = x_1! \cdot \binom{x_2}{x_1} \wedge x_2 > x_1.$$

Similarly as before we show that this is a diophantine relation.

Now, we may pass to the proof of Lemma 24.9. Let f be a polynomial, c—the sum of the absolute values of the coefficients of f, d—the degree of f. We shall show that

24.21

$$\begin{aligned} &\forall z < x \exists y_0, \ldots, y_k (f(z, x, x_1, \ldots, x_n, y_0, \ldots, y_k) = 0) \\ \text{iff} \quad &\exists Y_0, \ldots, Y - k, Z, e, K[K \ge c \cdot (x^2 x_1 \cdot \ldots \cdot x_n e)^d \wedge Y_0 > e \\ &\cdots \wedge Y_k > e \wedge e \ge 1 \wedge 1 + (Z+1)K! = \prod_{z=0}^{x-1} (1 + (z+1)K!) \\ &\wedge f(Z, x, x_1, \ldots, x_n, Y_0, \ldots, Y_k) \equiv 0 \bmod (1 + (Z+1)K!) \\ &\wedge \bigwedge_{i=0}^{k} \prod_{0 \le j < e} (Y_i - j) \equiv 0 \bmod (1 + (Z+1)K!)]. \end{aligned}$$

First we shall show the proof of the implication $\rightarrow$.

Proof of the implication $\rightarrow$. Assume

$$\forall z < x \exists y_0, \ldots, y_k (f(z, x, x_1, \ldots, x_n, y_0, \ldots, y_k) = 0).$$

As before, let $y_{z0}, \ldots, y_{zk}$ denote the sequence corresponding to z, that is, the sequence satisfying

$$f(z, x, x_1, \ldots, x_n, y_{z0}, \ldots, y_{zk}) = 0.$$

Let $e = \max\{x, y_{zi}\colon\ i = 0, \ldots, k,\ z = 0, \ldots, x-1\}$ and let K be such that $K \ge c(x^2 x_1 \ldots x_n e)^d$. Let, as before, Y_i be the code for the sequence $\langle y_{0i}, \ldots, y_{x-1,i} \rangle$ found by the Chinese remainder theorem, that is, Y_i satisfies

$$Y_i \equiv y_{zi} \bmod \big(1 + (z+1)K!\big).$$

Moreover, assume that $Y_i > e$. [As in Chapter 18 we may show that the numbers $1 + (z+1)K!$ for $z = 0, \ldots, x-1$ are relatively prime, hence there is a number Y_i

with the required property] Let

$$Z = \frac{\prod_{z=0}^{x-1}(1+(z+1)K!) - 1}{K!} - 1.$$

Then

$$1 + (Z+1)K! = \prod_{z=0}^{x-1}(1+(z+1)K!).$$

We have

$$Y_i - y_{zi} \equiv 0 \bmod (1+(z+1)K!).$$

Hence, $\prod_{j=0}^{e-1}(Y_i - j)$ is divisible by $1 + (z+1)K!$ for $z = 0, \ldots, x-1$, and thus, by $\prod_{z=0}^{x-1}(1+(z+1)K!)$. It remains to show that

$$f(Z, x, x_1, \ldots, x_n, Y_0, \ldots, Y_k) \equiv 0 \bmod (1+(Z+1)K!).$$

We have

$$Z \equiv z \bmod (1+(z+1)K!) \quad \text{for} \quad z = 0, \ldots, x-1,$$

since

$$(Z - z)K! = (1+(Z+1)K!) - (1+(z+1)K!) \equiv 0 \bmod (1+(z+1)K!).$$

Also we have,

$$Y_i \equiv y_{zi} \bmod (1+(z+1)K!) \quad \text{for } z = 0, \ldots, x-1.$$

Hence,

$$f(Z, x, x_1, \ldots, x_n, Y_0, \ldots, Y_k) \equiv 0 \bmod (1+(z+1)K!) \quad \text{for} \quad z = 0, \ldots, x-1.$$

By the fact that the numbers $1 + (z+1)K!$ are relatively prime for different z, we infer

$$f(Z, x, x_1, \ldots, x_n, Y_0, \ldots, Y_k) \equiv 0 \bmod (1+(Z+1)K!),$$

which completes the proof of the implication $\rightarrow$.

Proof of the implication $\leftarrow$ *in 24.21.* Assume that $Y_0, \ldots, Y_k, Z, e, K$ are given satisfying the right hand side of the equivalence 24.21.

Let y_{zi} be the remainder of the division of Y_i by $1 + (z+1)K!$ in the case where

$1 + (z+1)K!$ is a prime; in the case where $1 + (z+1)K!$ is not prime let us fix an arbitrary prime factor p_z of the number $1 + (z+1)K!$ and let y_{zi} be the remainder of the division of Y_i by p_z. Then we have

$$f(z, x, x_1, \ldots, x_n, y_{z0}, \ldots, y_{zk}) \equiv 0 \bmod p_z,$$

since

$$f(Z, x, x_1, \ldots, x_n, Y_0, \ldots, Y_k) \equiv 0 \bmod(1 + (Z+1)K!),$$

and so also

$$f(Z, x, x_1, \ldots, x_n, Y_0, \ldots, Y_k) \equiv 0 \bmod p_z.$$

We shall show that $f(z, x, x_1, \ldots, x_n, y_{z0}, \ldots, y_{zk}) < p_z$.We have $p_z | \prod_{j=0}^{e-1}(Y_i - j)$ for $i = 0, \ldots, k$, since $\prod_{j=0}^{e-1}(Y_i - j) \equiv 0 \bmod(1 + (Z+1)K!)$. Hence, for every $i = 0, \ldots, k$ there is a $j = j(i)$ such that $p_z | Y_i - j$, since p_z is prime. Thus, Y_{zi}, as the remainder of the division of Y_i by p_z, is $\leq j(i)$. Hence, $y_{zi} < e$, and thus

$$f(z, x, x_1, \ldots, x_n, y_{z0}, \ldots, y_{zk}) \leq c(x^2 x_1 \ldots x_n e)^d \leq K < p_z.$$

Hence, it follows immediately $f(z, x, x_1, \ldots, x_n, y_{z0}, \ldots, y_{zk}) = 0$ for $z = 0, \ldots, x-1$. Thus, we have shown

$$\forall z < x \exists y_0, \ldots, y_k\, f(z, x, x_1, \ldots, x_n, y_0, \ldots, y_k) = 0,$$

which completes the proof of the equivalence 24.21.

To present the right-hand side of the equivalence 24.21 in a diophantine form it is enough to introduce new variables w_1, w_2, w_3 to denote $K!$, $\prod_{z=0}^{x-1}(1 + (z+1)K!)$, and $\prod_{0 \leq j < e}(Y_i - j)$, respectively, to use the diophantine definitions of the relations

$$w_1 = K!, \qquad w_2 = \prod_{z=0}^{x-1}(1 + (z+1)K!), \qquad w_3 = \prod_{0 \leq j < e}(Y_i - j)$$

and to apply Example 24.4.

Thus, we have completed the proof of Lemma 24.9 and of Theorem. 24.1 □

EXERCISES

24.1. Define the set of primes by a formula of the form $\exists y_1, \ldots, y_k\ F$, where F is a conjunction of equations and of a formula defining the graph of the function $x!$.

24.2. Find an upper bound for the number of variables and for the degree of the polynomial corresponding to a diophantine formula defining the set of primes (see Exercise 24.1).

24.3. Show that every diophantine formula is equivalent in $\mathbb{N}$ with a diophantine formula for which the occurring polynomial is of degree ≤ 4.

24.4. Show that for any formula $F \in \Sigma_1$ there is a diophantine formula G such that we have $\mathrm{PA} \vdash (F \equiv G)$. In other words, prove the Matiyasevich theorem in PA.

24.5. Show that if $\mathbb{M}_1, \mathbb{M}_2 \in \mathrm{Mod}(\mathrm{PA})$, $\mathbb{M}_1 \subseteq \mathbb{M}_2$ and F is a bounded formula, then we have

$$\mathbb{M}_1 \models F[a_1, \ldots, a_n] \quad \text{if and only if} \quad \mathbb{M}_2 \models F[a_1, \ldots, a_n]$$

for any $a_1, \ldots, a_n \in M_1$.

24.6. Prove the following theorem of Gaifman: *if* $\mathbb{M}_1, \mathbb{M}_2 \in \mathrm{Mod}(PA)$, $\mathbb{M}_1 \subseteq \mathbb{M}_2$, *then* $\mathbb{M}_1$ *is equivalent with an initial segment of the model* $\mathbb{M}_2$.

24.7. Show that there is a diophantine formula $F(x, y_1, \ldots, y_n)$ such that for any diophantine formula $G(y_1, \ldots, y_n)$ we have $\mathrm{PA} \vdash \big(F(\Lambda_G, y_1, \ldots, y_n) \equiv G(y_1, \ldots, y_n)\big)$.

24.8. Show that there is a diophantine formula F such that for any model $\mathbb{M} \in \mathrm{Mod}(\mathrm{PA})$ we have $F^M = \{a \in M\colon\ \mathbb{M} \models F[a]\} \notin \Pi_1(\mathbb{M})$.

GUIDE TO FURTHER READING

The reader interested in such topics as recursion theory, decidability or other systems of logic is advised to read the following titles.

H. D. Ebbingaus, J. Flum, and W. Thomas. *Mathematical Logic*. Springer, New York, 1984.

H. Enderton. *A Mathematical Introduction to Logic*. Academic Press, New York, 1972.

E. Mendelson. *Introduction to Mathematical Logic*. Wadsworth & Brooks-Cole, Monterey, CA, 1987.

J. D. Monk. *Mathematical Logic*. Springer, New York, 1976.

The reader who wishes to deepen his knowledge in model theory is advised to consult the following works.

C. C. Chang, and H. J. Keisler. *Model Theory*. North-Holland, Amsterdam, 1973.

J. L. Bell, and A. B. Slomson. *Models and Ultraproducts*. North-Holland, Amsterdam, 1969.

G. E. Sacks. *Saturated Model Theory*. Benjamin, Menlo Park, CA, 1972.

Peano arithmetic has been vividly developed during the last twenty years. An extensive account on this subject can be found in the titles listed.

P. Hajek, and P. Pudlak, *Mathematics of First Order Arithmetic*. Springer, Berlin, 1993.

R. Kaye. *Models of Peano Arithmetic*. Oxford University Press, 1991.

The following is the most comprehensive text covering all parts of mathematical logic.

Handbook of Mathematical Logic. (J. Barwise, ed.), North-Holland, Amsterdam, 1978.

The forcing technique introduced by P. J. Cohen in the early 1960s is a powerful method of construction of models of set theory. It has been perhaps the most important achievement in the foundations of mathematics. For fascinating results obtained with the help of the forcing method, the reader is referred to the following works.

K. Kunen. *Set Theory: An Introduction to Independence Proofs.* North-Holland, Amsterdam, 1979.

T. J. Jech. *Set Theory.* Academic Press, New York, 1978.

REFERENCES

[B1] E. W. Beth. On Padoa's method in the theory of definitions. *Indag. Math.* **15** (1953), 330–339.

[B2] G. Birkhoff. On the structure of abstract algebras. *Proc. Camb. Phil. Soc.* **31** (1935), 433–454.

[B3] G. Boole. *An Investigation of the Laws of Thought.* Cambridge, 1854.

[C1] E. A. Cichon. A short proof of the recently discovered independence results using recursion theoretic methods. *Proc. Amer. Math. Soc.* **87** (1983), 704–706.

[C2] P. J. Cohen. Decision procedures for real and p-adic fields. *Comm. Pure Appl. Math.* **22** (1969), 131–151.

[C3] W. Craig. Three uses of the Herbrand–Gentzen theorem in relating model theory and proof theory. *J. Symb. Logic* **22** (1957), 269–285.

[DPR] M. Davis, H. Putham, and J. Robinson. The decision problem for exponential Diophantine equation. *Ann. of Math.* **74** (1961), 425–436.

[E1] H. Enderton. *A Mathematical Introduction to Logic.* Academic Press, New York 1972.

[E2] E. Engeler. A characterization of theories with isomorphic denumerable models. *Notices Amer. Math. Soc.* **6** (1959), 161.

[EGH] P. Erdös, L. Gillman, and M. Henriksen. An isomorphism theorem for real-closed fields. *Ann. Math.* 6 (1955), 542–554.

[F1] S. Feferman. Arithmetization of metamathematics in general setting. *Fund. Math.* **XLIX** (1960), 35–92.

[F2] A. Fraenkel. Zu den grundlagen der Cantor–Zermeloschen Mengenlehre. *Math. Annalen* **86** (1922), 230–237.

[FMS] T. E. Frayne, A. C. Morel, and D. S. Scott. Reduced direct products. *Fund. Math.* **51** (1962), 195–228.

[G2] K. Gödel. Die vollständigkeit der axiome des logischen funktionkalküls. *Monatsh. Math. Phys.* **37** (1930), 349–360.

[G3] K. Gödel. Über formal unentscheidbare Sätze der Prinzipia Mathematica und verwander Systeme, *Monatsh. Math. Phys.* **38** (1931), 173–198.

[G1] R. L. Goodstein. On the restricted ordinal theorem. *J. Symb. Logic* **9** (1944), 33–41.

[GMR] A. Grzegorczyk, A. Mostowski, and Cz. Ryll-Nardzewski. Definability of sets in models of axiomatic theories. *Bull. Acad. Polon. Sci.* **9** (1961), 163–167.

[HK] S. Hayden and J. F. Kennison. *Zermelo–Fraenkel Set Theory.* Mervill, Columbus, Ohio, 1968.

[H1] L. A. Henkin. The completeness of the first order functional calculus. *J. Symb. Logic* **14** (1949), 159–166.

[HA] D. Hilbert and W. Ackermann. *Grundzüge der Teoretischen Logik.* Springer, Berlin, 1928.

[HB] D. Hilbert and P. Bernays. *Grundlagen der Mathematik*, Vol 1. Springer, Berlin, 1934; Vol. 2. Springer, Berlin, 1939.

[H2] L. Hörmander. *The Analysis of Linear Partial Differential Operators*. Springer, Berlin, 1983.

[H3] E. V. Huntington. Sets of independent postulates for the algebra of logic. *Trans. Am. Math. Soc.* **5** (1904), 288–309.

[K2] H. J. Keisler. Theory of models with generalized atomic formulas. *J. Symb. Logic* **25** (1960), 1–26.

[K1] J. L. Kelley. *General Topology*. Van Nostrand, Princeton, 1955.

[K5] A. Khovanskii. On a class of systems of transcendental equations. *Soviet Math. Dokl.* **22** (1980), 762–765.

[KP] L. A. S. Kirby and J. Paris. Accessible independence result for Peano Arithmetic. *Bull. London Math. Soc.* **14** (1982), 285–293.

[K3] S. C. Kleene. Recursive predicates and quantifiers. *Trans. Amer. Math. Soc.* **53** (1943), 41–73.

[KPS] J. F. Knight, A. Pillay, and C. Steinhorn. Definable sets in ordered structures, II. *Trans. Amer. Math. Soc.* **295**, 593–605.

[Ł1] J. Łoś. Quelques remarques, théorèms et problèmes sur les classes définissables d'algèbres. In *Mathematical Interpretations of Formal Systems*, 98–113 North-Holland, Amsterdam, 1955.

[L2] L. Löwenheim. Über Möglichkeiten im Relativkalkül. *Math. Annalen* **76** (1915), 447–470.

[M1] A. I. Malcev. Untersuchungen aus der Gebiete der Mathematische Logik. *Matem. Sb.* **5** (1936), 323–336.

[M2] Y. I. Manin. *A Course in Mathematical Logic*. Springer, New York, 1977.

[M3] Yu. Matiyasevich. Enumerable sets are Diophantine. *Dokl. Akad. Nauk SSSR* **191** (1970), 279–282.

[MS] R. McDowell and E. Specker. *Modelle der Arithmetik*. In *Infinitistic Methods*, 257–263. Pergamon Press, London, 1961.

[M4] J. D. Monk. *Introduction to Set Theory*. McGraw-Hill, New York, 1969.

[M5] A. Mostowski. On definable sets of positive integers. *Fund. Math.* **34** (1947), 81–112.

[PS] A. Pillay and C. Steinhorn. Definable sets in ordered structures, I. *Trans. Amer. Math. Soc.* **295**, No 2, (June 1986), 565–592.

[RS] H. Rasiowa and R. Sikorski. A proof of the completeness theorem of Gödel. *Fund. Math.* **37** (1951), 193–200.

[R1] A. Robinson. A result on consistency and its applications to the theory of definitions. *Indag. Math.* **18** (1956), 47–58.

[R2] A. Robinson. On ordered fields and definite functions. *Math. Annalen*, **130** (1955), 257–271.

[R3] J. B. Rosser. Extensions of some theorems of Gödel and Church. *J. Symb. Logic* **1** (1936), 87–91.

[R-N] Cz. Ryll-Nardzewski. On the categoricity in power $\aleph^0$. *Bull. Acad. Polon. Sci* **7** (1959), 545–548.

[S1] A. Seidenberg. A new decision method for elementary algebra. *Ann. Math.*, **60** (1954), 365–374.

[S2] R. Sikorski. On an analogy between measures and homomorphisms. *Ann. Soc. Pol. Math.* **23** (1950), 1–20.

[S3] T. Skolem. Logisch-kombinatorische Untersuchungen über die Erfüllbarkeit oder Beweisbarkeit mathematischer Sätze nebst einem Theorem über dichte Mengen. *Skrifter Vidensk., Kristania I. Mat. Naturv. Kl.* **4** (1920), 1–36.

[S4] T. Skolem. Über die Nicht-Charakterisierbarkeit der Zahlenreihe mittels endlich oder abzählbar unendlich vieler Aussagen mit ausschlieslich Zahlenvariabeln. *Fund. Math.* **23** (1934), 150–161.

[S5] E. Steinitz. Algebraische Theorie der Körper. *J. Reine Angew. Math.* **137** (1910), 167–309.

[S6] M. H. Stone. The theory of representations for Boolean algebras. *Trans. Am. Math. Soc.* **40** (1936), 37–11.

[S7] M. H. Stone. Postulates for Boolean algebras and generalized Boolean algebras. *Am. J. Math.* **57** (1935), 703–732.

[S8] L. Svenonius. $\aleph^0$-categoricity in first-order predicate calculus. *Teoria (Lund)* **25** (1959), 82–84.

[T3] A. Tarski. A decision method for elementary algebra and geometry, 2nd ed. Univ. of California Press, Berkeley and Los Angeles, 1951.

[T1] A. Tarski. Some notions and methods on the borderline of algebra and metamathematics. *Proc. Int. Congr. Math. Cambridge USA 1950.* **1** (1952), 705–720.

[T2] A. Tarski. Der Wahrheitsbegriff in den formalisierten Sprachen. *Studia Philos.* **1** (1936), 261–405.

[TV] A. Tarski and R. L. Vaught. Arithmetical extensions of relational systems. *Composito Math.* **13** (1957), 81–102.

[vdD] L. van den Dries. Remarks on Tarski's problem concerning (R, +, ·, exp). In *Logic Colloquium '82* (G. Lolli, G. Longo and A. Marcja, eds.), 97–121. North-Holland, Amsterdam 1984.

[V1] R. L. Vaught. Denumerable models of complete theories. In *Infinitistic Methods*, Pergamon Press, London, 1961, 303–321.

[V2] R. L. Vaught. *Set Theory*. Birkhäuser, Boston, 1985.

[W1] S. S. Wainer. A classification of the ordinal recursive functions. *Arch. Math. Logik Grundlagenforsch.* **13** (1970), 136–153.

[W2] S. S. Wainer. Ordinal recursion and a refinement of the Grzegorczyk hierarchy. *J. Symb. Logic* **37** (1972), 281–292.

[W4] A. Wilkie. Model completeness results for expansions of the real field by restricted Pfaffian functions and the exponential function. *J. Amer. Math. Soc.* **9**:4 (October 96), 1051–1095.

[Z] E. Zermelo. Untersuchungen über die Grundlagen der Mengenlehre I. *Math. Ann.* **65** (1908), 261–181.

INDEX

PURE AND APPLIED MATHEMATICS

A Wiley-Interscience Series of Texts, Monographs, and Tracts

ADÁMEK, HERRLICH, and STRECKER—Abstract and Concrete Catetories
ADAMOWICZ and ZBIERSKI—Logic of Mathematics
AKIVIS and GOLDBERG—Conformal Differential Geometry and Its Generalizations
*ARTIN—Geometric Algebra
AZIZOV and IOKHVIDOV—Linear Operators in Spaces with an Indefinite Metric
BERMAN, NEUMANN, and STERN—Nonnegative Matrices in Dynamic Systems
BOYARINTSEV—Methods of Solving Singular Systems of Ordinary Differential Equations
*CARTER—Finite Groups of Lie Type
CHATELIN—Eigenvalues of Matrices
CLARK—Mathematical Bioeconomics: The Optimal Management of Renewable Resources, Second Edition
*CURTIS and REINER—Representation Theory of Finite Groups and Associative Algebras
*CURTIS and REINER—Methods of Representation Theory: With Applications to Finite Groups and Orders, Volume I
CURTIS and REINER—Methods of Representation Theory: With Applications to Finite Groups and Orders, Volume II
*DUNFORD and SCHWARTZ—Linear Operators
Part 1—General Theory
Part 2—Spectral Theory, Self Adjoint Operators in Hilbert Space
Part 3—Spectral Operators
FOLLAND—Real Analysis: Modern Techniques and Their Applications
FRÖLICHER and KRIEGL—Linear Spaces and Differentiation Theory
GARDINER—Teichmüller Theory and Quadratic Differentials
*GRIFFITHS and HARRIS—Principles of Algebraic Geometry
GUSTAFSSON, KREISS and OLIGER—Time Dependent Problems and Difference Methods
HANNA and ROWLAND—Fourier Series, Transforms, and Boundary Value Problems, Second Edition
*HENRICI—Applied and Computational Complex Analysis
Volume 1, Power Series—Integration—Conformal Mapping—Location of Zeros
Volume 2, Special Functions—Integral Transforms—Asymptotics—Continued Fractions
Volume 3, Discrete Fourier Analysis, Cauchy Integrals, Construction of Conformal Maps, Univalent Functions
*HILTON and WU—A Course in Modern Algebra
*HOCHSTADT—Integral Equations
JOST—Two-Dimensional Geometric Variational Procedures
*KOBAYASHI and NOMIZU—Foundations of Differential Geometry, Volume I
*KOBAYASHI and NOMIZU—Foundations of Differential Geometry, Volume II
LAX—Linear Algegra
LOGAN—An Introduction to Nonlinear Partial Differential Equations
McCONNELL and ROBSON—Noncommutative Noetherian Rings
NAYFEH—Perturbation Methods

NAYFEH and MOOK—Nonlinear Oscillations
PANDEY—The Hilbert Transform of Schwartz Distributions and Applications
PETKOV—Geometry of Reflecting Rays and Inverse Spectral Problems
*PRENTER—Splines and Variational Methods
RAO—Measure Theory and Integration
RASSIAS and SIMSA—Finite Sums Decompositions in Mathematical Analysis
RENELT—Elliptic Systems and Quasiconformal Mappings
RIVLIN—Chebyshev Polynomials: From Approximation Theory to Algebra and Number Theory, Second Edition
ROCKAFELLAR—Network Flows and Monotropic Optimization
ROITMAN—Introduction to Modern Set Theory
*RUDIN—Fourier Analysis on Groups
SENDOV—The Averaged Moduli of Smoothness: Applications in Numerical Methods and Approximations
SENDOV and POPOV—The Averaged Moduli of Smoothness
*SIEGEL—Topics in Complex Function Theory
Volume 1—Elliptic Functions and Uniformization Theory
Volume 2—Automorphic Functions and Abelian Integrals
Volume 3—Abelian Functions and Modular Functions of Several Variables
STAKGOLD—Green's Functions and Boundary Value Problems
*STOKER—Differential Geometry
*STOKER—Nonlinear Vibrations in Mechanical and Electrical Systems
*STOKER—Water Waves: The Mathematical Theory with Applications
WESSELING—An Introduction to Multigrid Methods
WHITMAN—Linear and Nonlinear Waves
ZAUDERER—Partial Differential Equations of Applied Mathematics, Second Edition

*Now available in a lower priced paperback edition in the Wiley Classics Library.